T0234404

Aerospace Manufacturing Processes

Aerospace Manufacturing Processes

Pradip K. Saha

The Boeing Company
Seattle, Washington, USA

CRC Press
Taylor & Francis Group
Boca Raton   London   New York

CRC Press is an imprint of the
Taylor & Francis Group, an **informa** business

CRC Press
Taylor & Francis Group
6000 Broken Sound Parkway NW, Suite 300
Boca Raton, FL 33487-2742

First issued in paperback 2020

© 2017 by Taylor & Francis Group, LLC
CRC Press is an imprint of Taylor & Francis Group, an Informa business

No claim to original U.S. Government works

ISBN-13: 978-1-4987-5604-4 (hbk)
ISBN-13: 978-0-367-73692-7 (pbk)

Library of Congress Cataloging-in-Publication Data

Names: Saha, Pradip K., author.
Title: Aerospace manufacturing processes / Pradip K. Saha.
Description: Boca Raton : CRC Press, 2017. | Includes bibliographical references and index.
Identifiers: LCCN 2016013744 | ISBN 9781498756044 (hardback : alk. paper)
Subjects: LCSH: Airplanes--Design and construction. | Space vehicles--Design and construction. | Aerospace industries.
Classification: LCC TL671.28 .S245 2017 | DDC 629.1--dc23
LC record available at https://lccn.loc.gov/2016013744

Visit the Taylor & Francis Web site at
http://www.taylorandfrancis.com

and the CRC Press Web site at
http://www.crcpress.com

In loving memory of my parents—Sushil Kumar Saha and Debrani Saha

and my parents-in-law—Durgadas Saha and Hemnalini Saha.

I would also like to dedicate this book to my lovely wife Ranu and my beautiful

daughters Rima and Priya without whom this book would not be possible.

Contents

Preface

To drive the content and vision of this book on aerospace manufacturing processes, the author takes the opportunity to present manufacturing processes with a proper balance of fundamentals of an aircraft's structure, along with analysis and application of the manufacturing process. Current books on manufacturing processes of engineering materials are used as textbooks mainly for undergraduate engineering programs in universities. Their content is generally very broad and consists of technologies applied toward a vast range of industries, from automotive to ship building to aerospace. This book is a focused attempt to present manufacturing process technologies exclusively for fabricating major aircraft components. Further, it is an effort to initiate a new undergraduate and/ or graduate-level course material for mechanical, manufacturing, aerospace engineering, or material science students who may be interested in pursuing careers in the aerospace industry. This book should be equally valuable for structural design and material manufacturing engineers and other technical personnel, as well as managers in the aerospace industry. While most schools have instituted courses in manufacturing or metal working processes, there is a great diversity in the material covered and in the background of the students taking these courses. The information being covered by this book will not only help students understand basic concepts of major aircraft components, but also understand the processes and technologies involved in manufacturing before they plan to look for a career in the aerospace industry. With the initial exposure to the information from this book, the student will be better prepared to explore future research opportunities in aerospace manufacturing processes. Topics covered in successive chapters are presented with a balanced perspective on the relevant fundamentals and examples from real-world practices. It is hoped that the breadth and scope of this book will provide comprehensive information about manufacturing processes of aerospace components, and will stimulate the development of new courses that cover the total subject matter.

The content of this book has been arranged to start with the fundamental features of aerospace vehicles covering the introduction of an aircraft. This is covered along with the associated aircraft terminology, function of different components of an aircraft, aerodynamic principles, and explanation of three major aerospace vehicles, including commercial airplanes, military aircraft, and space craft. Chapter 2 is designed to discuss the fundamentals of building an aircraft, starting with three major requirements, including market, design, and engineering materials. These topics are then followed by manufacturing planning, manufacturing technologies, design for manufacturing and finishing with structural, functional, and qualification tests. Chapter 3 focuses mainly on material selection, major aircraft metals and their properties, application of metals, nonmetals, and composites in an aircraft. Chapter 4 provides manufacturing principles and processes of major aircraft metal products used for fabrication of aircraft components. The metal product manufacturing processes include casting, extrusion, forging, rolling, tube drawing, tube pilgering, and rotary piercing of tube. Chapter 5 provides an introduction of composite (non-metallic) materials for aerospace. This chapter illustrates the fundamentals of composite materials, mechanics, and manufacturing technologies of primary composite materials. Before introducing any manufacturing processes to fabricate any metal and composite material for components of an aircraft, the author in Chapter 6 provides an overview of major structural components and operating system components of an aircraft.

The author has divided manufacturing technologies in two different categories: metals and composite materials. Accordingly, the author begins Chapter 7 as an introduction to manufacturing processes of metal components of an aircraft with additional knowledge of tribology applicable to manufacturing processes. Metal forming technology has wide applications in making aircraft parts of various sizes and complex geometries from different input raw materials received in various product forms. Metal forming technologies covered in Chapters 8 through 13 mainly include: cold forming of flat sheet, plate, and extrusion products; hot forming of flat sheet, plate, and extrusion products, high energy forming and joining, and finally tube and duct forming.

Welding and joining technologies are not commonly used to join metal parts or to make near-net-shaped pre-machining stock for aerospace parts. Research is being continued to explore various welding technologies in the field of aerospace manufacturing and to qualify the process for part acceptance criteria. Fundamentals of welding technology for aerospace have been discussed in Chapter 14. Metal cutting and machining technology is the most common and necessary technology that enables the aerospace industry to manufacture a series of small, medium, and larger components of an aircraft. Chapter 15 covers fundamentals of metal cutting and machining technology as applied to machining of various metals: aluminum alloys, titanium, and steel. This chapter also covers some fundamentals of machining of composite materials. Abrasive metal removal processes are generally used as post-machining operations to remove burs and sharp edges, and even for polishing the surface and improving the aircraft metal component to meet the engineering surface finish requirements. Chapter 16 illuminates the fundamentals that draw on the principles of tribology to explain how the abrasive particle interacts on the metal surface during a metal removal process. This chapter also covers an overview of abrasive water-jet cutting, which is widely used in the aerospace industry. A chemical metal removal process known as chemical milling is commonly used by the aerospace industry for selectively removing metal to reduce weight and produce parts of varying gage. It is particularly beneficial on structure where the removal of small amounts of metal over large surface areas is required, or where the part geometry has difficult access for cutting tools. Chapter 17 covers the fundamentals of the chemical milling process and also major chemical cleaning and coating processes required for fabricating aluminum and hard metal alloy parts, including plating of hard metals.

Composites with glass and carbon fibers were steadily gaining a very high level of importance within the aerospace industries as of the 1970s. Continued technology development brought a new revolution to the commercial aircraft industries, including Boeing and Airbus, for introducing wide body aircraft to the world airline customers in the twenty-first century. Worldwide research and development on composite materials and their manufacturing processes mitigated high costs of these materials to some extent. Reducing the cost of materials and their manufacturing costs helped aircraft industries to produce more highly fuel-efficient aircraft. After discussing various manufacturing technologies to fabricate metal aircraft parts, the author introduces Chapter 18 to cover manufacturing processes of composite materials to fabricate structural, nonstructural, and interior components for an aircraft.

Measurement techniques play a very important role in every step of manufacturing processes, starting from the input raw materials to part manufacturing to meet the basic engineering dimensions and tolerances. It is also important to ensure the proper detailed part geometry, including simple to complex contours to maintain the highest quality of the aerospace products. Chapter 19 introduces various standard measurement systems, including the automation of some measurements. The chapter also includes both destructive and nondestructive inspection/testing to verify material soundness and integrity of the part, which will thereby be free from any surface and subsurface flaws including cracks. Finally, Chapter 20 discusses certain directions of research and development in the area of developing high-strength lighter aircraft materials, and cost-effective manufacturing processes.

Acknowledgment

It is a great pleasure to express my deep gratitude to The Boeing Company where I made my research career in aerospace manufacturing technology and obtained my working experience with various technologies. That experience has given me the opportunity and courage to write this book to introduce the field of aerospace manufacturing for the enlightenment of future aerospace engineers. Congratulations Boeing for achieving 100 years (1916–2016) in aerospace manufacturing.

It has been a unique privilege to be associated with my colleague Billy L. Small who provided constant support and valuable suggestions during his careful review of the book. I am equally grateful to many of my colleagues, subject matter experts of various technologies, for their critical review of individual chapters and also providing valuable articles, photographs, and more: James D. Cotton, Jeffrey D. Morgan, Rodney R. Boyer (retired), David H. Gane (retired), Larry D. Hefti, Peter K. Hwang, Liangji Xu, Richard G. Wire, Colin F. Wright, David W. Waugh, McKay A. Kunz, Melissa A. Johnson, Mark E. Bice, and Robert A. Kisch.

I am grateful to Mike Russell and Richard Freeman of TWI, UK, and also Simon Jones of Thompson, UK, for careful review of the chapter on Welding Technology in Aerospace and their support to provide valuable information and permission to use many photographs.

I am thankful to a very large number of people who helped me on many aspects to overcome the big challenge of completing this book by way of providing fundamental information and granting permission to use many photographs. Dean Malejan of Universal Alloy Corporation, Robert T. Johnson of Sandvik Special Metals, Gregor Kaustraeter of Aluminiumwerk Unna AG, Germany, Glynn Phillips of KGS Diamond International, The Netherlands, Nikolas Lehrke of IMM Maschinenbau GmbH, Germany, Christopher Rollag of Despatch Industries, Joe Jansen of Magnum Venus Plastech, Clarissa Hennings of Ingersoll Machine Tools, Michael Standridge of Sandvik Coromant Co, Industry Specialist-Aerospace, Gerry Filipski of RTI International, Brian A. Cheney of Kaiser Aluminum–Trentwood Works, Rob Specht and Dom Cimino of Metal Improvement Company, Jim Ingram (retired) of Weber Metals, Marcel Oud of 3D-Metal Forming B.V., The Netherlands, Dave Reischman of 3M Abrasive division, Conrad Bollinger of RTI International Metals, Inc., Jeff Sieck of University Swaging Corporation.

I would like to express my special thanks to a number of companies and educational institutions for providing technical information and photographs: Hexagon Metrology, Simplytoolsteel, The L.S. Starrett Company, Mitsubishi Rayon, Techsouth, Inc., McMaster Manufacturing Research Institute (MMRI), Buffalo Machines, Inc, Spincraft Inc., US Composites, MTS Systems Corporation, FARO, and also The Boeing Company.

I would also like to acknowledge the dedication and continued help and cooperation of my editorial manager/senior editor Gagandeep Singh at CRC Press, Taylor & Francis Group, and the editorial staff at Taylor & Francis, including Jennifer Ahringer and Kyra Lindholm. I would especially like to thank Rahul Mitra, professor at IIT Kharagpur, India, Jyhwen Wang of Texas A&M University, and Billy Small of the Boeing Company for their valuable time to review the book proposal and their recommendations to CRC Press for its acceptance for publication.

Finally, I would like to thank my family for their help, understanding, and patience during the seven years that it took to write and produce this book.

Author

Pradip K. Saha is a technical fellow in the Boeing Research and Technology organization and is engaged heavily in research in metallic fabrication technology in Seattle, Washington since 1997. He earned his bachelor's (1975) and master's (1978), both in mechanical engineering from the BE College, Shibpur under Calcutta University, India, and earned his PhD (1993) in mechanical engineering from Northwestern University, Evanston, Illinois. He is a recognized technical expert in wide range of metal forming and joining technologies including (1) cold forming technologies of aluminum flat sheet and extrusion products, (2) aluminum extrusion technology, (3) tribology in metal forming and other manufacturing technologies, and (4) high energy (electromagnetic and explosive) forming and joining technology at the Boeing Enterprise, and at national and international levels. He has developed new processes, part designs, specifications for measurement systems, and has resolved critical quality issues. Dr. Saha has provided creative and innovative solutions, is a self-motivated problem solver and well capable of conducting, managing, and collaborating both applied and fundamental research. He received multiple meritorious invention cash awards and various other cash awards for technology breakthroughs from Boeing, including three US patent awards. He also bagged a gold medal for academic excellence in master's program from Calcutta University, India. He has authored a book *Aluminum Extrusion Technology* published by ASM International (2000), widely used in the aluminum extrusion industries, universities, research institutes, and more. He has published 24 research papers in the national and international proceedings and journals.

1

Fundamentals of Aerospace Vehicles

1.1 Introduction

The term aerospace is derived from the words aeronautics and spaceflight. An aircraft is a vehicle which is able to fly through the earth's atmosphere. The vehicles connected with or driven by rockets are not aircraft because they are not supported in flight by the surrounding air. All the activity including the design, development, production, and the operation of an aircraft is called aviation. Aircraft can either be heavier than air or lighter than air. Lighter than air vehicles include balloons and airships whereas heavier than air vehicles are powered aircraft, gliders, helicopters, and space craft. The first form of an aircraft was the kite, designed in the fifth century BC. Later on in the thirteenth century, Roger Bacon, an English monk, performed studies which later gave him the idea that air could support a craft just like water supports boats. In the sixteenth century, Leonardo da Vinci studied birds' flight, and later produced the airscrew and the parachute. The airscrew, leading to the propeller later on, and the parachute were tremendously important contributions to aviation.

In the early 1900s, two American brothers, Orville and Wilbur Wright from Dayton, Ohio began to experiment with gliders. The gliders were built using data from Otto Lilienthal in Europe. Most of these flights turned out to be failures. In 1901, the Wright brothers decided to gather their own wing data by conducting systematic experiments on different type of wing configurations. In 1902, their Glider had a wingtip to wingtip measurement of 32 ft. and wing width of 5 ft. This was the first aircraft with three-axis control, the aircraft could go up or down, left or right, and could also roll about its longitudinal axis. At Kitty Hawk, they performed over 800 flights and resolved early problems of the aircraft (Figure 1.1).

From 1903 to today there have been remarkable developments in the aviation technology. The technology development in various areas, including aerospace structure and propulsion systems, during the last few generations has brought the aerospace industry to a very solid foundation. The aerospace industry has reached a much matured stage capable of designing and manufacturing very large capacity, fuel efficient commercial vehicles, global positioning system/laser-guided bombers, high-technology fighter jets as well as a wide variety of satellites and space vehicles. The aerospace vehicles are mainly classified in the following three categories based on the purpose of each vehicle:

- Commercial airplanes—airplanes used for flying passengers/freight goods in the air from one point to another point in any part of the world and also taking business executives for business meetings using business jets.

FIGURE 1.1
Wright brother's first flight.

- Military aircraft—aircraft used as a bomber, fighter jet, refueling tanker or to transport armies and military vehicles and equipment, gunship helicopters, and many more type of military hardware.
- Spacecraft—spacecraft used for space research and most useful space communication satellite systems. A satellite is an object that moves around, or orbits, a larger object, such as a planet.

This chapter introduces the essential parts of an aircraft, along with their function, underlying aerodynamic principles, and basic structural features.

1.2 Aircraft Basics

To provide a wide-scale overview of an aircraft, the fundamental information about the forces acting on aircraft is presented, followed by brief discussion the aerodynamic principles allowing the aircraft to fly [1].

1.2.1 Forces Acting on Aircraft

Figure 1.2 shows the schematic representation of an aircraft with its major functional components for providing high lift and the flight control system. Three axes with four major forces act on the aircraft. Each axis is perpendicular to other two axes. All three axes intersect at the center of gravity (CG) of the aircraft. The axis which extends lengthwise through the fuselage shown in Figure 1.2 from the nose to the tail is called the longitudinal axis. The axis transverse to the longitudinal from the left to the right wing of the aircraft represents lateral axis. The axis which passes through the fuselage at the CG perpendicular to the longitudinal plane is called the vertical axis. The rotation of aircraft about its longitudinal axis is called roll, rotation about its lateral axis is pitch, and finally rotation

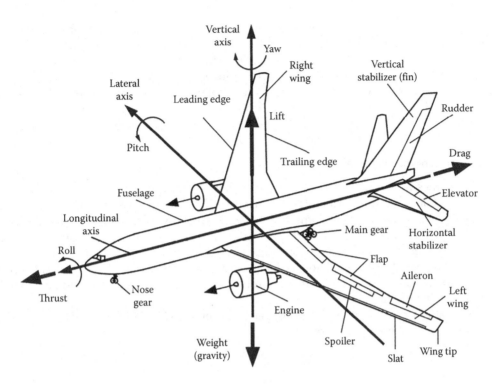

FIGURE 1.2
Aircraft terminology and four major forces.

about its vertical axis is yaw. Those three terms are always precisely handled by the pilot to control the flight at any flying stage.

The four major forces acting on an aircraft during the flight through the air as shown in Figure 1.2 are identified as

1. Weight or gravity force
2. Lift
3. Thrust
4. Drag

Weight or gravity force is directed toward the center of the earth. The magnitude of the total takeoff weight of the aircraft, W_T, is the summation of the following components:

$$W_T = W_G + W_F + W_P \tag{1.1}$$

where W_G is the gross weight of the aircraft, W_F the weight of fuel at the takeoff, and W_P the weight of payload on board (passenger, baggage, freight, etc.).

The weight is actually distributed throughout the aircraft. But it could be considered that W_T is acting through a single point which is the CG of the aircraft. In flight, the aircraft has three axes of rotation as shown in Figure 1.2 about the CG. During the flight, the weight of an aircraft decreases due to fuel burned. As a result, the distribution of weight and the CG also changes; it is required to adjust the flight control devices to keep the aircraft balanced.

Lift and drag discussed in the next section are considered as aerodynamic forces, since the forces are generated due to the movement of the airfoil-shaped wing and stabilizers of an aircraft with the air flow. Thrust is the mechanical force generated by the engines attached normally to the wing (Figure 1.2) to move the aircraft through the air. Thrust opposes drag caused by air resistance to the airplane. During takeoff, thrust must be greater than drag and lift must be greater than weight so that the airplane can become airborne.

1.2.2 Aerodynamic Principles

An aircraft flies in accordance with the aerodynamic principles. In this section, the fundamentals of aerodynamic principles are discussed. Aero is derived from a Greek word meaning air whereas dynamics come from another Greek word meaning power. Air currents are movement of the air with respect to the object. Aircraft wings, horizontal, and vertical stabilizers are built with airfoil-shaped cross sections as shown in Figure 1.3.

Figure 1.4 shows the type of airfoil and its terminology. The wing with an airfoil-shaped cross section used to produce lift force to lift the aircraft from the ground and also to balance the aircraft in flight.

Angle of attack is the acute angle measured between the chord of an airfoil and the relative wind flow as shown in Figure 1.5.

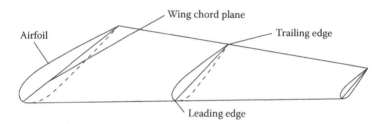

FIGURE 1.3
Airfoil-shaped wing and stabilizer.

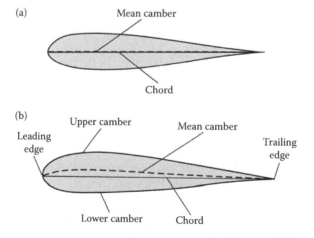

FIGURE 1.4
Type of airfoil and its terminology: (a) symmetric, (b) non-symmetric.

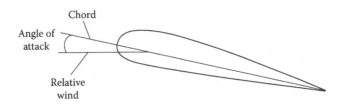

FIGURE 1.5
Angle of attack.

The lift of an aircraft is produced by the differential pressure between the lower pressure created on the upper surface and the higher pressure at the lower surface of the wing. The lift has to be sufficient to overcome the gravity force (total weight of the aircraft) before it leaves the ground during takeoff (Figure 1.6). The special geometry of the aircraft wing (airfoil) is required for the air flow to travel a greater distance and at faster speed on the upper side of the wing to create low-pressure area according to Bernoulli's principle. The principle states that as the air velocity increases, the pressure decreases and when the velocity decreases, the pressure increases as shown in Figure 1.6. The lift force will be equal to the lift pressure multiplied by the area of the wing surface. The magnitude of the lift depends on several factors including:

- Angle of attack
- Lift devices such as slats and flaps
- Surface area of the wing
- Shape of the wing
- Density of the air
- Velocity of the aircraft

As shown in Figure 1.2, drag acts in a direction that is opposite to the motion of an aircraft. Drag is an aerodynamic resistance to the motion of an aircraft through the air. Drag

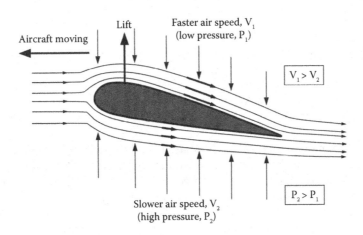

FIGURE 1.6
Lift pressure on the wing.

acts along and opposite to the direction of flight. The following factors affect the magnitude of the drag force:

- Shape and size of an aircraft
- Mass, viscosity, and compressibility of air
- Velocity of the aircraft

The shape of a wing greatly influences the performance of the aircraft, including speed, maneuverability, and its handling qualities. The wing shape includes mainly:

- Straight—small low-speed airplane
- Sweep (backward and forward)—high-speed airplane
- Delta—supersonic airplane (retired Concorde)

When an aircraft flies through the air, the air molecules near the aircraft are physically disturbed and moved around the aircraft. If the aircraft flies at a low speed, typically less than 250 miles per hour (mph), the density of the air remains unchanged. When the speed increases further, some of the energy of the aircraft contributes to compress the air and locally changes the density of the air. The effect becomes more important as speed increases and approaches to near and beyond the speed of sound, about 330 m/s or 760 mph at sea level. The ratio of the speed of the aircraft to the speed of sound in the gas determines the magnitude of many of the compressibility effects. Because of the importance of this speed ratio, aerodynamicists designated it with a special parameter called the Mach number in honor of Ernst Mach in the late nineteenth century as follows:

$$\text{Mach number, m} = \frac{\text{Speed of Aircraft}}{\text{Speed of Sound}}$$

Based on the Mach number, the aircrafts are classified as follows:

- Subsonic, $m \leq 1$
- Transonic, $m = 1$
- Supersonic, $m \geq 1$
- Hypersonic, $m \geq 5$

1.2.3 Major Components of an Aircraft and Their Function

1.2.3.1 Fuselage

The fuselage is the main structural body of an aircraft which carries the crew members, passengers, cargo, instruments, and other essential equipment. The fuselage structures of two commercial aircraft shown in Figure 1.7 could be fabricated by joining a few large sub-assemblies called sections that are built separately with aluminum metal structure shown in Figure 1.7a or more recent development of composite material structure. The fuselage is built in sections to ease production and handling problems. Figure 1.7b shows composite fuselage sections ready to join together. The fuselage has two major compartments. The upper compartment is designated for the passengers, and the lower compartment is

(a)

(b)

FIGURE 1.7
Fuselage structure: (a) Aluminum, (b) composite. (From The Boeing Company.)

structured for cargo storage and other equipment for use by the aircraft. Depending on the model of the aircraft, the fuselage structure is designed and fabricated to meet the need of the customer. The pilot cabin is located to the front of the fuselage.

1.2.3.2 Wing

As shown in Figure 1.4, the wing is an airfoil-shaped cantilever structure attached to each side of the fuselage as shown in Figure 1.2. Wings provide the main lifting surface from the bottom of the wing and support the entire weight of the aircraft in flight. Wing design, size, and shape are determined by the manufacturer for each model of the aircraft to satisfy aerodynamic need of the aircraft. An example of both the wing structures of an aircraft is shown in Figure 1.8.

FIGURE 1.8
Wing structures ready for installation to the fuselage. (From The Boeing Company.)

Based on the type of aerospace vehicle, there are a variety of wing geometries, as viewed from the side and looking onto the wing as shown in Figure 1.9.

The end of the wing is called the wing tip and the total distance from one wing tip to the other wing tip is called the wing span, S_w, as shown in Figure 1.10. The wing area, A_w, is the projected area of the wing viewed from above looking down onto the wing. Aspect ratio is a measure of how long and slender a wing is from tip to tip. The aspect ratio of a

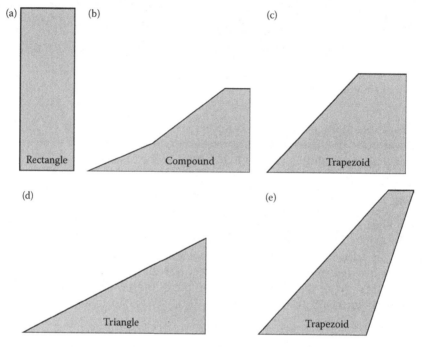

FIGURE 1.9
Wing geometries of various aerospace vehicles: (a) Wright Brothers, (b) Space Shuttle, (c) F-18, (d) Concorde, and (e) commercial airplane.

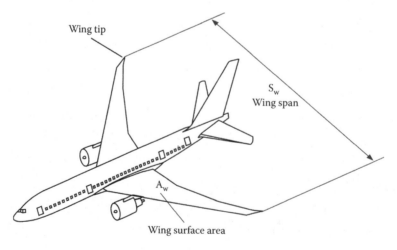

FIGURE 1.10
Wing span of an airplane.

wing is defined to be square of the span (S_w) divided by the projected wing area, A_w. High aspect ratio wings have long span like commercial aircraft, while low aspect ratio wings have short span like military fighter jets.

$$\text{Aspect ratio} = \frac{S_W^2}{A_W}$$

Figure 1.11 shows wing-to-fuselage configurations based on how the wing is attached to the fuselage, above or below the CG of the airplane. Wing dihedral is the upward angle of an airplane wing from the wing root where the wing joins with the fuselage to the tip of the wing as shown in Figure 1.11. The dihedral angle determines the inherent stability of the airplane along the roll axis. Most of the commercial airplanes utilize the low-wing configuration.

On low-wing aircraft, the CG is above the wing and roll stability is less pronounced. This factor requires the use of greater *dihedral* angles in low-wing airplanes (Figure 1.11a). On high-wing aircraft, the CG is below the wing, so less *dihedral* is required (Figure 1.11b). Many military aircraft, including fighter jets and transport planes, are utilizing high-wing design.

The main wing structure is connected to the leading and trailing edge flight control surfaces including slats and flaps as shown in Figure 1.12. In the leading edge some of the aircraft have attached slats. Slats are the movable auxiliary airfoils attached to the

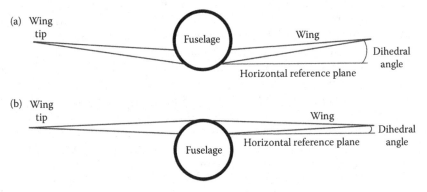

FIGURE 1.11
Wing to fuselage configurations: (a) Low-wing and (b) high-wing.

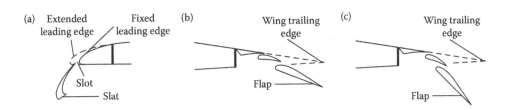

FIGURE 1.12
Example of leading and trailing edge control surfaces of a commercial airplane wing: (a) Leading edge slat (takeoff), (b) trailing edge flap (takeoff), and (c) trailing edge flap (landing).

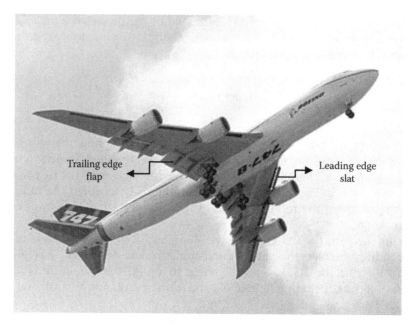

FIGURE 1.13
Wing leading edge slat and trailing edge flap positions during takeoff. (From The Boeing Company.)

leading edge of the wing (Figure 1.12a) of some of the aircraft models. The ailerons are the movable control surface attached to the trailing edge of the outboard end of the wing. In addition to the aileron there are two other flight control devices including spoilers and flaps, which are connected to the trailing edge of the wing as shown in Figure 1.2. In certain aircraft designs, the spoilers are mounted on the upper surface of each wing. Flaps are hinged at the trailing edge of the wing from the inboard end of the aircraft. Flaps are controlled by the pilot according to the need during takeoff and landing of the aircraft. The trailing edge of the wing is hinged with flight control flaps that can be lowered and extended within a definite curvature range as required during takeoff and landing of the aircraft. When lowered, the flaps increase airplane lift by extending the surface area of the airfoil (Figure 1.12b). When the flaps are extended all the way down, it worked like air brake increasing air drag (Figure 1.12c). During taking off, the leading edge is kept open and it forms a slot and increases lift. The slot is an elongated passage through wing to improve the airflow over the wing at high angles of attack. Figure 1.13 shows the position of leading edge slats and the trailing edge flaps during an actual takeoff of an aircraft.

1.2.3.3 Empennage

The empennage includes the entire tail portion of the airplane, consisting of vertical and horizontal stabilizers as shown in Figure 1.2. The rudder is connected with the fixed vertical stabilizer. The rudder is used to control yaw, the horizontal movement of the aircraft. The elevators are connected to the two horizontal stabilizers to control the vertical rotation, pitch of the aircraft. In some aircraft, the entire horizontal stabilizer can be rotated as a complete unit about its own horizontal axis to control the pitch in flight. The empennage of an airplane in flight is shown in Figure 1.14. The function of each component of the wing and the stabilizers are shown in Table 1.1.

FIGURE 1.14
Empennage of 787. (From The Boeing Company.)

TABLE 1.1

Function of Different Components of Wing and Empennage

Wing or Stabilizer Component	Function
Aileron	The aileron can be moved upward or downward to provide roll, the rotation of the aircraft along its longitudinal axis. The ailerons work simultaneously in opposite direction to bank (roll) the aircraft (Figure 1.2).
Elevator	Hinged airfoil section to the trailing edge of the horizontal stabilizer designed to apply a pitching movement to the airplane. A pitching movement is to rotate the airplane about the lateral axis (Figures 1.2 and 1.14), that is nose up or nose down.
Rudder	Hinged or movable auxiliary airfoil-shaped section attached to the vertical fin or stabilizer to apply yawing movement to the airplane so that it turn to the left or right about the vertical axis (Figures 1.2 and 1.14).
Flap	Hinged sections mounted on the trailing edge of the wing and that can be lowered or extended. When lowered, flaps increase airplane lift during takeoff by increasing the camber of the wings (Figures 1.2 and 1.13).
Slat	Movable auxiliary airfoil attached to the leading edge of the wing. When opened, it forms a slot and increases lift by improving the airflow over the wing at high angle of attack. When closed, it forms the total contour of the leading edge of the wing (Figures 1.2 and 1.13).
Spoiler	The function of the spoiler is to spoil or disrupt the smooth flow of air over the wing to reduce the lift force of the wing. This allows the pilot with a means of increasing the decent rate without increasing the aircraft speed. Left and right spoilers can be raised alternately for high-speed lateral control or can be raised together as speed brakes during landing (Figure 1.2).

1.2.3.4 Engine

The aircraft is powered by the engines attached most commonly to the leading edge of both the wings as shown in Figure 1.13. The engines provide the thrust to move the airplane forward after overcoming the total drag of the aircraft. The engines also provide power to other units to generate electricity, hydraulic, and pneumatic power to run instruments and

FIGURE 1.15
Engine nacelle components ready for installation to the wing. (From The Boeing Company.)

move other devices including landing gear, flight control wing surfaces, etc. The nacelle is a cover housing that holds the engine of an aircraft. The design of the nacelle involves both the external shape and the inlet internal geometry. The design of the engine inlet is generally the job of the airframe manufacturer, not the engine manufacturer and is of great importance to the overall efficiency of the aircraft. Figure 1.15 shows the inlet and nacelle components of an engine prior installation.

1.3 Commercial Aircraft Product and History

Owing to business globalization in the world market, the demand for air travel is gradually increasing every year across cities all over the world. As a result, the demand for commercial airplanes for the next 20 years is growing faster all over the world, especially in the Asian and Middle Eastern countries. To meet the customer's demand, aircraft manufacturers are constantly coming up with continuous improvements in aircraft design, materials, and manufacturing technologies. A result is the launching of new airplane models, or introduction of the derivatives of the existing models with increased passenger carrying capacity and/or longer range and better comfort. Commercial aircraft manufacturers are exploring to a large degree the use of high-performance lightweight materials to design and manufacture fuel efficient modern aircraft to remain viable in a competitive aircraft manufacturing business.

Commercial aircraft can be classified according to the following categories:

- Single aisle (narrow body)
- Double aisle (wide body)

Examples of seating arrangements of both single aisle and double aisle aircraft are shown in Figure 1.16.

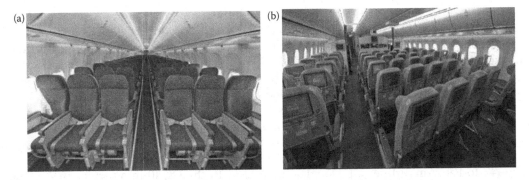

FIGURE 1.16
Examples of aisles in an aircraft: (a) Single aisle, Boeing 737 and (b) double aisle, Boeing 787. (From The Boeing Company.)

In the United States, an aircraft manufacturing company, named The Douglas Company, developed passenger aircraft models DC-1, DC-2 (Figure 1.17), and DC-3, which were built in high numbers throughout the 1930s and 1940s. Thousands of DC-3s were built. During World War II, there was a huge increase in airplane production.

In the 1950s, Douglas and Boeing both planned to develop the first American commercial jet airliner. Boeing took the technological initiative to develop a narrow body, medium size B707 model introduced into service in 1958 by Pan Am. The Boeing 707 was developed from the Boeing 367–80 or Dash–80, a prototype jet aircraft, in 1954. Douglas was not far behind with the DC-8. The DC-8 was put into service in the fall of 1959, offered by United and Delta airlines. In the 1960s, Douglas suffered from the financial burdens from the enormous costs of developing both the DC-8 and DC-9. Boeing introduced a mid-size narrow body three-engine aircraft 727 in February, 1964 first in service with Eastern

FIGURE 1.17
DC-2 from Douglas.

Airlines. In order to ensure a more viable future for the company, Douglas Aircraft merged with military and aerospace manufacturer McDonnell Aircraft in 1967 to become the McDonnell Douglas Corporation. Boeing introduced yet another narrow body short-to-medium range two-engine aircraft the 737–100 in February 1968 with Lufthansa. Boeing continued advancing aircraft technology to come up with the next generation 737NG, numbered 600, 700, 800, and 900 models. Boeing completed first flight on January 29, 2016 of a new 737MAX family of aircraft, and is scheduled to introduce 737 MAX 8 to the launch customer Southwest Airlines in 2017. This family of aircraft will be powered by very efficient new engines and will also feature some modification of airframe structures.

In December 1969, Boeing delivered the first large wide body jet, dubbed a jumbo jet, the 747 to Pan Am airlines [2]. Douglas brought forth the DC-10 into service in August 1971. McDonnell Douglas ended up having to split the domestic airline market for mid-range wide bodies with its competitor, Lockheed, which produced the similar sized L-1011. The first L-1011 went into service for Eastern Airlines in 1972. L-1011 production ended in 1983, ending Lockheed's involvement in the commercial aircraft industry. McDonnell Douglas continued to produce the DC-10 throughout the 1980s, while introducing a new narrow body passenger jet, the MD-80 (modified from the DC 9), which entered service in October 1980 by Swiss Air. Boeing designed its first twin-jet wide body aircraft the B767 (smaller than preceding aircraft such as the 747) and introduced in September, 1982 in service with United Airlines. In January, 1983 Boeing introduced another mid-size narrow body aircraft the 757 in service with Eastern Airlines. The 757-300 was the largest among all the narrow body aircrafts, which was built and could carry from 200 to 289 passengers. The 757 has since been discontinued from production status in 2005. In the 1990s, McDonnell Douglas introduced two new passenger jets—the MD-90, an updated version of the MD-80 was introduced in service by Delta Airlines in 1995, and the MD-11, a larger version of the DC-10 was introduced in December, 1990 flown by Finn Air. Further financial strain in the 1990s took its toll on McDonnell Douglas. Boeing introduced the world's largest twin engine wide body aircraft the 777 in June, 1995 in service with United Airlines. Boeing acquired McDonnell Douglas in 1997, which made the combined company the largest aerospace firm in the world. After the merger, Boeing introduced another narrow body aircraft MD-95 as B717 in October, 1999 to Air Tran for the 100 seat market. The B717 was discontinued after 2006. Boeing officially launched 777X a new series of the Boeing 777 family in November, 2013 and scheduled to be introduced in 2020. The 777X will have two variants; the 777-8 and the 777-9. The 777X will feature new engines, new composite wings, and technologies from the Boeing 787. Boeing's most recent revolutionary composite fuel efficient aircraft the 787-8, introduced into service in October, 2011 with All Nippon Airways (ANA). Later, added two larger stretched versions, 787-9 in service from 2014 and 787-10 is currently in production.

Airbus, a subsidiary of European Aeronautic Defense and Space (EADS) aircraft manufacturing company was founded in the European city Toulouse, France in 1970 in an attempt to compete with American aircraft manufacturers. Airbus introduced its first wide body medium size passenger jet, the A300, in May, 1974 with Air France. The A300's two engines stood in contrast to its three-engine rivals: the Lockheed L-1011, the Douglas DC-10, and the Boeing 727.

After going through tough initial competition with U.S. aircraft manufacturers, in 1978, Airbus broke through. Eastern Airlines agreed to purchase 23 A300s, marking the first major American airline to select an Airbus jet. Airlines around the world soon were placing their orders for Airbus jets, making the company a viable alternative to the U.S. aircraft manufacturers. Over the last couple of decades, Airbus introduced several aircraft models, which were derived from basic design elements developed with the A300. In March, 1988 the single-aisle two engine A320 was introduced in service with Air France to serve short-distance routes. Airbus continued developing new aircraft models, introducing the wide body aircraft A340 in March, 1993 with Lufthansa and the A330 in January, 1994 with Air Inter.

Airbus has now offered the largest commercial airliner ever built, the A380. The A380 exceeds the Boeing 747 model by offering two full-length passenger decks. The A380 was introduced in service in October, 2007 with Singapore Airlines. The most recent development of Airbus is dubbed the extra wide body A350-900, introduced to Qatar Airways in January, 2015. Figure 1.18 shows the most recent Airbus commercial jet models in service.

Similarly, Figure 1.19 shows the most recent Boeing commercial jet models that are in service. Table 1.2 shows some Airbus and Boeing commercial aircraft models and their operating capacities [3,4].

FIGURE 1.18
Airbus commercial aircraft recent models in service: (a) A320, (b) A330–300, (c) A340–600, (d) A380 and (e) A350 XWB. (From commons.wikimedia.org.)

FIGURE 1.19
Boeing commercial aircraft recent models in service: (a) B737–800, (b) B767–300ER, (c) B777–300ER, (d) B787–8 and (e) B747–8I. (From The Boeing Company.)

1.4 Military Aircraft Product and History

Military aircraft can be classified in the following categories:

1. Bomber
2. Fighter
3. Cargo transport
4. Refueling tanker
5. Helicopter or rotorcraft
6. Airborne early warning (AEW) and control

TABLE 1.2

Operating Capacities of Most Recent Commercial Aircraft Models in Service from Airbus and Boeing

Model	Aisle	Passenger Capacity (2–3 Class Config.)	Maximum Takeoff Weight, lb (kg)	Nautical Distance, Nautical Miles (km)	Maximum Fuel Capacity, U.S. Gallon (L)	Cruise Speed (35,000 ft) Mach
A320	Single	150	162,000 (73,500)	3294 (6100)	6395 (24,210)	0.82
A330–300	Double	295	507,000 (230,000)	6425 (11,900)	25,764 (97,530)	0.86
A340–600	Double	380	811,300 (368,000)	7883 (14,600)	51,650 (195,520)	0.86
A350–800	Double	270	546,700 (248,000)	8504 (15,750)	36,455 (138,000)	0.89
A380	Double	525	1,234,600 (560,000)	8477 (15,700)	84,535 (320,000)	0.89
B737–800	Single	162	174,200 (79,010)	3115 (5,765)	6875 26,020	0.785
B767–300ER	Double	218	412,000 (186,880)	5990 (11,070)	23,980 90,770	0.80
B787–8	Double	210	502,500 (227,930)	8200 (15,200)	Not Available	0.85
B777–300ER	Double	365	775,000 (351,530)	7930 (14,685)	47,890 (181,280)	0.84
B747–8I	Double	467	987,000 (447,700)	8000 (14,815)	64,055 (242,470)	0.855

Note: A: Airbus, B: Boeing.

In May, 1944, the B-29 Superfortress, a four-engine propeller-driven heavy bomber, was introduced by Boeing to the United States Air Force (USAF). In 1946, the United States Army Air Forces used B-29s in World War II. Two years later, in 1948 after World War II, the B-50, and a revision of the B-29 was introduced. In 1949, the B-36 "Peacemaker," a strategic bomber, was introduced by Convair. In 1951, six jet-engine B-47 medium size bombers were introduced by Boeing. A revolutionary aircraft was introduced by Boeing in 1955, a subsonic jet-powered strategic bomber called the B-52. Further additions to the U.S. bomber fleet included the supersonic B-58 Hustler, designed and built by Convair and introduced in 1960. In 1967, General Dynamics introduced the F-111 medium range tactical strike bomber to the USAF and the Royal Australian Air Force.

Rockwell developed the B-1, Lancer; a four-engine variable-sweep wing strategic bomber to replace the Boeing B-52 bomber in the early 1970s. Due to the change of needs by the USAF, the new version of the B-1 became the B-1B. The first operational B1-B bomber joined the service in 1986. The main purpose of the B-1B was to focus on low-level penetration bombing using stealth technology. Stealth technology provides a unique ability to penetrate an enemy's most sophisticated defenses and threaten its most valued, and heavily defended, targets. Its capability to penetrate air defenses and threaten effective retaliation provides an effective deterrent and combat force well into the twenty-first century. The B-1B has supported U.S. and NATO military forces in both the Iraq and Afghanistan wars. The B-2 Spirit, a Stealth bomber, was developed by Northrop Grumman and introduced to the USAF in April, 1997. Along with the B-52 and B-1B, the B-2 provides the penetrating

FIGURE 1.20
Examples of some earlier Bombers in operation: (a) B-52 Bomber and (b) B-1B Bomber. (From The Boeing Company.)

flexibility and effectiveness inherent in manned bombers. Figure 1.20 shows examples of a couple of earlier bombers which are still in operation.

Among all the fighter jets developed in the world, the F-14 was the first of the American "teen-series" fighters designed by Grumman to compete against Russian-made MIG fighters during the Vietnam War. The F-14 first flew in 1970, and was introduced in 1974 to the U.S. Navy. McDonnell Douglas (currently Boeing) introduced a twin engine all-weather tactical fighter, the F-15 Eagle in January, 1976. Later, the F-15E was developed and entered in service in 1989. General Dynamics (currently Lockheed Corporation) came up with the F-16 Falcon all-weather multirole fighter jet in 1978. Over 4500 fighter jets were produced since the first flight in 1974. The F-18 "Hornet" is a supersonic, single seat twin engine, all-weather, night, combined fighter and attack aircraft, which entered into service in January, 1983. The F/A-18 multi-mission aircraft can operate from either aircraft carriers or land bases. The resilient F/A-18 Hornet was the first aircraft to have carbon fiber wings and the first tactical jet fighter to use digital fly-by-wire flight controls. Compared to the Hornet, the Super Hornet is larger, heavier, and has improved range and payload. The Boeing Super Hornet F/A-18E/F was originally proposed as an alternative to an all-new aircraft to replace existing dedicated attack aircraft such as the A-6. The larger variant was also directed to replace the aging F-14, thus serving a complementary role with Hornets in the U.S. Navy, and serving a wider range of roles including use as a refueling tanker. Figure 1.21 shows a couple of recent fighter jets in service by the USAF and other nations.

FIGURE 1.21
Recent fighter jets: (a) F-15 fighter jet and (b) F/A-18 fighter jet. (From The Boeing Company.)

In the category of fixed-wing military transport aircraft, many varieties of transport models have been introduced, starting with the C-47 in the 1940s, to the most recent popular Boeing C-17, and also including the new A400M, in flight testing by Airbus. Among all the transport aircrafts to have been built so far, the Lockheed C-130, a four-engine turboprop military transport aircraft introduced with USAW in December, 1957, has been used the most. The C-130 was originally designed for transporting troops, medical evacuation and cargo transport. Later Lockheed has developed many variants of the C-130 to support a variety of other roles including gunship, airborne assault, search and rescue, aerial refueling, maritime patrol and aerial firefighting, and many more. Over 2300 different variants of C-130 aircraft have been built so far. The Boeing made C-17 is the newest and the most efficient airlift aircraft to enter the USAF inventory since 1993. The C-17 is also operated by many other nations, including the United Kingdom, Australia, Canada, Qatar, the UAE, and most recently, India. C-17 is a four-engine turbofan aircraft capable of airlifting large payloads over long intercontinental ranges without refueling. Its design is intended to allow delivery of outsize combat cargo and equipment directly into airfields, strategic airlift, airdrop, and many more special operations missions. A couple of model of transport aircraft are shown in Figure 1.22.

An aerial refueling, also called air refueling, is the process of transferring fuel from one aircraft (tanker) to another aircraft (receiver) during flight. Tankers play a major role in providing combat support air refueling during war time operations. Air refueling has major advantages since the receiver aircraft will be receiving fuel while still in the air, such as:

- Take off with greater payload
- Shorter take off roll
- Reduced fuel consumption on long distance flights

There are two main refueling systems including probe-and-drogue and the flying boom being fitted to existing aircraft. The KC-97, aerial refueling tanker, based on the Boeing C-97, was introduced by the USAF in 1950. Over 800 KC-97 were built before it was retired in 1978. The Lockheed AC-130 is also used by the USAF for aerial refueling. The KC-135 was originally derived from the Boeing 367–80, which was later developed to become the

FIGURE 1.22
Military cargo transport aircrafts: (a) Lockheed C130, transport and (b) Boeing C17, transport. (From The Lockheed Martin and the Boeing Company.)

Boeing 707. The KC-135 has a similar outward appearance to the B707. The KC-135 was introduced in 1957 to the USAF. Little over 800 KC-135s were built till 1965. Now the KC-135 has been supplemented by the larger version KC-10. The McDonnell Douglas (currently Boeing) KC-10 tanker is the military version of the three-engine DC-10 airliner and was introduced in 1981 to the USAF. A total of 60 KC-10s were built till 1987. The KC-46 is the latest aerial refueling tanker, derived from the B767 jetliner. In 2011, Boeing received a contract from the USAF to build the KC-46, which will replace it, the oldest KC-135s. Figure 1.23 shows a couple of models of aerial refueling tankers.

A helicopter (or also called rotorcraft) is an aircraft whose rotors are driven by engines to allow the helicopter to take off vertically as well as land vertically, along with other forward, backward, and lateral movement with hover. Helicopters have several different basic configurations, with one or more number of rotors. Figure 1.24 shows a couple of very popular rotor craft widely used by U.S. Army. The Chinook was initially designed and built by Vertol in the early 1960s, and was introduced to the U.S. Army in 1962. Over 1200 Chinook rotor craft have been built so far, and are still in production for the Army. The AH-64 Apache is an all-weather day–night military attack helicopter with a four-bladed main and tail rotor, and a crew of two pilots who sit in tandem. The Apache was first designed and manufactured by Hughes Helicopters in 1975, and entered in U.S. Army

FIGURE 1.23
Refueling tankers: (a) KC 10 refueling tanker and (b) KC 767 refueling tanker. (From The Boeing Company.)

FIGURE 1.24
Example of helicopter or rotor craft: (a) CH-47, Chinook and (b) AH-64, Apache. (From The Boeing Company.)

FIGURE 1.25
AEW vehicles: (a) 737 Wedgetail and (b) 767 AWACS. (From The Boeing Company.)

service in 1986. McDonnell Douglas purchased Hughes Helicopters and continued the development of the AH-64, resulting in the AH-64D Apache Longbow, which is currently produced by Boeing. Over 1200 Apaches have been built so far since 1983, and are still in production for the U.S. Army.

AEW is the detection of enemy air or surface units by radar or other equipment carried in an airborne vehicle, and the transmitting of a warning to friendly units. Airborne early warning and control system (AWACS) is air surveillance at a high altitude, and control is provided by AEW aircraft. These aircrafts are equipped with search and height-finding radar and communications equipment to detect aircraft, ships, and vehicles at long ranges, and provide control and command of the battle space in an air engagement by directing fighter and attack aircraft strikes. The E-3 Sentry (AWACS) was developed by Boeing in the early 1970s and was introduced to the USAF in 1977. The E-3 was derived from the Boeing 707. A total of 68 aircraft were built until 1992 for the USAF, NATO, the Royal Air Force, the French Air Force, and the Royal Saudi Air Force. Figure 1.25 shows the most recent Wedgetail, derived from the Boeing 737, and an AWACS derived from the Boeing 767.

1.5 Space Craft Product and History

A spacecraft is a vehicle designed to fly in space. Spacecrafts are generally used for a variety of purposes, including earth observation, communications, meteorology, navigation, planetary exploration, and transportation of humans and cargo to outer space. Sputnik 1 was the first artificial satellite launched by Soviet Russia in 1957. The first manned spacecraft was Vostok 1 that carried a Soviet Cosmonaut into space in 1961, and completed a full earth orbit. This was followed by Freedom 7 carrying an American astronaut, Alan Shepard. After that there were five other manned missions carried out using Mercury spacecraft. Reusable space vehicles, designed for carrying man to space, are often called space planes. The first example was the North American X-15 space plane which conducted a couple of manned flights and reached a height of over 100 km in the 1960s. The X-15 was the first air-launched on a suborbital trajectory in 1963.

The first partially reusable orbital non-capsule winged spacecraft called the space shuttle was launched by the United States by NASA in 1981. NASA's Space Shuttle, officially called the Space Transportation System, is a space craft used for human space flight missions, and operated by the U.S. government. The reusable shuttle enables regularly scheduled transportation for people and cargo between Earth and low Earth orbit for space research. Five space shuttles including the Columbia, Challenger, Discovery, Atlantis, and Endeavour have travelled to and from the space. Columbia was the first to go into the space. There was a tragedy involving the Columbia, which disintegrated during reentry in February, 2003. The space shuttle was finally retired in 2011 due to aging vehicles, and the very high cost of each mission. Figure 1.26a shows a space shuttle in the launch pad ready for lift off. Prior to launch (Figure 1.26b), the entire weight of the space shuttle is supported on the launch pad

FIGURE 1.26
Journey of space shuttle: (a) At launch pad (1. External fuel tank, 2. Booster rocket, and 3. Shuttle), (b) launch, (c) shuttle at international space station, (d) shuttle reenters and lands and (e) shuttle moved to Kennedy Space Center. (From The Boeing Company.)

by two solid booster rockets along with external tank. After deploying the non-reusable main tank and two reusable booster rockets, the payload of the spacecraft is carried by the shuttle, operating the onboard scientific instruments, and working in the international space station (Figure 1.26c). After the mission, the shuttle reenters the atmosphere and lands (Figure 1.26d), normally at either Kennedy Space Center in Florida or Edwards Air Force Base in California. When the shuttle lands in California, the shuttle rides atop a specially modified Boeing 747 aircraft back to Kennedy Space Center (Figure 1.26e) completing the end of a mission of a space shuttle.

A couple of other manned Soviet spacecraft included the Voskhod and Soyuz. Similarly, other manned American spacecraft included Gemini, Apollo, the Skylab space station, and the Space Shuttle. The international space station has been manned since November, 2000, and is a joint venture among Russia, the United States, Canada, and several other countries. China recently developed the Shenzhou space craft and completed its first mission in 2003. The role of transporting humans to space by the shuttle by NASA is planned to be replaced by the partially reusable Crew Exploration Vehicle in the near future. Heavy cargo transport to the space station is expected to be replaced by another launch vehicle, which could be derived from the Space Shuttle.

A satellite (Figure 1.27) is a space vehicle that goes around, or orbits, a large object, such as a planet. There are hundreds of satellites that are orbiting the Earth on a regular basis since; Russia launched its first satellite "Sputnik." Since then many countries developed their own satellites for their specific needs or applications. Today, satellites have become common place tools of technology. Most satellites serve one or more functions:

1. Communications
2. Navigation

FIGURE 1.27
Example of a satellite. (From The Boeing Company.)

3. Weather forecasting

4. Environmental monitoring

5. Manned platforms

There are three major components of satellite that including:

- Communication devices (antennas, radio receivers, and transmitters)
- Power sources
- Control systems

Communication devices enable the satellite to communicate messages with one or more ground stations, called the command centers. Messages sent to the satellite from a ground station are uplinked, whereas messages sent from the satellites to the earth are down linked.

References

1. www.grc.nasa.gov
2. Bauer, E.E., *BOEING: The First Century*, TABA Publishing, Inc., WA, 2000.
3. www.airbus.com
4. www.boeing.com

2

Fundamentals of Building an Aircraft

2.1 Introduction

Continuing development of new materials and state-of-the-art manufacturing technologies for building structural components of aircraft are being applied to new aircraft design. Among other desirable features, this produces more fuel-efficient commercial aircraft entering into service for airline customers across the world for the transport of passengers and cargo from point to point or hub to hub. For each new aircraft, every new material needs to be tested, modeled, and analyzed before being considered in the design configurations. Testing is conducted to establish the performance of new materials in order to analyze how well they satisfy all the design requirements. Those requirements often include simultaneously making the aircraft more fuel efficient, cost effective, safe, reliable, and comfortable for passengers.

The aircraft design process starts with the feasibility phase—the development of an idea for the project. This phase is heavily dependent on marketing requirements to identify the concept and specific features. Sincere attention and efforts are then applied to the concept and definition phase to minimize overall design costs and development risks. The aircraft design process integrates input from various disciplines including aerodynamics, payload, structure, materials and process, flight control, research and technology, etc. The aerodynamic definition of the aircraft will determine the least amount of air drag to make the aircraft more fuel efficient. Improved materials will also help to reduce the total weight of the aircraft, and improve fuel efficiency as well.

After the geometric definition and the final shape of the aircraft is determined from the aerodynamic analysis, the material and process technology and structural group will assess the design to assure alignment with the best available light weight and high strength materials to make the airplane fuel efficient. The next step is the selection of the right type engine by the jet propulsion group to provide the required thrust designed for the aircraft model. The engine manufacturers work together with the aircraft manufacturer for matching the needed engine performance by running series of experiments on the aircraft manufacturer's test bed. The goal is to reach the maximum performance parameters, including maximum thrust, speed, and finally the most fuel-efficient airplane.

Once the concept design is complete, the next phase starts with detailed design of major aircraft components including fuselage, wing, empennage, engine, and their individual elements within the individual major components. This step is followed by materials selection and their producibility reviews, which determine the cost of manufacturing while maintaining the highest quality and reliability of the product. The primary goal is to investigate the initial concept design as a whole and come up with one geometric model of the aircraft that is completely analyzed by each of the internal divisions including structural, stress, payload, flight control, materials, and manufacturing. The aircraft manufacturing company is also updated with feedback from airline customers and outside partner companies who will produce the parts and assembly of the aircraft components. In this development process, the aircraft is broken into a few distinct major components, and those components are successively divided into subassemblies until the derived aircraft parts can be designated. From this level, manufacturing parameters will be included in the part drawing for manufacturing and assembly of the components. The final shape of the aircraft will be subjected to structural tests followed by functional and qualification tests. Figure 2.1 shows an outline of different phases involved in building an aircraft, starting from market requirements to the final delivery of an aircraft to the customer. This chapter discusses the function of each phase related to building aircraft.

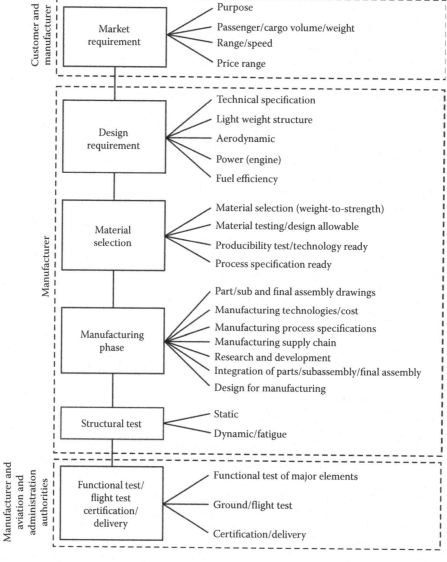

FIGURE 2.1
Various phases of building an aircraft.

2.2 Market Requirement

Market study, competition, and economical requirements demand that the aerospace industry continuously explore new materials and new manufacturing technologies. Aircraft manufacturers constantly look for new and proven technologies available in the market, and also stay in a research made for improving of manufacturing aircraft to satisfy customer needs. The customer needs consist of the following:

- Flying higher altitude—more comfort during flying—avoiding rough weather
- Higher cruise speed—shorter flight duration
- Longer range
- Less noise
- Low environmental impact
- Fuel efficiency

Market requirement analysis is the most critical phase of aircraft manufacturing. The major requirements include the purpose of the aircraft, number of passengers, weight of cargo, range, speed, and cost of manufacturing. The purpose of the aircraft at the highest level would be the use of the aircraft either for commercial or for military applications.

The total take-off weight of a commercial airplane will include the summation of gross weight of the airplane including fuel, total passenger capacity, and the total cargo weight. The overall size of an aircraft is a function of total number of passengers, and the cargo carrying capacity. The current world's largest in-service commercial aircraft, the A380–800, can carry 525 passengers in standard three-class configurations as compared to a maximum of 476 passengers carried by the second largest, B747-8I.

Flying range of an aircraft as shown by a line diagram in Figure 2.2 is a very important factor. It dictates the ability of an airline customer to fly passengers from point A to B or hub A to B, as determined by the airlines, based on the market study of travel flow of passengers. Any airport used should have sufficient length of runway to allow the aircraft to taxi and take off safely within the length of runway. Similarly, the same condition to be fulfilled for the landing procedure of the aircraft to have sufficient runway length from touching down and taxiing, to the final stop of the aircraft at the gate.

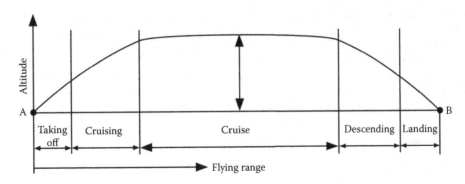

FIGURE 2.2
Altitude versus flying range.

The flying speed of an aircraft is an important design parameter to determine the flying time required to reach a destination. Any saving of flying time is a very big benefit for a passenger for any personal or business travel. Commercial airplanes for both passengers and cargo normally operate in the subsonic range (Mach number varies between 0.84 and 0.89).

2.3 Design Requirement

Technical specification of an aircraft is derived based on the marketing research performed with the world airline customers, before the designers can start focusing on determining the required shape and geometry of the aircraft. In the technical specifications the following factors are being considered by the aircraft manufacturers after going through series of reviews with many different airlines all over the world to satisfy their requirements:

- Number of passengers with different class configurations
- Cargo volume
- Engines maximum thrust
- Maximum fuel capacity
- Maximum take-off weight
- Maximum range (nautical distance)
- Typical cruise speed
- Basic dimensions (Figure 2.3)
  - Exterior wing span
  - Overall length
  - Tail height
  - Interior cabin width

The basic dimensions of some current models of two major aircraft manufacturers are shown in Table 2.1 [1,2]. The length and span of an aircraft is very much limited by the airport configuration, so the aircraft can get enough space to maneuver the aircraft in taxiing from the gate to the runway, and from runway to the gate.

The basic structural circumferential geometry of the fuselage and the wing span is mainly a function of total gross weight at take-off. When the circumferential geometry of the aircraft is large to allow for higher passenger and cargo carrying capacity, the aircraft is made wide body with a double aisle configuration, instead of a single aisle narrow body configuration.

2.4 Materials Selection

Materials selection for designing and manufacturing an efficient aircraft is one of the major challenges for aircraft manufacturers. R&D organizations continue research on various

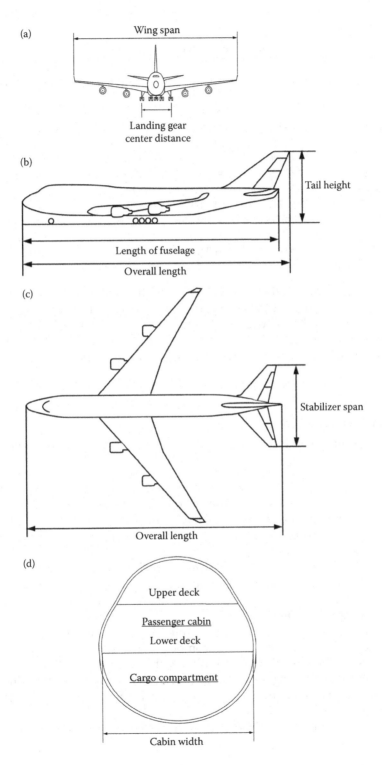

FIGURE 2.3
Example of basic dimensions of an aircraft: (a) Exterior wing span, (b) tail height, (c) overall length, and (d) interior cabin width.

TABLE 2.1

Basic Dimensions of Some Commercial Aircraft Models

Aircraft Model	Overall Length m (ft/in)	Wing Span m (ft/in)	Tail Height m (ft/in)	Interior Cabin Width m (ft/in)
A380	72.72 (238/7)	79.75 (261/8)	24.09 (79)	6.54 (21/6)
A340-600	75.36 (247/3)	63.45 (208/2)	17.22 (56/6)	5.28 (17/4)
A350-800	60.54 (198/7)	64.75 (212/5)	17.05 (55/11)	5.61 (18/5)
A330-300	63.69 (208/11)	60.30 (197/10)	16.83 (55/3)	5.28 (17/4)
A320	37.57 (123/3)	34.10 (111/11)	11.76 (38/7)	3.70 (12/2)
B747-8I	76.3 (250/4)	68.5 (224/9)	19.4 (63/8)	6.1 (20)
B777-300ER	73.9 (242/5)	64.8 (212/7)	18.5 (60/8)	5.86 (19/3)
B787-8	57 (187)	60 (196/10)	17 (56)	5.74 (18/10)
B767-300ER	54.9 (180/1)	47.6 (156/2)	15.8 (52)	4.7 (15/5)
B737-800	39.5 (129/7)	34.3 (112/6)	12.5 (41)	3.53 (11/7)

material technologies, especially determining the high strength and light weight material for major structural components, which will make the aircraft more efficient and attractive to the customers. Material properties need to satisfy the basic structural requirements and justify the weight-to-strength relationship to make it more fuel efficient. Materials selection for the design of aerospace structure depends on major critical requirements:

- Environmental resistance (temperature and corrosion)
- Stiffness
- Static and dynamic loads
- Durability (fatigue and damage tolerance including crack growth and residual stress)

Since the early 1920, the airframe structures have been built mainly with high strength aluminum alloys (2xxx, 7xxx series). The overall compositions started changing when high-performance polymer matrix composites were developed during mid-1960s and early 1970s. Military aircrafts were the first applications of composites materials. The earliest aircraft usages of composites were on the empennages of the F-14 and F-15 military fighter jets in 1975. The usage of composites, mainly carbon/epoxy, continuously increases in military applications from 2% to 25% approximately by the early 1980s. The areas of application include wing, forward fuselage and the horizontal stabilizer, with a typical weight savings of approximately 20%.

In the commercial aircraft industry, the application of composites has accelerated in pace. Airbus started earlier (1972), using composite materials for horizontal stabilizers and vertical fins for their A300 series aircrafts with great success. However, Boeing has recently made significant changes in the commercial aviation industry by using more than 50% composite materials in their new light weight fuel efficient 787 series airplanes rolled out in 2007 (Figure 2.4).

Major metals and materials used to manufacture aircrafts are as follows:

- Aluminum
- Magnesium
- Titanium

FIGURE 2.4
First commercial airplane with major composite structures. (From The Boeing Company.)

- High strength steels
- Super alloys (nickel, iron-nickel, and cobalt based)
- Composites (polymer, ceramic, and metal matrix)

Major structural metal components are made using various metal product forms including sheet, plate, extrusion, and forging. The fabrication costs of many lighter weight structural materials are more expensive as compared to aluminum, the most popular aircraft metal. The various product forms of input raw materials and also material breakdown for some commercial and military aircraft are discussed in Chapter 3.

2.5 Manufacturing Phase

Before making a plan for manufacturing aircraft components, the detailed design parameters defined engineering drawings of each part of the aircraft need to be ready for fabrication. In the first phase of the manufacturing process, the aircraft is divided into major aircraft components including:

- Fuselage
- Wing
- Wing box
- Stabilizer
- Engine
- Landing gear

Each major component consists of many parts and subassemblies, which are described with useful parameters provided by the design group. The above-mentioned major components are then divided into different parameter groups:

- Outer structure
- Inner structure
- Operating system
- Interior and exterior installations

Material selection for the efficient design for manufacturing is directly related to manufacturing technologies connected with research and development. Both material selection and manufacturing technologies are required to make individual part components for many detail parts, subassemblies, and final assembly. Manufacturing of individual parts and small assemblies of major components are done within the manufacturing facilities of the aircraft manufacturers. Some are also off-loaded to the outside suppliers or partners based on their technical expertise, and the relative cost of manufacturing. Due to business globalization, major large components of the aircraft are being manufactured by outside partners from all over the world. The aircraft manufacturers do the final integration and assembly in their own facility before the aircraft goes for flight test. Various materials and manufacturing technologies are involved in fabrication of aircraft components to build an aircraft. It starts with the design and drawing of the individual part. The design engineer will assign the proper material selection and manufacturing processes including the final finishes on the part.

The price of an aircraft offered to an airline customer may be calculated on the basis of total manufacturing cost as noted in the flow diagram as shown in Figure 2.5 plus some level of profit. The total manufacturing cost is associated with major cost components including:

- Design
  - Materials
  - Manufacturing technologies
  - R&D for both materials and manufacturing technologies
- Part components fabrication
- Subassemblies
- Total assembly

Cost of manufacturing individual aircraft parts is a very important factor toward the total manufacturing cost of an aircraft. Total cost of manufacturing can be optimized by introducing lean activities at each stage of manufacturing, starting from each part component to the total (or final) assembly of the aircraft. Lean practices are very powerful tools to apply in any manufacturing environment, and each functional organization is required to establish manufacturing targets and plans. Each manufacturing unit may apply lean activities to help meet these plans, providing tools for streamlining processes, reducing flow time, and distributing resources efficiently to enable continuous improvements in quality and productivity. Costs can be broken into direct and indirect categories. Direct costs are cost of materials and labor, and cost of

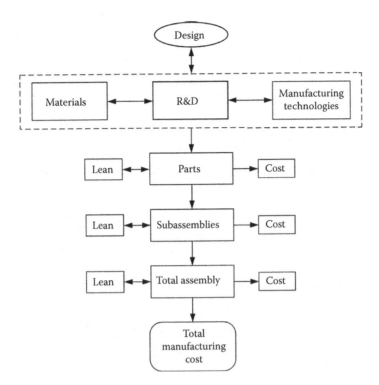

FIGURE 2.5
Cost of manufacturing.

manufacturing. Indirect costs include overhead, shipping, and handling of part components and subassemblies from the suppliers to the final assembly location. The base price of the aircraft fundamentally varies with the aircraft model because of its size and shape. Price further varies with different options and configurations requested by the airlines customers. Configurations and options include the choice of engine manufacturer, interiors, avionics, and more.

Cost is involved in each block, starting from the design to the final integration of the subassembly, and final assembly of the aircraft. The total manufacturing cost of the airplane is the integration of costs at each part level. It is essential to optimize the price of each part in the aircraft. Any variation of price in manufacturing of any aircraft part based on a particular design input and choice of material will add up quickly when assembling millions of parts to make the final assembly of an aircraft.

Design engineers may not be always aware of the manufacturing technologies available for fabrication of the parts. There should be very consistent understanding between the design engineers, material engineers, and the manufacturing engineers to manufacture the aircraft parts in the most efficient manner, including minimum cost, along with the highest quality to meet the need of the airlines customer. For successful part design and manufacturing, there should be a direct feedback system for the three major elements including part design, material selection, and manufacturing technologies applied toward the quality and final cost of the part. Figure 2.6 shows a design for manufacturing concept.

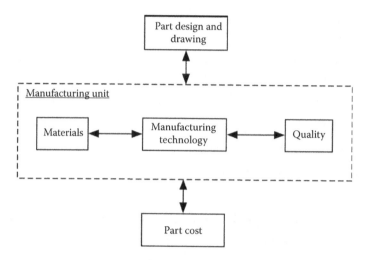

FIGURE 2.6
Design for manufacturing concept.

2.6 Structural Test

During the material selection process, considerable testing was carried out to determine the design allowable of the individual materials used for structural elements of the large aircraft structure. The structural material forms used include mainly sheet, plate, extrusion, forging, and tubular product shapes. The aircraft structure is designed and built in such a way that the aircraft should be able to withstand (greater than) all the major forces acting on the aircraft in normal flight. The major forces shown in Figure 2.7 are listed below:

W_G = total gross weight

F_L = lift force

F_T = thrust

F_D = drag force

R_W = vertical reaction at the wing tip (L stands for left and R stands for right)

T_W = torque at the wing

M_W = bending moment at the wing tip

R_S = vertical reaction at the stabilizer tip

T_S = torque at the stabilizer

M_S = bending moment at the stabilizer tip

Based on the dynamic loading on the aircraft in flight, some simulated structural static and dynamic tests are required to complete as outlined in Figure 2.8, from coupon and component level testing to subassembly and full-scale testing.

The purpose of the structural tests is to validate the internal state of stress and associated strain predictions, validate deflections, and validate the models that were used

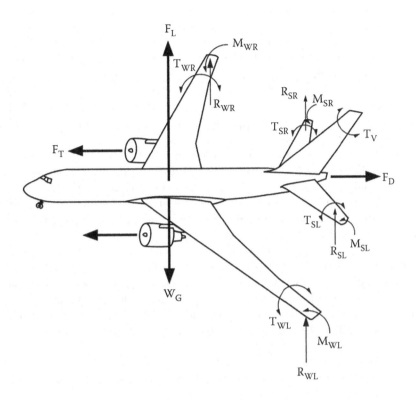

FIGURE 2.7
Forces acting on an aircraft in flight.

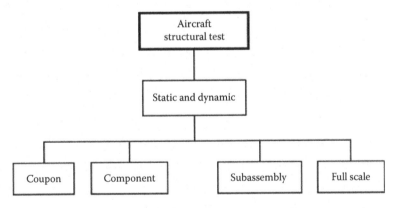

FIGURE 2.8
Fundamentals of structural test.

during designing the aircraft. Basically, there are three types of tests designed for various components and finally on the complete aircraft:

- Static
- Fatigue
- Rupture

Test plans are coordinated closely with the Federal Aviation Administration (FAA) to meet all the design and safety requirements, before the aircraft is ready for the flight tests.

Based on the nature of the cyclic loading system in the aircraft for each take-off and landing for each flight cycle, fatigue tests are very critical from coupon to full-scale testing. One important fatigue test is conducted by pressurizing and depressurizing the aircraft to simulate take-off and landing cycles. Fatigue cycle test programs are carried out for a three-year period. "A Building block approach" is used to test each area of the airplane.

A high-pressure test also known as a "high blow" test is being conducted as part of the static airframe test. This test is one of the three static tests that must be completed prior to the first flight of a new aircraft. As seen during pressurization of an aircraft in service, the internal pressure reaches the level of around 15 psi, which is about 1.5 times normal pressurization during flight. When the pressure is slowly increased into the fuselage (flying pressure vessel), the fuselage tends to expand, so it is very important to ensure the integrity of the aircraft fuselage structure.

According to aircraft inspection regulations, every aircraft needs to be inspected after a certain number of years (FAA regulations, 14 years), to look out for any fatigue cracks in the fuselage. These begin as small cracks that develop on a plane's fuselage components over time. The cracks are caused by the stress resulting from repeated pressurization and depressurization of the aircraft. Pressurization causes the fuselage to expand as oxygen is pumped into the cabin during flight to support the comfort of the crew and passengers. The aircraft compresses upon returning to ground level as the exterior atmospheric pressure increases.

For uniform internal pressure p applied to a cylindrical pressure vessel shown in Figure 2.9a, the membrane stresses σ_1 and σ_2 (Figure 2.9b) for the cylinder radius, r, and the wall thickness, t, can be calculated as [3].

Uniform longitudinal stress

$$\sigma_1 = \frac{pr}{2t} \tag{2.1}$$

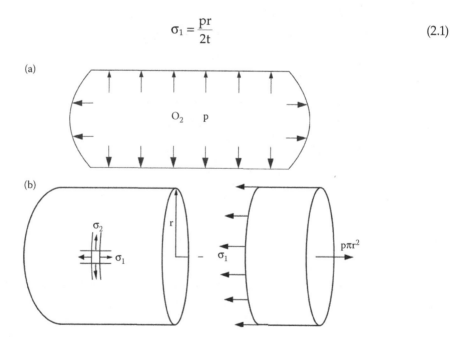

FIGURE 2.9
Fundamentals of pressurizing the cylindrical pressure vessel: (a) Pressurizing with oxygen and (b) principal stresses.

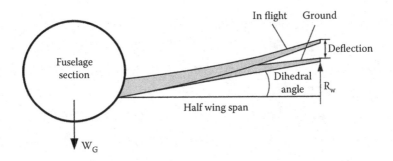

FIGURE 2.10
Deflection of wing in flight.

and hoop stress

$$\sigma_2 = \frac{pr}{t} \tag{2.2}$$

Dihedral angle is the upward angle of the aircraft wing from the wing root to the wing tip as shown in Figure 2.10. Most commercial aircraft wings are designed with some amount of dihedral angle. When the aircraft takes off from the runway rising into the sky, the wings flex or deflect by several feet along the wing span. Figure 2.10 shows schematically the deflection of the wing in flight. The amount of deflection depends on a number of factors, mainly including:

- Gross weight of the aircraft, W_G
- Gravity loads on the wing
- Wing span
- Altitude of the aircraft

As required by FAA, all commercial aircraft must be able to withstand for at least 3 s 150% of the expected maximum load the aircraft is ever expected to encounter during normal operation. A structural loading test is used to demonstrate a safety margin for the design and is part of the certification process to show the aircraft can withstand extreme loads. Figure 2.11 shows the wing ultimate up-bending tests performed on one of the Boeing aircraft models. Ultimate wing load testing is a standard procedure for any new aircraft design. The test structures shown in the figure flex the wings to apply the necessary loads.

2.7 Functional Test/Flight Test/Certification/Delivery

The final phase of building an aircraft is the series of functional tests designed by the manufacturer, with the agreement with the government aviation authorities, to make sure the airplane is ready for flight testing.

FIGURE 2.11
Static test of maximum wing deflection. (From The Boeing Company.)

Taxi testing is one of the final steps before the new airplane makes its first flight. The taxi test provides data about the airplane's braking performance and handling characteristics on the ground, and ensures all systems perform as designed while the airplane is in motion. During the flight tests there are certain requirements planned by the aviation authorities required to be fulfilled before the airplane gets its certification for delivery to the airline customers to commence services with passengers.

Some of the flight tests include:

- Fuel consumption
- Reliability
- Directional stability control functions
- Interior systems including heating, venting and air conditioning, smoke detections, and galleys

References

1. www.airbus.com
2. www.boeing.com
3. Timoshenko, S. and Young, D.H., *Elements of Strength of Materials*, 5th Edition, Van Nostrand Reinhold Co., New York, NY, 1968.

3

Major Aircraft Materials and Its Classification

3.1 Introduction

There has been a long history of development of materials for aircraft manufacturing since the first aircraft was developed in the early 1900s by the Wright Brothers. The choice of aircraft materials available has changed since World War II when aluminum alloys were exclusively used for the body frame, along with some steel in selected areas where strength and stiffness was required; for example, in the landing gear and engine support structural components. In 1950, titanium was introduced for aerospace applications, when extraction of titanium from its ore had been economically developed for production. Titanium has now become a key material for both airframe and engine structures, due to its high specific strength and corrosion resistance compared to aluminum alloys. During the early to mid-1960s, composite materials with boron-based fibers impregnated with various polymers were introduced into aerospace applications. The boron fiber composites were costly, and difficult to machine. Composite technologies continued to develop successfully into the present day carbon fiber composites, commonly referred to as carbon fiber reinforced plastic (CFRP). The composite manufacturing industries are constantly working on various matrices with graphite fibers, along with the process technologies to produce cost effective and more attractive and efficient materials with lower density, high strength and stiffness, and excellent fatigue characteristics. Thermoset and thermoplastic epoxy-based resin systems have widely been used in making various aircraft components satisfying the engineering requirements for these applications.

Major aircraft materials are classified according to their specific application in the aircraft. These classifications could be structural, semi-structural, or non-structural, and also for interior applications. The primary aircraft structure can be broken into the following major components as shown in Figure 3.1:

1. Fuselage
2. Wing
3. Vertical stabilizer
4. Horizontal stabilizer
5. Engine
6. Landing gears

In each major component of the airplane, the selection of materials is a critical part to ensure the best design advantage of high strength-to-weight ratio. To satisfy the design

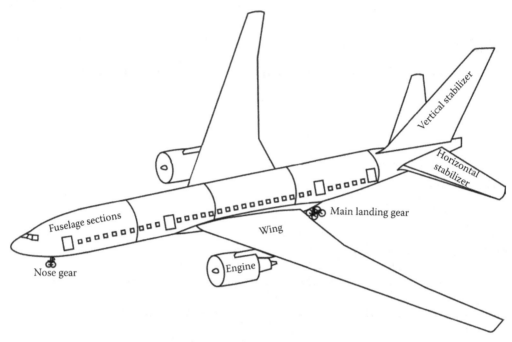

FIGURE 3.1
Major structural components of an aircraft.

requirements for any part of the major components, the type of raw material, basic raw material manufacturing process, and finally the producibility aspects of the materials are considered to meet the customer requirements. As mentioned in Chapter 2 materials selection for the design of aerospace structure is the most important factor to consider to satisfy the critical design factors including structural stiffness, static and dynamic load-bearing capacity, and durability (environmental resistance, fatigue, and damage tolerance including crack growth and residual stress).

Owing to the continuous evolution of aerospace raw materials, design, materials, and manufacturing, engineers in the aerospace industry constantly engage with primary material manufacturers to look for lighter and stronger materials for both the airframe structure and the propulsion systems. Structural durability would necessarily be a major consideration, since the service life of both commercial and military aircrafts is being extended far beyond their designed lives.

3.2 Major Aircraft Materials

The major raw materials commonly used to build aircraft structure are classified into two distinct categories—metal structures and non-metal structures. Primary metals are classified into two groups including non-ferrous and ferrous alloys, whereas non-metals like composite materials are classified mainly into polymer, metal, and ceramic matrices as explained in the tree diagram shown in Figure 3.2:

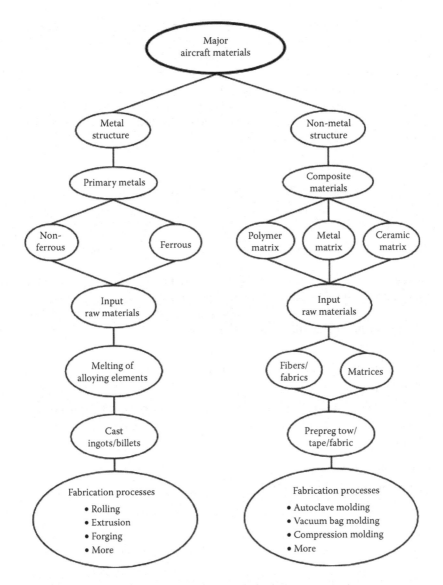

FIGURE 3.2
Classification of major aircraft materials.

- Non-ferrous metals
  - Aluminum alloys
  - Titanium alloys
  - Nickel alloys
- Ferrous metals
  - High strength alloy steel
  - Stainless steel
- Non-metals
  - Composite materials

Major structural metal components of an aircraft are made out of various shapes of primary raw material as shown in Table 3.1. Fundamentals of manufacturing processes of each form of metal shapes are discussed in Chapter 4.

Proper selection of metal alloys for producing an aircraft component is critical. The primary metal producers may need to work together with the aircraft manufacturers to satisfy both product and business requirements of both manufacturers. A primary metal producer's goal is to develop the material properties to satisfy the following basic design requirements of aircraft manufacturers:

- Higher mechanical properties
- Higher fracture toughness and damage tolerance
- Improved corrosion resistance
- Lower density
- Product mix including sheet, plate, extrusion, forging, and tube

Similarly, aerospace manufacturers will look to satisfy the following major requirements before introducing the metal in the airplane structure:

- Meet engineering design allowable values
- Cost of raw materials
- Meet producibility and quality requirements in terms of:
  - Formability
  - Cold and hot forming trials of extrusion, plate and sheet products
  - Effect of working strain on the final properties of metal structural component
  - Machinability
  - Chemical finish
  - Corrosion and environmental effects

TABLE 3.1

Various Shapes of Metal for Aircraft Major Components

Aircraft Major Components	Major Structural Elements	Various Metal Shapes
Fuselage	Skin, frames, stringers	Flat sheet Extrusion
Wing	Skin, chords, webs, stringers, channel vent, terminal fittings, flap track	Extrusion Plate Flat sheet Forging
Empennage	Skin, spars, webs, stringers, fittings, bracket	Extrusion Plate Flat sheet Forging
Engine	Pylon, bracket, fittings, nacelles, lip skin, tail cone, heat shield	Flat sheet Forging
Landing gear	Struts, beams, axles, braces, links	Extrusion Forging

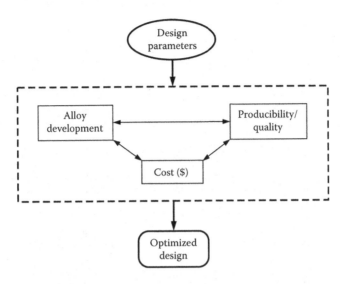

FIGURE 3.3
Relationship between alloy development and producibility.

Alloy development by the primary metal producers and the producibility study by the aircraft manufacturers are closely related. They need to satisfy the design parameters set by the design engineers and to meet the final product costs as shown by a block diagram in Figure 3.3. Aircraft structural design engineers provide their material design parameters to satisfy the technical need of the structural component of the aircraft and work together with the primary metal producers to optimize the design need. At the same time, the manufacturing facilities to produce aircraft components look for the producibility aspects of the material to make sure the structural part could be produced with the highest quality using the existing manufacturing technology with optimized product design and cost.

Table 3.2 compared the advantages and disadvantages between different materials used for making various components of an aircraft. There is a wide application of materials used in various location or various components of major metal aircraft. Table 3.3 shows some major materials as applied in major primary structural and secondary structural applications of an aircraft as shown in Figure 3.4. Aluminum alloys make up the major share of the traditional aircraft structure including fuselage and wing of the aircraft. Figure 3.5 shows the application of steel and titanium in the main landing gear and Figure 3.6 shows the application of titanium in the empennage and tail cone structure.

3.2.1 Aluminum

Aluminum alloys have been the backbone of manufacturing aircraft just prior to 1920. Most of the advanced alloys are simply variants of 2024, which was introduced in 1921, and 7075 that was introduced in 1943. The timeline of aluminum alloy development is shown in Figure 3.7. The 2024 alloy was first introduced in DC 3 aircraft in 1935. After a very long gap of nineteen years, the first 7xxx series alloy 7075 was introduced in the Boeing 707 jetliner. Since then, the continuous development of both 2xxx and 7xxx series alloys by the primary metal producers continued to meet the design needs of the aircraft. Refinements in the chemistry, processing, and heat treatment have led to the evolution of properties of

TABLE 3.2

Advantages and Disadvantages of Different Materials

Material	Advantages	Disadvantages
Aluminum alloys	Low density Good mechanical properties Low cost Low manufacturing cost-forming at room temperature	Corrosion sensitive High strength alloys not wieldable Used up to 200°F Not compatible with composite
Titanium alloys	High strength-to-weight ratio Corrosion resistance High operating temperatures Super plastic forming/diffusion bonding capability Compatibility with composites	High initial cost High machining costs Difficult to form at room temperature High notch sensitivity
Low-alloy steel	High strength Moderate toughness Low raw material cost	High density Corrosion sensitive Requirements of rigorous heat treatment and machining for highest strengths
Corrosion-resistant steels	Corrosion resistance Moderate-to-high strength Good fracture toughness Reduced manufacturing costs/flow time	High density Low strength relative to low-alloy steels
Nickel and Heat Resistant Alloys	Moderate-to-high strength Excellent corrosion resistance High operating temperature	High density High cost Difficult to form and machine
CFRP	Light weight High stiffness High strength High fatigue strength High corrosion resistance	High cost of manufacturing Poor compatibility with aluminum

Source: Reprinted from *Encyclopedia of Materials: Science and Technology*, 1, Boyer, R., Airframe Materials, 66–73, Copyright 2001, with permission from Elsevier.

2024 with emphasis on improving fracture toughness, fatigue crack initiation, and crack growth resistance. Emphasis has been given on 7075 type alloys to improve primarily the compressive strength and fracture toughness, without sacrificing the other properties in either case. Further emphasis has also been given on the critical factor of improving corrosion resistance. Developments have been accomplished on the heat treatment processes for both 2xxx and 7xxx series alloys to improve corrosion resistance, while sacrificing just a few points in their strength capabilities. As shown in the time line (Figure 3.7), Al–Cu–Li alloys like 2098 and 2099, which are roughly 10% lighter in density, have been introduced in Boeing's new 787 aircraft. These are some typical aluminum alloys and tempers used for some primary structures of an aircraft [1–3]:

- Fuselage skin: 2024-T3, 2524-T3, 2524-T4, 2524-T36
- Wing skin: 7075-T7751 (upper), 2324-T39 (lower)
- Stringers: 7150-T77511 (crown), 7075-T6 (lower surface)
- Primary forging alloys: 7075-T73 and 7050-T74

The general requirements of major structural elements of metal aircrafts are determined by the design engineers based on the loading and stresses applied in the aircraft in normal

TABLE 3.3

Examples of Major Materials in Major Primary Structural and Secondary Structural Applications

Aluminum Alloys	Titanium	Steel	Graphite Composites	Fiberglass/Norex Panels
Fuselage skin	Main landing gear beam	Steering yoke	Floor panels	Radom
Wing spar chords	APU Fire wall	Truck position bell crank	Main deck side wall panels	Ailerons
Wing spar web	Torque link	Wing landing gear lower side strut	Main deck ceiling panels	Wing-to-body fairing
Fuselage stringers	Hydraulic tubing	Torsion link	Overhead stowage bins	Vertical and horizontal stabilizer tips
Wing stringers	Inboard flap fitting	Flap linkages	Winglet ribs and panels	Rudder
Wing channel vents	Inboard carriage	Flap tracks and flap carriage		Elevator
Wing skin	Lavatory attach fittings	Actuator arms		Upper and lower trailing edge panels
Wing-to-body chords		Landing gear		Trailing edge flap segments
		Aft engine mount		Floor panels
				Spoilers
Fuselage chord				
Wing ribs				

Source: Reprinted from *Encyclopedia of Materials: Science and Technology*, 1, Boyer, R., Airframe Materials, 66–73, Copyright 2001, with permission from Elsevier.

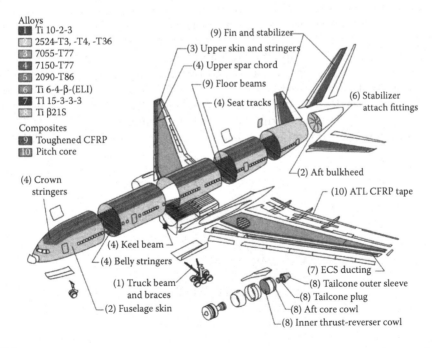

FIGURE 3.4

Schematic illustration of application of different metals in a commercial aircraft. (Adapted from Boyer, R., Titanium applications on Boeing Aircraft, *Proceedings of International Titanium Association*, CO, 115–129, 1996.)

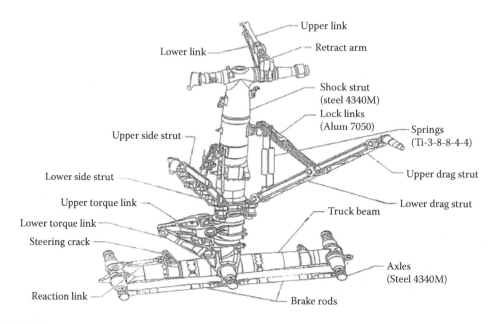

FIGURE 3.5
Major applications of titanium and steel in main landing gear. (Adapted from Boyer, R., Titanium applications on Boeing Aircraft, *Proceedings of International Titanium Association*, CO, 115–129, 1996.)

and extreme service scenarios. Some of the critical requirements for a transport aluminum metal aircraft component are summarized in Table 3.4.

One of the 6000 series alloys 6013 is used for formed parts of secondary structures, due to its higher formability than that of 2024 alloy. The 6013 alloy can be formed in T4 temper and artificially aged to T6, whereas 2024 alloy must typically be formed in either "O" or "W" depending on the severity of deformation. Solution heat treatment to make "W" temper from original "O" temper creates distortion problems in the parts formed in "O" temper prior to solution heat treatment. The 2024 and 7075 alloys can be produced in the "hi-form" condition, as developed by a primary metal producer providing a fine grain size and higher formability. The high form material can be formed in either "O" or "W" condition in order to tighter bend radii in comparison with the regular 7075 and 2024 alloys.

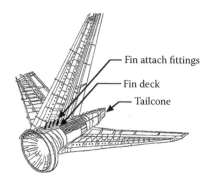

FIGURE 3.6
Application of titanium in the empennage and tail cone of an aircraft. (Adapted from Boyer, R., Titanium applications on Boeing Aircraft, *Proceedings of International Titanium Association*, CO, 115–129, 1996.)

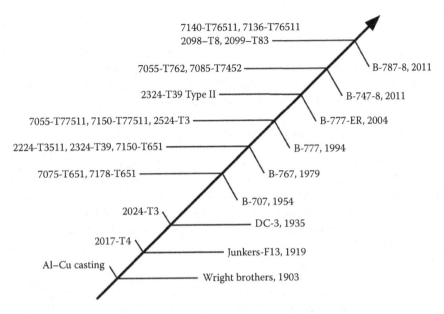

FIGURE 3.7
Timeline of aluminum alloy development for aerospace. (After Axter S. et al., Evolution of an Integrated Approach to Material Development, Invited paper for American Institute of Aeronautics and Astronautics, Special Edition, 2004.)

TABLE 3.4

Critical Requirements for the Design of Major Aluminum Aircraft Components

Major Component	Subcomponent/Metal Form	Critical Requirement
Fuselage	Skin (Al 2xxx sheet)	Fatigue, damage tolerance, and corrosion resistance
	Stringer (7xxx sheet)	Fatigue and compression strength
	Frames (7xxx sheet)	Stiffness, fatigue, compression strength, and damage tolerance
	Floor beams (extrusion and sheet)	Static strength
	Seat tracks (extrusion)	Corrosion resistance, static strength, fatigue, and damage tolerance
Wing	Lower skin (2xxx plate)	Damage tolerance and fatigue
	Lower stringer (2xxx extrusion)	Fatigue, damage tolerance, and tension strength
	Upper skin (7xxx plate)	Compression strength (damage tolerance for heavy gage skins)
	Upper stringer (7xxx extrusion)	Compression strength
	Ribs shear tie (7xxx plate)	Shear strength
	Intermediate (7xxx sheet)	Stiffness and shear strength
Horizontal Stabilizer	Lower skin (7xxx plate)	Compression strength
	Lower stringer (7xxx extrusion)	Compression strength
	Upper skin (2xxx plate)	Tension strength, damage tolerance, and fatigue
	Upper stringer (2xxx extrusion)	Tension strength
Vertical fin	Skin	Compression strength and damage tolerance
	Stringer	Compression strength

Source: Reprinted from *Encyclopedia of Materials: Science and Technology*, 1, Boyer, R., Airframe Materials, 66–73, Copyright 2001, with permission from Elsevier.

The aluminum alloys used for military aircraft are similar to those on commercial aircraft, primarily 7075, 7175, and 2124 alloys. The 2124-T81851 thick plate materials are used to make major bulkheads of F-16 and F-22 aircraft. In addition to that, 7075-T76, 7475-T761 and 2024-T81, 2124-T81 are used for sheet applications. The 2124-T81 plate product is also used for heavy structural applications. The 6013 sheet also has potential for future applications because of its improved formability as mentioned earlier, excellent exfoliation corrosion resistance, and lower cost.

Aluminum–lithium (Al–Li) alloys were developed primarily for reducing the weight of aircraft and aerospace structures [3]. Al–Li alloys have decreased density because of the very low density of metallic lithium. Addition of 1% lithium results in decreasing density of the alloy by 3%, and increases modulus by 6%. In addition to the main element lithium (Li), copper (Cu), magnesium (Mg), zirconium (Zr), and silver (Ag) are used as other alloying elements. Like other aluminum alloys, Al–Li alloys are heat treatable. Al–Li alloys possess increased modulus of elasticity, high specific stiffness, increased fatigue strength, and cryogenic strength. Alloys containing silver also have good weld ability. Addition of zirconium to the alloy controls grain structure during heat treatment. The major development work started in the 1970s, when aluminum producers accelerated the development of Al–Li alloys as replacements for traditional airframe 2xxx and 7xxx alloys. The development work led to the introduction of commercial alloys 8090, 2090, and 2091 in the mid-1980s. Primary aluminum alloy producers are continuously improving various aspects of the alloys including mechanical properties, formability, and machinability to manufacture airframe structures. Several Al–Li alloys including 2055, 2060, 2098, 2195, 2198, and 2099 are currently being used in modern aircraft. The cost of Al–Li alloys is typically three to five times higher than that of the conventional aerospace alloys due to relative high cost of lithium and high processing and handling cost. It is a challenge to the aerospace industry to make proper trade study comparisons with the Al–Li alloys to compete with the recent trend of using light carbon fiber composite materials in the airframes. There is a positive trend to increase using Al–Li alloys in the fuselage and wing structures of new commercial aircraft.

3.2.2 Titanium

The use of titanium in aerospace applications is growing due to its high strength, excellent corrosion resistance, and elevated temperature properties and also having lower density by 40% than that of steel and nickel-based alloys. Owing to initial high cost of raw materials followed by costly forming and machining operations, the use of titanium is generally limited. But some primary benefits must be considered to justify on a part-by-part basis against the added cost. The primary benefits for the gradually increasing use of titanium in the aerospace industry include [4]:

- Weight savings: Primarily as a replacement for steel. Using lower density titanium alloy compared with high strength steel permits significant weight savings. There is an example of a Boeing 777 aircraft truck beam; it saved about 350 Lb by replacing 4340 M (tensile strength 275 ksi) with Ti-10V-2Fe-3Al (160 ksi strength). The use of titanium on the main landing gear saved a total weight of over 1000 Lb/airplane.

- Space limitations: Replacement of aluminum alloys. Volumetric constraints provide the opportunity for some of the largest titanium components such as landing gear beams. The choice of material would have been aluminum alloys such as 7075 or 7050 with this lower cost option. But the size of the aluminum part to carry the specific load would not fit within the available space within the wing. Again steel would have been an option, but higher density steel incurs higher weight, which is not acceptable.

- Operating temperature: Replacement of aluminum, nickel, and steel alloys. High strength conventional aluminum alloys have a maximum operating temperature on the order of 270°F, whereas titanium alloys can be used when the operating temperature is much higher than this. Steel and nickel-based alloys could be used to satisfy temperature requirements, but again the weight will be higher due to their higher density.

- Corrosion resistance: Replacement of aluminum and low-alloy steel. Corrosion is a very important issue for the life of an aircraft component. Titanium and its alloys are corrosion-resistant in an aerospace environment without non-protective coatings or paint. The application areas include wheel wells, galleys, and lavatories.

- Composite compatibility: Replacement of aluminum alloys. Compatibility of light weight CFRP while joining with the metal parts has become a significant issue. Titanium is highly compatible with the carbon fibers in the CFRPs, while there is galvanic reaction taking place between the carbon fibers and aluminum or low-alloy steel leading to corrosion to the metallic portion of the airframe structure. Although there are corrosion protection methods that could be used to prevent corrosion, the integrity of the barrier coating over the life time of the aircraft is questionable. Thus, titanium may be used safely when the component is critical and it is difficult to inspect and replace.

Titanium alloys used in aircraft cover the entire range of commercially produced conventional alloys including commercially pure (CP) Ti, Ti-3Al-2.5V, and Ti-6Al-4V, Ti-6Al-2Sn-4Zr-2Mo, Ti-10V-2Fe-3Al, Ti-15V-3Cr-3Al-3Sn, and more. CP titanium with formability and corrosion resistance has driven its use from α or super-α type alloys such as Ti-6Al-2Sn-4Zr-2Mo for high temperature applications to α/β alloys such as Ti-6Al-4V, which is widely used in the aerospace industry, and β alloys such as Ti-10V-2Fe-3Al. Ti-6Al-4V was developed in the late 1950s and still represents approximately 80% of the aerospace market. It is a moderate strength alloy with minimum ultimate strength (UTS) of about 130 ksi, and can be utilized for every product shape including sheet metal, extrusion, forging, and also in tube form. Since the application of titanium is growing faster, the aerospace industries are exploring more to manufacture wide variety of parts used for aerospace vehicles including fixed wing aircrafts, rotary aircrafts, and also jet engines.

3.2.3 Steel

The structural steel alloys used in the commercial aircraft are mainly the high strength low-alloy steels such as 4340, 4330, and 4340 M. These alloys are used when very high strength (275–300 ksi) requirements are needed to satisfy the working stress of the parts such as landing gear and flap tracks. But the protective finishes are required to address general corrosion and stress corrosion issues. More recent commercial models and their derivatives have started using high strength corrosion-resistant steels such as 15-5PH wherever possible. Relative to high strength low-alloy steel, 15-5PH steel reduces manufacturing flow time and costs in production. The only limitation of 15-5PH steel is the range of the maximum UTS (180–200 ksi).

3.2.4 Composite Materials

Gross weight of an aircraft is a big concern to make it fuel-efficient. Weight-to-power ratio needs to be optimized to make the aircraft more efficient. The demand for low-density composites and other non-metals have continued to go up to make the aircraft lighter. Carbon fiber composites have been used in applications for some structural and mostly semistructural parts for quite some time. Continuous improvements in the technology of fabricating large structural carbon fiber composite parts have given a new direction to the aircraft manufacturers to explore many different designs to make the aircraft more fuel-efficient as delivered to the airline or military customers. Molding technology helps to make an integral or one-piece part to avoid many joining processes with many metal fasteners. Figure 3.8 shows the timeline for composite materials used mainly in the commercial aircraft. The most revolutionary commercial aircraft to date, the Boeing 787, introduced in 2011, is made with 50% composite materials as shown in Figure 3.9. There is also a continuous growth of use of composite materials in military applications, as shown in Figure 3.10. Chapter 5 discusses mainly the introduction of composite materials, and Chapter 18 discusses various manufacturing processes of composite materials useful to make aircraft components.

The materials breakdown for some commercial and military aircraft is shown in the pie chart displayed in Figure 3.11. It is clear from Figure 3.11 that the trend is toward increased usage of composites and titanium as compared to the traditional aluminum

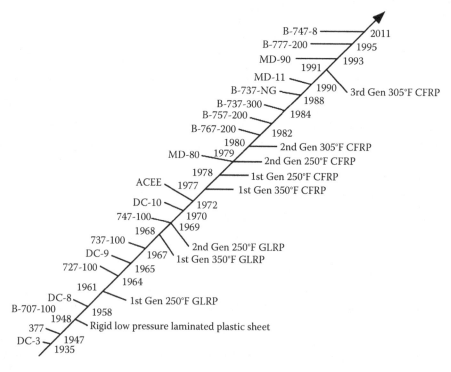

FIGURE 3.8
Timeline of composite materials in aircrafts. (Adapted from Axter, S. et al., Evolution of an Integrated Approach to Material Development, Invited paper for American Institute of Aeronautics and Astronautics, Special Edition, 2004.)

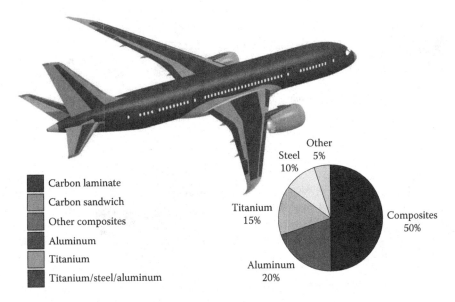

Carbon laminate
Carbon sandwich
Other composites
Aluminum
Titanium
Titanium/steel/aluminum

Other 5%
Steel 10%
Titanium 15%
Aluminum 20%
Composites 50%

FIGURE 3.9
The 787 with 50% composite materials. (Adapted from Kisch, R.A., Composites processing technology and innovation at Boeing, Key Note Presentation for the *Society of Plastics Engineers Automotive Composites Conference & Exhibition*, September, Troy, MI, 2007.)

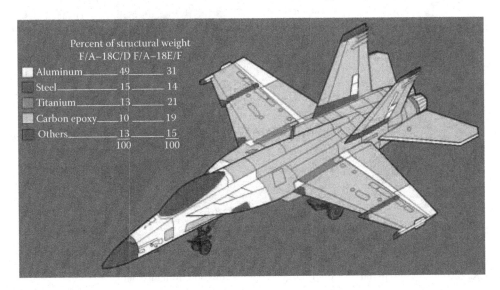

Percent of structural weight	F/A–18C/D	F/A–18E/F
Aluminum	49	31
Steel	15	14
Titanium	13	21
Carbon epoxy	10	19
Others	13	15
	100	100

FIGURE 3.10
Schematic illustration of application of different materials in a military aircraft. (From The Boeing Company.)

alloys, especially as witnessed in the new Boeing 787. The trend of increased use of composite materials in the airframe structure is also being pursued by the other prime aerospace manufacturing companies including Airbus for their new A350. Further considerations are being discussed to use composite wings for future Boeing 777X model.

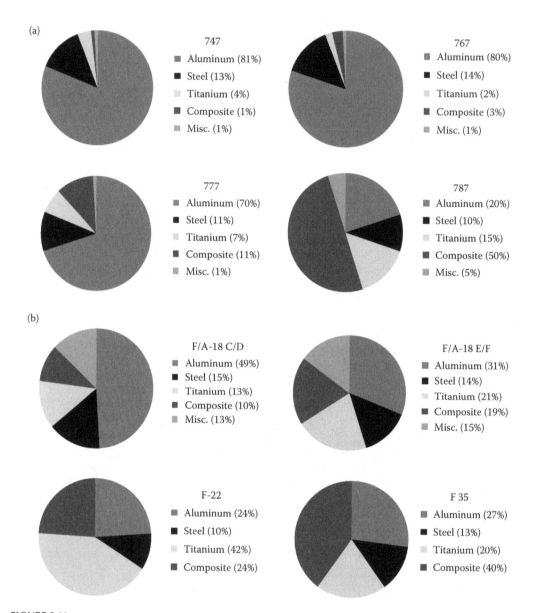

FIGURE 3.11
Examples of structural materials used in various aircraft, showing gradual increase of composites: (a) commercial aircrafts and (b) military aircrafts. (Reprinted from *Encyclopedia of Materials: Science and Technology*, 1, Boyer, R., Airframe Materials, 66–73, Copyright 2001, with permission from Elsevier.)

3.3 Fundamentals of Mechanical Behavior and Testing of Metals

It is important to know the fundamentals of mechanical behavior of metals [6,7] for better understanding of forming of metals for making structural components of an aircraft. The most fundamental data on the mechanical properties of materials are obtained from a tension test, using the ASTM standard test coupon shown schematically in Figure 3.12a.

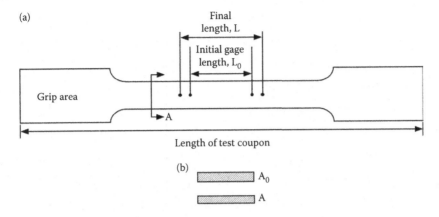

FIGURE 3.12
Tensile test coupon nomenclatures: (a) tensile test coupon and (b) cross-sectional area.

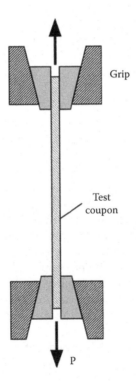

FIGURE 3.13
Schematic representation of tensile testing.

The test coupon has two shoulders for gripping in the tensile test machine and a reduced gauge section in between. The test specimen is subjected to an applied axial load, P, as shown in Figure 3.13 until it fractures. The load and elongation are measured at frequent intervals during the test and are expressed as average stress and strain data. The stress and strain data are generally plotted as a stress–strain diagram as shown schematically in Figure 3.14.

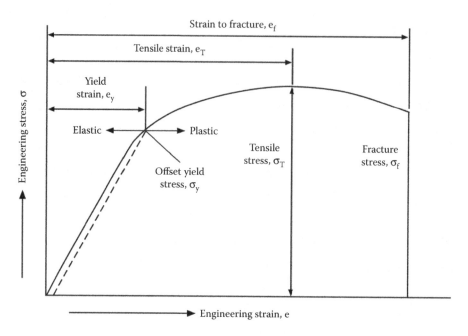

FIGURE 3.14
The engineering stress–strain curve.

The initial linear portion of the curve is the elastic region that obeys Hooke's law. The slope of the stress–strain curve in the elastic region is the modulus of elasticity. The yield strength is defined as the stress that will produce a small amount of permanent deformation as shown by σ_y. The yield strain is shown by e_y, or offset. Plastic deformation begins after exceeding the material's elastic limit. As the plastic deformation of the test coupon increases, the metal becomes stronger due to strain hardening and as the load required continue extending the test coupon increases with further straining. Eventually, the stress reaches a maximum value. The maximum stress is called ultimate tensile strength, σ_T. For a ductile material, the cross-section of the test coupon begins to decrease from the initial area, A_0, to the final area, A (Figure 3.12b), rapidly until the test coupon fractures.

The stress–strain curve obtained by uniaxial loading in the tension test provides the engineering stress–strain curve. In plastic deformation of metals, a true stress–strain curve is plotted.

Engineering stress or nominal stress is defined as the ratio of the applied load, P, to the original cross-sectional area of the gage section, A_0, of the test specimen.

$$\sigma = \frac{P}{A_0} \tag{3.1}$$

The engineering or nominal strain is defined as

$$e = \frac{\Delta L}{L_0} = \frac{L - L_0}{L_0} \tag{3.2}$$

where, ΔL is the change in gage length, L_0 is the initial gage length, and L is the final length

$$e + 1 = \frac{L}{L_0} \tag{3.3}$$

The modulus of elasticity, or Young's modulus, E is defined as

$$E = \frac{\sigma}{e} \tag{3.4}$$

This linear relationship between stress and strain is known as Hooke's law.
Ultimate tensile stress is calculated by

$$\sigma_T = \frac{P_{max}}{A_0} \tag{3.5}$$

The ductility of the metal is commonly quantified by elongation or reduction of area. Elongation is defined by

$$\frac{L - L_0}{L_0} \cdot 100\% \tag{3.6}$$

Reduction in area is given by

$$\frac{A_0 - A}{A_0} \cdot 100\% \tag{3.7}$$

where, A_0 is the initial gage area and A is the final gage area.
True strain or natural strain can be defined as

$$\varepsilon = \ln \frac{L}{L_0} = \ln \frac{A_0}{A} \tag{3.8}$$

Relation between true or natural strain with the engineering or nominal strain

$$\varepsilon = \ln \frac{L}{L_0} = \ln(e + 1) \tag{3.9}$$

True stress is defined as where A_i is the actual or instantaneous area supporting the load, P

$$\sigma = \frac{P}{A_i} = \frac{P}{A_0} \cdot \frac{A_0}{A_i} \tag{3.10}$$

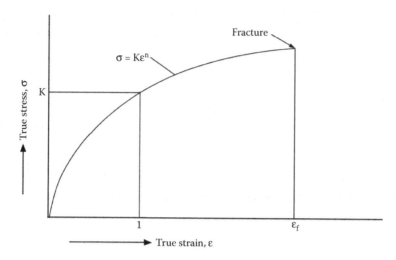

FIGURE 3.15
True stress–strain relationships.

From the volume constancy relation,

$$\frac{A_0}{A_i} = \frac{L}{L_0}$$ (3.11)

True stress can be calculated from

$$\sigma = \frac{P}{A_0}(e+1)$$ (3.12)

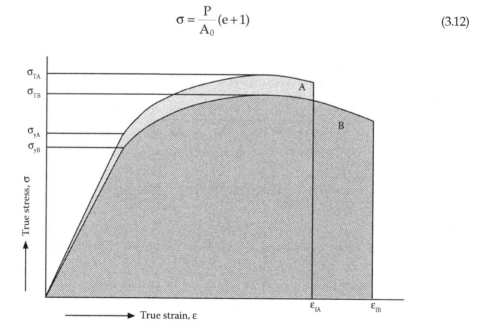

FIGURE 3.16
Variation of toughness between two alloys A and B.

TABLE 3.5

Typical Mechanical Properties of Common Aircraft Structural Metals

Alloy and Temper	Yield Strength ksi (MPa)	Tensile Strength ksi (MPa)	Elongation (%)
Bare 2024-T3	50 (345)	70 (485)	18
Bare 2024-T4, T351	47 (325)	68 (470)	20
Alclad 2024-T3	45 (310)	65 (450)	18
Alclad 2024-T4, T351	42 (290)	64 (440)	19
Bare 7075-T6, T651	73 (503)	83 (572)	11
Alclad 7075-T6, T651	67 (462)	76 (524)	11
CP-2 Ti	55 (379)	65 (448)	18
Ti-3Al-2.5V	80 (551)	90 (620)	17
Ti-6Al-4V	126 (869)	134 (924)	10
Ti-6Al-2Sn-4Zr-2Mo	136 (938)	143 (986)	10

Source: Adapted from Davis, J.R., *ASM Specialty Handbook of Aluminum and Aluminum Alloys*, ASM International, 1998; Welsch, G., Boyer, R. and Collings, E.W., *Materials Properties Hand Book Titanium Alloys*, ASM International, Ohio, 2007; Metallic Materials Properties Development and Standardization (MMPDS-06), Battelle Memorial Institute, Ohio, 2011; www.grantadesign.com

A true stress–strain curve shown in Figure 3.15 is frequently called a flow curve because it gives the stress required to cause the metal to flow plastically to any given strain. The common mathematical power equation of the form

$$\sigma = K\varepsilon^n \qquad (3.13)$$

when, K is the stress at $\varepsilon = 1.0$ and n is the strain hardening coefficient.

Toughness is another important parameter to be considered designing aircraft structural components. Toughness is defined by the area under the true stress–strain curve as shown in Figure 3.16.

$$\text{Toughness} = \int_0^{\varepsilon_f} \sigma d\epsilon \qquad (3.14)$$

Table 3.5 shows some typical tensile properties of some common aerospace aluminum and titanium alloy wrought products used for making various structural components of an aircraft.

References

1. Boyer, R., Airframe materials, In *Encyclopedia of Materials: Science and Technology*, Buschow, K.H.J., Cahn, R.W., Flemings, M.C., Ilschner, B., Kramer, E.J., and Mahajan, S., eds. Elsevier, Oxford, Vol. 1, 66–73, 2001 (invited manuscript).

2. Axter, S., Boyer, R., Burford, I., Davis, K., Giuffre, M., Mohaghegh, M., and Pryor, J., Evolution of an Integrated Approach to Material Development, Invited paper for American Institute of Aeronautics and Astronautics, Special Edition, 2004.
3. Davis, J.R., *ASM Specialty Handbook of Aluminum and Aluminum Alloys*, ASM International, Ohio, 1998.
4. Boyer, R., Titanium applications on Boeing Aircraft, *Proceedings of International Titanium Association*, CO, 115–129, 1996.
5. Kisch, R.A., Composites processing technology and innovation at Boeing, Key Note Presentation for the *Society of Plastics Engineers Automotive Composites Conference & Exhibition*, September, Troy, MI, 2007.
6. Dieter, G.E., *Mechanical Metallurgy*, International Student Edition, McGraw-Hill Kogakusha, Ltd., Tokyo, 1961.
7. Kalpakjian, S. and Schmid, S.R., *Manufacturing Processes for Engineering Materials*, Pearson Education, Inc., publishing as Prentice Hall, NJ, 2008.
8. Welsch, G., Boyer, R., and Collings, E.W., *Materials Properties Hand Book Titanium Alloys*, ASM International, Ohio, 2007.
9. Metallic Materials Properties Development and Standardization (MMPDS-06), Battelle Memorial Institute, Ohio, 2011.
10. www.grantadesign.com

4

Manufacturing Principle and Processes of Major Aircraft Metal Products

4.1 Introduction

Input raw materials and their shapes required for fabricating different components of an aircraft are produced by using various manufacturing processes. Specific processes used depend on the material, geometry, size, and allowable properties to satisfy various engineering needs for each component. Initial raw materials are classified mainly by various product shapes including flat sheet, bars, plates, beams, pipes and tubes, solid, and hollow extrusions. Different manufacturing principles and processes are involved to make the input product shapes. As discussed in Chapter 3, a high volume of aluminum alloy products have been commonly used in building aircraft for many years, since the development of the first passenger aircraft model DC-3 by the Douglas Company in 1935 in the United States. Aluminum alloy products mostly in the form of flat sheet, plate, extrusion, forging, and drawn tubes satisfied the engineering requirements and the economics of aircraft manufacturing for many decades. This changed when Boeing introduced high volume usage of carbon fiber composite materials in commercial aircraft manufacturing, with the all-new 787-8, which entered service in October, 2011. Figure 4.1 shows a pie chart of usage of different aluminum wrought products in a typical pre-787 airplane model.

The manufacturing processes for producing input raw materials are generally classified into two basic categories based on the type of forces applied to produce the product size and shapes [1]:

1. Direct compression process
2. Indirect compression process

In a direct compression process, the force is applied by the die to the surface of the primary ingot to induce plastic deformation, and the metal flows at right angles to the direction of compression, for example, as applied to a forging process (Figure 4.2a). Another example of a direct compression process is rolling, where the forces are applied by the rolls plastically deforming metal by passing it between rolls (Figure 4.2b).

Indirect compression processes mainly include two bulk plastic deformation processes: extrusion and drawing (Figure 4.3a and b), the primary applied forces in drawing are frequently tensile (Figure 4.3b), but high indirect compressive stresses are developed by the reaction of the primary ingot material with the die surface. As a result, the metal flows under the action of a combined state of stress that includes high compressive stress in one of the principal directions.

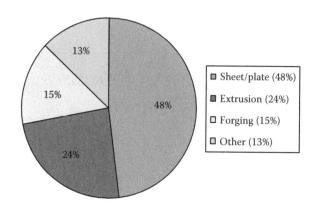

FIGURE 4.1
Usage of aluminum wrought products in an airplane.

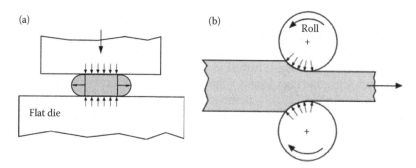

FIGURE 4.2
Direct compression processes, (a) forging and (b) rolling.

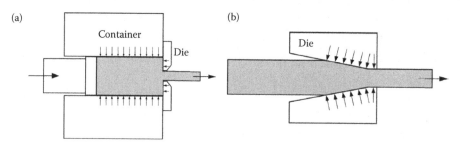

FIGURE 4.3
Indirect compression processes, (a) extrusion and (b) drawing.

The main purpose of the above-mentioned bulk deformation processes is to break down a cast ingot or billet structure to produce various wrought products. The plastic deformation processes shown in Figures 4.2 and 4.3 could be performed either hot or cold or as a combination of both. A hot working is the initial step of bulk deformation process for most of the metals and alloys for making various wrought products. Hot working process decreases the energy required to deform the high strength material, and also increases the ability of material to flow without cracking under different states of stress. Blowholes and porosity are being eliminated by welding together any cavities at high compressive stress. The coarse columnar grains of the cast ingot structure are broken down and refined into smaller equiaxed recrystallized grains. These changes allow the deformed metal structure to obtain an increase in ductility and toughness over the cast billet or ingot structure.

The fundamental knowledge of how to produce primary metal stocks for use in fabricating thousands of small, medium to very large metal components of an aircraft using various manufacturing technology is very important to learn for students of aerospace and manufacturing engineering and also the practicing engineers of aerospace industries. This chapter will provide mainly the basic principles, processes, and fundamental aspects of the following manufacturing technologies used to produce the major portion of the input wrought products required to manufacture the major structural components of an aircraft:

1. Casting of billets and ingots
2. Extrusion
3. Forging
4. Rolling
5. Drawing (aluminum tubing)
6. Pilgering (titanium tubing)

The fundamentals of each of the above-mentioned manufacturing technology will be covered mainly with examples of aerospace aluminum alloys, which are in high volume usage, since the aerospace industry started producing a variety of airborne vehicles. Besides aluminum wrought products, titanium alloy extrusion, forging, and rolled products are also in very high demand. This is because these materials have very high strength to weight ratios for many applications, and are compatible with composite aircrafts.

4.2 Casting of Ingot or Billet

Cast ingot or billet is the input raw material before transforming the cast ingot structure into desired wrought product shapes. Casting of ingot or billet is the first step in converting the primary input material and other alloying elements from the liquid molten stage to the solid cast structure. Two basic casting processes including static casting and continuous casting are mainly used to make ingots and billets for producing various wrought products. In static casting, the typical practice is to pour molten metal into a vertical open-ended split type metal (cast iron) mold. After solidification, the cast ingot can be easily withdrawn. Static casting is a batch process not ideal for volume production. However in continuous casting, molten metal is poured into the water-cooled mold and solidification starts before entering through the spray water for complete solidification. Continuous casting is a high-volume production process. Fundamentals of continuous casting of aluminum billet will be discussed in the next section. Ingots made from either static or continuous casting are generally hot worked and plastically deformed mainly by extrusion, forging, and rolling to make near-net-shaped wrought products. The products are in turn used to fabricate various structural components of an aircraft.

4.2.1 Casting of Aluminum Billet for Extrusion/Forging

The direct chill (D.C.) continuous casting process developed in 1933 by W.T. Ennor is the method used today to cast aluminum extrusion billets. The major process steps of billet making are shown in Figure 4.4. There are many steps involved between input raw materials and the final casting process [2].

Figure 4.5 shows the flow diagram of the basic steps of the melting and the billet casting processes. Figure 4.6 illustrates the principle of the D.C. casting process. Molten aluminum is poured into a shallow water-cooled mold normally of a round cross-sectional shape. When the metal begins to freeze in the mold, the false bottom in the mold is lowered at a controlled speed and water is sprayed on the surface of the freshly solidified billet as it comes out of the mold.

Figure 4.7 shows continuous casting of aerospace aluminum alloy extrusion billets at a raw material supplier. The principal variables that mainly influence the production performance of aluminum billet casting are

- Pouring molten metal temperature
- Casting speed
- Type of mold
- Metal head
- Rate of water flow

There are several advantages of the D.C. casting system especially for hard aerospace alloys, like 2xxx and 7xxx series systems, compared to other techniques such as in tilt-mold casting:

- Has minimum metal segregation
- Can produce large ingots
- Flexible to cast with varying speeds
- Minimizes cracking in hard alloys

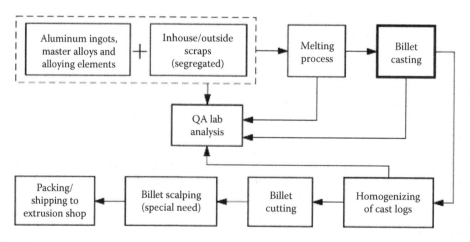

FIGURE 4.4

Major process steps of billet making for extrusion shop.

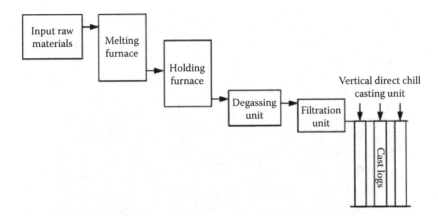

FIGURE 4.5
Functional block diagram of melting and casting processes.

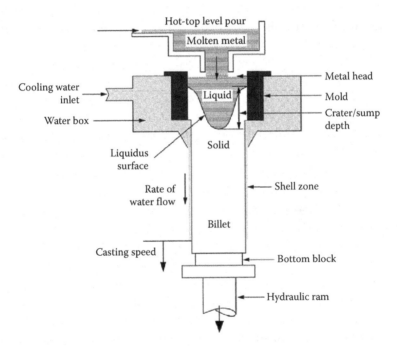

FIGURE 4.6
Principle of D.C. billet casting.

- Transfers molten aluminum slowly, uniformly with a relatively low temperature to avoid many problems.

4.2.2 Billet Geometry

Aluminum extrusion billets are cast in round-shaped continuous long lengths (usually called logs). These billets are then cut from the cast log into the desired length as shown in Figure 4.8a. Round billets have two dimensions including diameter (D) and length (L). The length of the billet is determined from the product shape and geometry and the extrusion press capacity.

FIGURE 4.7
Casting of aerospace aluminum alloy billet. (From Universal Alloy Corporation.)

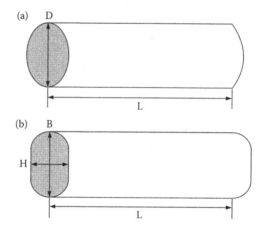

FIGURE 4.8
Shapes of aluminum alloy billets, (a) round and (b) rectangular/oval.

For very wide-shaped extrusion, the rectangular/oval-shaped billet (Figure 4.8b) is used. Rectangular billets have three dimensions such as major axis (B) and minor axis (H) of oval cross section and length (L). The ratio of B and H is normally kept 2:1 or 2.5:1 as a maximum.

4.2.3 Billet Homogenizing

Cast aluminum billet logs are usually homogenized before extrusion. The as-cast condition gives a product of unsatisfactory quality and lower extrudability. These effects can be partly or completely eliminated by homogenization, a kind of heat treatment of the cast billet logs. However, the purpose of the billet heat treatment varies according to quality and economic requirements and can involve dissolution, precipitation, or uniform distribution of the alloying components. The homogenization process has three major steps, including heating billet logs at a particular rate, holding at a constant temperature for a

TABLE 4.1

Typical Values for Billet Homogenizing of Some Aerospace Aluminum Alloys

Alloy	Homogenizing Temperature °F (°C)	Holding Time (h)
2014–2024	896–914 (480–490)	12
7075, 7079	878–896 (470–480)	12

Source: Adapted from Saha, P.K., *Aluminum Extrusion Technology*, ASM International, Materials Park, Ohio, 2000.

certain time, and cooling using a proper cooling rate. Table 4.1 shows homogenizing cycle parameters of some common aerospace alloys.

4.2.4 Billet Scalping

Scalping of the billet means the machining of the cast liquated skin (Figure 4.9) of the billet, generally for harder aluminum alloys. For harder alloy aerospace extrusion, especially for the indirect process where there is no relative displacement between billet and container, the machined billets are primarily being used. Generally, the skin surfaces of harder alloys are much harder than those of softer grade alloys. It is always recommended to use machined billets for harder alloy extrusions, to obtain the best-quality extrusion, especially for the aerospace industries.

4.2.5 Common Casting Defects

The objective of the D.C. casting process is to produce quality aluminum billets of uniform chemistry, fine metal structure, and strength. The common problems or defects associated with D.C. casting aluminum billets are as follows. The defects are explained with examples in the aluminum extrusion technology book [2]:

- Cracking and splitting
- Segregation
- Bleeding
- Cold shutting
- Porosity
- Grain growth

FIGURE 4.9
Cast versus scalped billet surface. (From Universal Alloy Corporation.)

4.3 Extrusion

In an airframe structure, extrusions of different profile geometries and sizes are used to manufacture many different components. According to the pie chart shown in Figure 4.1, the usage of extrusion in an aircraft is around 25% of the total metals used in the aircraft. Extrusion is a plastic deformation process in which a block of metal called a billet is forced to flow by compression through a die opening of a smaller cross-sectional area than that of the original billet as shown in Figure 4.10. Extrusion of aerospace aluminum alloys is generally produced by hot extrusion. In hot extrusion, the billet is preheated to facilitate plastic deformation. The fundamentals of aluminum extrusion technology are well covered in the book [2].

The two basic types of extrusion are direct and indirect, which are commonly used in the aluminum extrusion industry. The fundamental differences between the two processes are shown in Figure 4.11.

In direct extrusion, the direction of metal flow will be in the same direction as ram travel. During this process, the billet slides relative to the walls of the container. The resulting frictional force at the billet–container interface increases the needed ram pressure considerably. In indirect extrusion, the die at the end of the hollow stem moves relative to the container, as a result there is no relative sliding between billet and container. Therefore, this process is characterized by the absence of friction between the billet and the container interface.

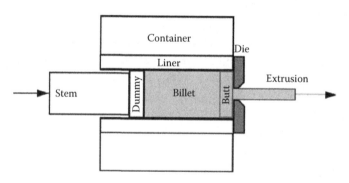

FIGURE 4.10
Definition of extrusion.

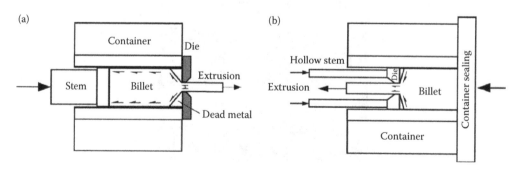

FIGURE 4.11
Difference between direct and indirect extrusion, (a) direct and (b) indirect.

4.3.1 Principal Variables

The principal variables that influence the force required to cause extrusion and quality of material exiting from the die are shown in Figure 4.12:

- Extrusion ratio, ER
- Working temperature (billet, container, and die)
- Speed of deformation
- Alloy flow stress

A dead metal zone builds up in the corners of the die, and the material shears along this face. The material may continue to extrude over this generated zone, which acts like a conical die surface. The surface and subsurface defects are likely to occur on the extruded product if a sufficient amount of butt is not kept. According to industry practice, the standard butt thickness for direct extrusion is 10%–15% of the billet length. This is more important for harder alloy extrusion, especially for aircraft manufacturing. The extrusion should be stopped within a safe margin zone, preventing oxide and other metallic or nonmetallic inclusions from flowing into the extrusion. The extrusion speed could be calculated for any extrusion die by using a volume constancy relation, which means that the volume of metal in the container becomes equal to the volume of extrusion coming out of the die because there is no loss of metal during extrusion. To determine the volume constancy as shown in Figure 4.12, it is given by

$$V_R A_C = V_E A_E \tag{4.1}$$

where V_R is the ram speed, A_C the area of container bore, V_E the extrusion speed, and A_E the area of the extruded shape. If the die is multi-hole with the same cross-sectional area of each hole, the volume constancy relation can be written as

$$V_R A_C = V_E(n\, A_E) \tag{4.2}$$

where n is the number of symmetrical holes in the die.
The extrusion speed is given by

$$V_E = V_R \frac{A_C}{n(A_E)} \tag{4.3}$$

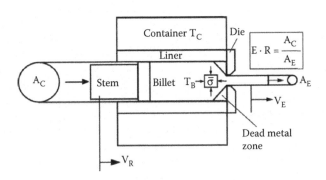

FIGURE 4.12
Extrusion variables.

The extrusion speed can also be written as

$$V_E = V_R(E \cdot R)$$ (4.4)

where the extrusion ratio, $E \cdot R$ is defined as

$$\frac{A_C}{n(A_E)}$$ (4.5)

4.3.2 Extrusion Pressure

Figure 4.13 shows schematically the variation of load or pressure with the ram travel between direct and indirect extrusion. This plot has three distinct regions:

I The billet is upset, and pressure rises rapidly to its peak value
II The pressure decreases, and what is termed "steady state" extrusion proceeds
III The pressure reaches its minimum value followed by a sharp rise as the "butt" is compacted

The parameter that determines whether extrusion will proceed or fail to extrude is the magnitude of the maximum pressure, which must be within the extrusion press capacity. The factors that influence successful extrusion are

- Extrusion temperature
- Temperature of container, die, and associated tooling
- Extrusion pressure
- Extrusion ratio
- Extrusion speed
- Billet length
- Chemistry of the alloy

In the direct extrusion process, pressure reaches a maximum at the point of breakout at the die. The difference between the maximum and minimum pressures can be attributed to the force required in moving the billet through the container against the frictional force.

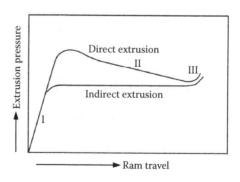

FIGURE 4.13
Variation of load or pressure with ram travel.

The actual pressure exerted on the ram is the total pressure. The total extrusion pressure required for a particular extrusion ratio is given by

$$P_T = P_D + P_F + P_R \tag{4.6}$$

P_D is the pressure required for the plastic deformation of the material, which is given in the functional form as

$$P_D = f(\bar{\sigma}, \bar{\varepsilon}) \tag{4.7}$$

where the flow stress, $\bar{\sigma}$ is defined by

$$\bar{\sigma} = f(\bar{\varepsilon}, \dot{\bar{\varepsilon}}, T) \tag{4.8}$$

strain and strain rate are defined by

$$\bar{\varepsilon} = \ln \frac{A_C}{A_E} \tag{4.9}$$

$$\dot{\bar{\varepsilon}} = \frac{d\bar{\varepsilon}}{dt} \tag{4.10}$$

and T is the temperature of the material.

Figure 4.14 shows the friction components in direct solid die extrusion. P_F is the pressure required to overcome the surface friction at the container wall friction, dead metal zone friction, and die bearing friction, which is given in the functional form

$$P_F = f(p_r, m, m', m'', D, L, L') \tag{4.11}$$

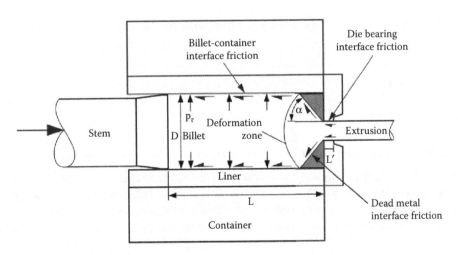

FIGURE 4.14
Friction components in direct solid die extrusion.

where p_r is the radial pressure applied to the container wall by the billet, m is the friction factor between billet and container wall, m' is the friction factor at the dead metal zone-flowing metal interface, m" is the friction factor between extruded material and die bearing, D is the billet diameter, L is the length of the billet, and L' is the die bearing length of a solid die.

P_R is the pressure required to overcome redundant or internal deformation work, which is given in the functional form

$$P_R = f(\overline{\sigma}, \alpha) \tag{4.12}$$

where α is the semi-dead metal zone angle. This work expenditure, which is not related to the change in dimensions from the billet to the extrusion, is called redundant work [2]. The redundant work is mainly responsible for the large difference between the actual extrusion pressure and the calculated pressure on the basis of uniform plastic deformation.

4.3.3 Extrusion Force

The force required for extrusion depends on the flow stress of the billet material, the extrusion ratio, friction condition at the billet-container interface and friction condition at the die material interface, and the other process variables such as initial billet temperature and the speed of extrusion. The required extrusion force F_r is given by

$$F_r = P_T A_c \tag{4.13}$$

where P_T is the extrusion pressure and A_c is the area of the container bore.

The force term is essential in determining the capacity of the extrusion press. The external force given by the extrusion press will determine the press capacity. For successful extrusion, the force balance has to be satisfied as follows:

$$F_p > F_r \tag{4.14}$$

Force (compression power) applied by the press is given by

$$F_p = p\,A_1 + p(2A_2) \tag{4.15}$$

where A_1 is the area of the main cylinder, A_2 the area of each side cylinder, and p the applied hydraulic pressure to the cylinders as shown in Figure 4.15.

4.3.4 Tribology and Thermodynamics in Extrusion

The thermodynamic and tribological relationship [3] plays a very critical role in aluminum extrusion technology. Knowledge of the fundamentals of tribology (friction, lubrication, and wear) is essential in dealing with the field of metal working processes. During the hot extrusion of aluminum, the friction at the die/material interface has a considerable influence on the accuracy of the shape and surface quality of the extrusion including die wear and life of dies. Figure 4.14 shows the friction force components in direct extrusion, which are the sources of heat energy to the direct extrusion process. Most of the work of

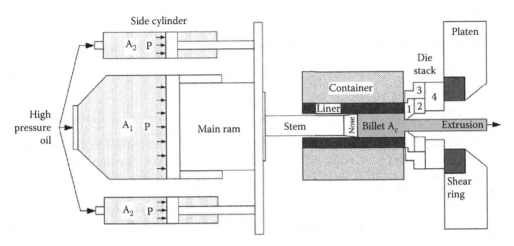

FIGURE 4.15
Schematic diagram of direct extrusion press for aluminum alloys, (1) die, (2) backer, (3) die ring, and (4) bolster.

deformation is transformed into heat. This temperature rise due to plastic deformation can be several hundred degrees. Friction forces acting in three different locations affect the overall temperature change in the billet as well as in the extruded product leaving the die.

Temperature is one of the most important parameters in the extrusion process. The flow stress is reduced if the temperature is increased, and deformation is, therefore, easier. At the same time, however, the maximum extrusion speed is reduced because localized temperature can lead to a reaching of the incipient melting temperature of the specific alloy. The temperature changes during extrusion depend on many factors, such as the following:

- Initial billet temperature
- Flow stress of alloy at given temperature, strain, and strain rate
- Plastic deformation (homogeneous and redundant work)
- Friction at billet container, dead metal flowing material, and die bearing flowing material interfaces
- Heat transfer (both conduction and convection)

Extrusion with constant temperature is referred to as *isothermal extrusion* and has practical interests due to significant major benefits including

- Improved dimensional stability
- Uniform surface quality
- Improved or consistent mechanical properties with uniform microstructure
- Faster extrusion speed to increase productivity

Extrusion press designers and manufacturers have developed solutions to improve the isothermal extrusion technology by introducing close loop feedback. Input variables are provided by a monitoring device for the exit temperature, producing benefits to extrusion manufacturers as well as the end users of the extrusion, mainly for aerospace applications.

4.3.5 Extrusion Press

The extrusion presses are classified mainly as [2]

1. Direct
2. Direct for seamless hollow
3. Indirect

The schematic diagram of a direct extrusion press is shown in Figure 4.16. Direct presses are used to make solid bars, rods, strips, and integrated sections. The press could also be used to extrude tubes and hollow sections from softer grade aluminum using solid billet through porthole or bridge dies.

Figure 4.17 shows the schematic diagram of a direct extrusion press for seamless hollow shapes and tubes. Tubes from harder aluminum alloys are generally produced from hollow short billets using floating or fixed mandrel on the stem.

The basic type of press developed for indirect extrusion consists of the same elements as the presses used for direct extrusion as shown in Figure 4.18. In the case of indirect extrusion, however, the die fixed on the front end of the hollow stem penetrates into the container.

With indirect extrusion, the circumscribed circle of a section is smaller than with direct extrusion and the stress on the stem is higher. The indirect method, however, affords the following advantages:

- Longer initial billets
- Higher extrusion speed for many materials
- Thinner butt ends
- More uniform structure over the extruded length

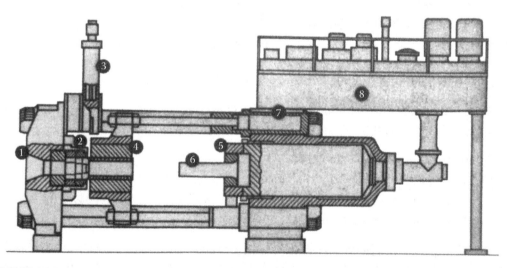

FIGURE 4.16
Schematic diagram of direct extrusion press ((1) counter platen, (2) die slide or rotary die head, (3) shear, (4) billet container, (5) moving crosshead, (6) stem, (7) cylinder crosshead, and (8) oil tank with drive and controls). (From Schloemann.)

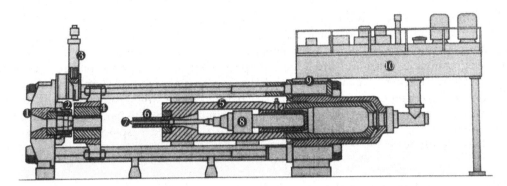

FIGURE 4.17
Schematic diagram of direct extrusion press for seamless hollow and tube ((1) counter platen, (2) die slide or rotary die head, (3) shear, (4) billet container, (5) moving crosshead, (6) stem, (7) mandrel, (8) piercer, (9) cylinder crosshead, and (10) oil tank with drive and controls). (From Schloemann.)

- Thinner sections
- Closer tolerances over the entire length of the product
- More uniform container and billet temperatures during extrusion
- Longer service life of the container and liner

Generally for hard alloy extrusion especially for aerospace industry, the metal flow properties obtained with indirect extrusion are much more favorable than in the case of the direct method. With the aid of the flow pattern occurring during direct extrusion, it is possible to decide for which materials and products the indirect extrusion method should be considered. Indirect extrusion is often more economical for the manufacture of rods, bars, sections, and tubes made from many aluminum alloys.

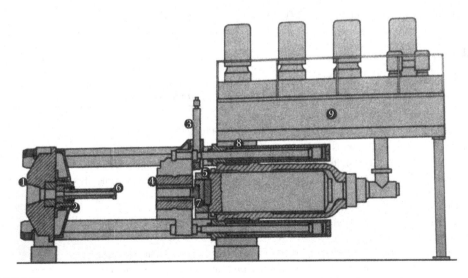

FIGURE 4.18
Schematic diagram of indirect extrusion press ((1) counter platen, (2) die slide, (3) shear, (4) billet container, (5) moving crosshead, (6) die stem, (7) sealing element, (8) cylinder crosshead, and (9) oil tank with drive and controls). (From Schloemann.)

Most modern presses have a relative position indicator to monitor the alignment of the container and the moving cross head during extrusion. They are also provided with press monitoring equipment to display press operation data on the monitor screen in real time. The programmable controllers are linked to operate billet heater, extrusion press, puller, cooling table, stretcher, saw table, saw, etc. Extrusion data are measured and stored using the controlled computer system for each die. That data are available and can be effective for designing and modifying dies and selecting optimum extrusion parameters including billet temperature, container temperature, billet size, ram speed/displacement, and extrusion speed/puller speed.

4.3.6 Extrusion Dies and Tooling

Performance of extrusion dies and tooling impacts product quality, productivity, recovery, and product design. That is why the aluminum extrusion industries have a huge demand on the extrusion die technology to make integrated shapes with critical tolerances. The functions of the individual tools are shown in Table 4.2.

4.3.7 Extrusion Die for Solid Shapes

This section will cover some basic knowledge of die and tooling for extrusion of solid shapes. Die design and die making are the most important and demanding aspects of the entire extrusion process. Die design is a vast subject and also can be considered an art of a die maker. The fundamental aspect of die design is mainly covered based on aluminum extrusion technology [2]. The basic consideration of die design is to determine the following factors:

- The number of die openings based on the shape and size of the profile, and also to check the existing tooling
- Location of die opening with respect to the billet axis

TABLE 4.2

Functions of the Individuals Tools

Tool	Function
Die	Makes the shape of extrusion
Die holder/ring	Holds the die with feeder plate and backer
Die backer	Supports the die to prevent collapse or fracture
Bolster	Transmits the extrusion load from die to the pressure ring
Pressure ring	Transmits the extrusion load from bolster to press platen and also prevents bolster to deflect
Die carrier/slide	Holds the complete die set (die ring and bolster) in the press
Bridge/spider/ porthole die	Special die to make a hollow shape with welding joint along the length of the shape
Feeder plate	Sits in front of die to balance the metal flow and also to make a continuous extrusion
Stem	Fitted with the main ram to push the billet through dummy pad
Dummy pad/nose	Protects the life of expensive stem fitted or floats in front of the stem
Liner	Protects the life of an expensive huge container from thermal and mechanical stresses

Source: Adapted from Saha, P.K., *Aluminum Extrusion Technology*, ASM International, Materials Park, Ohio, 2000.

- Orientation of the openings around their centroids to match the handling system
- Determination of the final die openings based on thermal shrinkage, stretching allowance, and die deflection (both die and deep tongues)
- Optimization of bearing lengths to increase productivity

To start with the design of die, the designer needs some fundamental information including geometry of the shape, alloy to be extruded, size of the press, billet size, extrusion length required by the customer, support tooling like backer or bolster to be used, weight of extrusion per unit length, and so on.

Basically, there are two major methods of layout for multi-hole dies; these are radial and flat layouts. In radial layout, the major axis of each shape lies along a radius, as shown in Figure 4.19a. The geometric layout of the openings within the die face is determined by a number of factors such as

- Proper clearance "A" between the die opening and the container wall and also the distance "B" between the openings (Figure 4.19b)
- Balanced metal flow to avoid any distortion of the shape
- Ease in die design and manufacture
- To avoid overlapping and scratching a particular part of the extrusion on the run-out table

A minimum clearance between the die opening and the container wall is required to avoid the flow of the oxide skin of the billet surface into the extrusion in the case of the direct extrusion. At the same time, the minimum distance between two openings of a multi-hole die must be adequate to have proper strength to withstand the pressure applied by the billet. Sufficient strength in the die may avoid cracking and deflection in the die. The number of holes in the die is determined by many factors and some of them are existing tooling to fit the number of holes, handling ease and facilities on the runout table, length, profile dimensions, and metal flow.

(a)

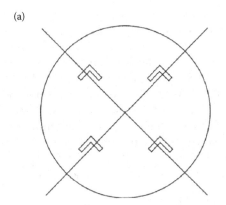

(b)

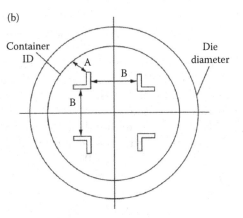

FIGURE 4.19
Die layouts, (a) radial and (b) flat.

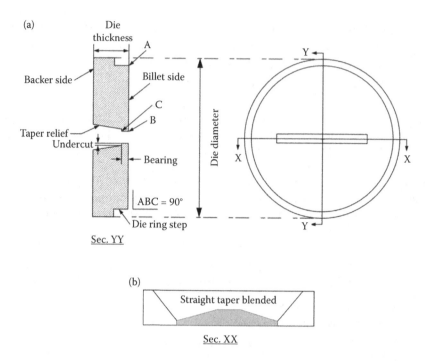

FIGURE 4.20
Solid flat-face extrusion dies, (a) die configuration and (b) bearing length.

The schematic diagram of a solid flat-face die configuration is shown in Figure 4.20a. The bearings of a die are of utmost importance. The function of the bearing is to control size and shape, finish, speed of extrusion, and also determine the life of the die. Friction at the die land is the controlling factor to retard the metal flow. The length of bearing at any location of the die opening depends upon the extent to which the metal flow has to be retarded at that point. Basically, there are three parameters that determine the dimensions of the die bearing to control the metal flow:

- Distance of the opening from the center of the billet
- Section thickness at that location
- Pocket shape and size—applicable to pocket/feeder plate die for softer alloy

In direct extrusion, the frictional resistance at the billet–container interface slows down the metal flow near the billet surface. The center of the billet thus moves faster than the periphery of the billet. To balance the flow, bearing length needs to be inversely proportional to its distance from the center of the billet. The thinner the section, the slower the flows due to passage through a smaller die opening. Similarly, to balance the flow in the thinner section, the bearing length needs to be smaller and vice versa. Sharp changes in bearing may cause streaks, due to uneven flow of the metal or to inadequate filling of the die opening. Variations in bearing lengths at the junction points must be properly blended to prevent streaking. An example of a straight taper blended bearing is shown in Figure 4.20b.

The treatment of bearing surfaces at the front and back of the die aperture is known as "choke" or "relief," respectively, as shown in Figure 4.21. For harder materials like 2xxx and 7xxx series alloy extrusion, the front of the bearing is generally choked (Figure 4.21a)

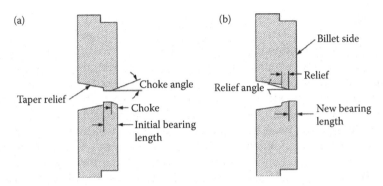

FIGURE 4.21
Choke and relief in die bearing, (a) choke and (b) relief.

at an angle up to 3°. This slows the metal flow and consequently fills out the die aperture to give better dimensional stability. Increasing the relief angle (Figure 4.21b) at the back or exit side of the bearing to as high as 7° increases the speed of metal flow by decreasing the original bearing length.

4.3.8 Extrusion of Hollow Shapes

Extrusion of tubing and hollow shapes is a complex process as compared to solid-shape extrusion. Hollow extrusion die and tooling systems are completely different from that of a solid-shape extrusion. There are two major ways of producing tubing and hollow extrusion shapes in the industry depending on the type of hollow. They are based on the requirements of the product, and the chemistry of the alloy being extruded. Tubing is classified as type I, seamless (tubing extruded from hollow billets using die and mandrel) and type II, seam (tubing extruded from solid billets using porthole or spider dies, or similar tooling). Seamless type I hollow extrusions are produced by using the direct extrusion press, having mandrel attached to the stem as shown in Figure 4.22, or a press with piercing ram or mandrel through a hollow stem, controlled independently by the main ram (Figure 4.17). This process is usually restricted to thick-walled round tube or very simple symmetrical hollow shapes. Tubes from harder alloys are generally made using hollow short billets. Extruded round tubes are then processed through a series of tube drawing operations with subsequent annealing processes to create the final dimension and shape

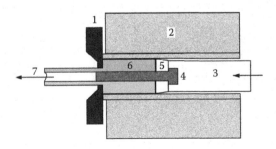

FIGURE 4.22
Schematic representation of seamless (type I) hollow extrusion using fixed-type mandrel ((1) die, (2) container with liner, (3) stem, (4) mandrel attached with stem, (5) hollow nose, (6) hollow billet, and (7) seamless hollow).

of the tube. Drawn seamless tubes are process intensive, incurring higher production cost. Manufacturing technology of drawn tubing will be discussed in Section 4.6.

Figure 4.23a shows the sectional view of type II hollow extrusion through a special type extrusion die. The die has two major components: a mandrel, that determines the inside shape; and a cap that determines the outer shape of the tube or hollow sections and section thickness. In type II hollow extrusion, the solid billet is forced to flow into separate ports and around the bridges that support the mandrel (Figure 4.23b). The separate streams of metal that flow through the ports are brought together in a welding chamber surrounding the mandrel, and the metal exits the die through the gap between mandrel and dies cap as a hollow extrusion, as shown in Figure 4.23a. Metal flow through a hollow die takes place in three different stages. In stage one, metal flows from the solid billet inserted into a container of cross-sectional area, A_C, to the die mandrel port. In stage two, the metal flows from the mandrel port into the weld chamber cavity in the die cap. In the final stage, metal

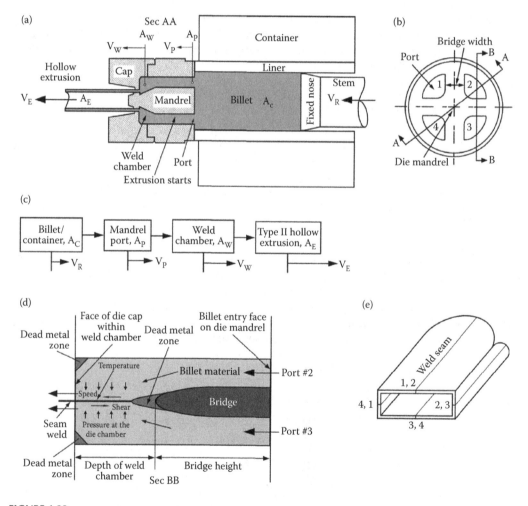

FIGURE 4.23
Schematic diagrams of hollow type II extrusion dies and metal flow mechanisms, (a) schematic representation of type II hollow extrusion, (b) front face of die mandrel with billet entry port, (c) control volume of a hollow die, (d) weld seam formation model, and (e) weld seam configuration with the mandrel port numbers.

flows from the weld chamber to the gap between the mandrel and cap to form the hollow extrusion of cross-sectional area, A_E. The weld chamber includes the total volume between the top of the bridge of the mandrel and the cavity in the die cap. The volume constancy relationship in the control volume as shown in Figure 4.23c is given by

$$A_C V_R = A_P V_P = A_W V_W = A_E V_E \qquad (4.16)$$

where A_P is the cross-sectional area of the port in the mandrel, V_P the material speed through the mandrel port, A_W the area of the weld chamber, and V_W the material speed though the weld chamber [2,4]. From this relationship, the extrusion speed, V_E, or the speed of metal flowing through the die can be calculated from a given ram speed, V_R.

Seam welds are formed in some specific locations within hollow extruded profiles based on the number of ports and design layout of a weld chamber-type die. The seam weld mechanism has been modeled schematically as shown in Figure 4.23d. The sheared streams of billet material entering through ports #2 and #3 (Figure 4.23b, Section BB) separate around the bridge and start joining together at the tapered end of the bridge at certain pressures and temperatures at the die welding chamber. This allowed a continuous seam welding mechanism before the metal finally flows through the gap between the die cap and mandrel bearing surfaces. Figure 4.23e shows schematic representation of a hollow rectangular shape extruded from a four-port weld chamber die as shown in Figure 4.23b. Figure 4.23e shows four seam welds along the longitudinal or extrusion direction. Each seam weld is formed by metal entering through two adjacent ports and forming seams according to the port numbers shown in Figure 4.23e.

Although the separate metal streams are joined within the die, where there is no atmospheric contamination, creating a perfectly sound weld at the joint is a challenge for harder aerospace alloy extrusion [4]. The following variables are critical to determine the quality of the hollow extrusion:

- Port and bridge geometry
- Depth of welding chamber
- Billet chemistry and temperature
- Extrusion speed
- Lubrication system at the die/container interface

4.3.9 Extrusion of Aluminum Alloys

Aluminum extrusion technology book [2] has covered almost all the branches of aluminum extrusion technology. Aluminum extrusions are generally classified in solid and hollow shapes as illustrated in Figure 4.24. Extrusion of solid shapes of varying cross sections (both heavy and light weight) is being used for different applications including primary and secondary airframe structures, interiors, hydraulic systems design, and many more. Extrusion geometry and the corresponding alloy are selected based on the engineering design requirements for each application. In addition to solid shape extrusions, there are many uses of hollow extrusions in aerospace applications including commercial airplanes, military aircraft, and spacecraft.

Extrusion of aluminum alloy shapes involves the most advanced technology in terms of alloy making, press design, tooling, and die design. The billet chemistry determines the type of alloy and the appropriate applications. In general, aluminum extrusions are

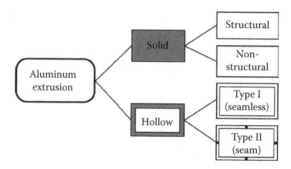

FIGURE 4.24
General classifications of extrusions.

categorized into two different areas based on the properties and application of the alloys in the commercial and industrial application:

- Soft and medium grade alloy—architectural, commercial, and industrial applications (heat treatable alloys)
- Harder alloys—structural, industrial, and especially aerospace applications (heat treatable alloys)

Extrudability, which can be measured by the maximum extrusion speed, is one of the most significant factors influencing cost and efficiency of the extrusion process. Temperature and speed parameters, together with the state of stress in the deformation zone mainly in the die region, play a significant role in improving extrudability of a given alloy. The alloys are classified into three different groups according to their extrudability [5,6]:

I. Alloys easy to extrude: pure aluminum, AlMn, AlMg1, AlMgSi0.5, AlMgSi0.8

II. Moderately difficult alloys: AlMg2-3, AlMgSi1, AlZnMg1

III. Difficult alloys: AlCuMg, AlCuMgPb, AlZnMgCu, AlMg > 3%Mg

Aluminum and aluminum alloys [7] under groups I and II are considered to be soft and medium grade alloys with extrudability rating between 40 and 100. Alloys under group III are considered to be harder alloys. The extrudability rating of some heat treatable harder alloys is given in Table 4.3. The extrudability as a limiting deformation under the given

TABLE 4.3

Extrudability Rating of Some Harder Alloys

Alloy	Major Alloying Elements	Relative Extrudability[a]
2014	Cu, Si, Mn, Mg	20
2024	Cu, Mg, Mn	15
7001	Zn, Mg, Cu, Cr	7
7075	Zn, Mg, Cu, Cr	10
7079	Zn, Mg, Cu, Mn, Cr	10

Source: Adapted from *Aluminum Extrusion Alloys*, Kaiser Aluminum and Chemical Sales, Inc., Oakland, California, 1964.

[a] Difficult to extrude: <30.

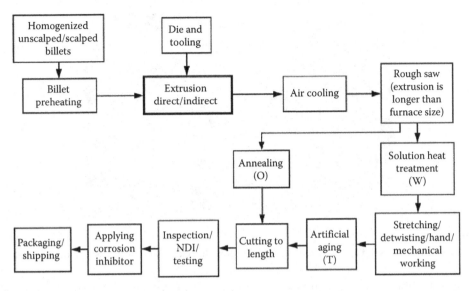

FIGURE 4.25
Functional block diagram showing the principal process steps involved in harder aluminum alloy extrusion plant.

process conditions is defined and the maximum extrusion speed without cracks is suggested as its measure.

Figure 4.25 shows the functional block diagrams of harder alloy extrusion plant with different heat treatment processes required satisfying customer's requirement.

Extrusion parameter, for example, billet temperature and extrusion speed are fully dependent on the chemistry of each alloy. The container temperature is controlled based on the initial billet temperature. The typical values of extrusion parameters of some heat treatable harder alloys are given in Table 4.4. Temperatures and extrusion speeds are dependent on the final shape and the extrusion ratio and it may be necessary to start with lower billet temperatures than mentioned in the table. Figure 4.26 shows an extrusion of solid-shaped profile coming out of the press. Figure 4.27 shows some harder alloy light and medium gage solid extrusion shapes for aerospace applications. Figure 4.28 shows extrusion of hollow (Type I and Type II) exiting the press. Figure 4.29 shows examples of some harder alloy hollow shapes.

Type I round tube extrusion from aerospace alloys are produced as an input raw stock for further tube drawing operations to produce tubes of varying OD and wall thickness. Extrusion of tubes and hollow shapes (type II) are generally produced from softer alloys using welding chamber-type dies [2]. These dies have a wide range of applications in the aluminum extrusion industry and are suitable for aluminum alloys such as AlMn, AlMgSi, and some AlMg and AlZnMg. Pressure welding properties of aluminum alloys

TABLE 4.4

Typical Values of Billet Temperature and Extrusion Speed of Harder Alloys

Alloy	Billet Temperature °F (°C)	Exit Speed ft/min (m/min)
2014–2024	788–842 (420–450)	5–11 (1.5–3.5)
7075, 7079	572–860 (300–460)	3–7 (0.8–2)
7049, 7150, 7178	572–824 (300–440)	2.5–6 (0.8–1.8)

Source: Adapted from Laue, K. and Stenger, H., *Extrusion*, ASM International, Materials Park, Ohio, 1981.

FIGURE 4.26
Aluminum extrusion of solid aerospace profile leaving through the die. (From Universal Alloy Corporation.)

for the production of tubes and hollow shapes are varied, compared with their billet chemistry. The welding chamber dies have been successfully used to produce hollow extrusion shapes in longer lengths with thin walls and complex geometry.

Type II hollow dies for aerospace alloys like AlZnMgCu are normally treated separately in the extrusion press with controlled variables including billet temperature, speed, and lubrication system to avoid extrusion defects like weld (seam) failure, tearing, poor surface finish, and flawed extrusion shape. Extrusion manufacturers explored different die technologies to develop cost-effective, harder alloy type II hollow extrusions to meet engineering requirements of products, mainly for aerospace and other industrial applications.

4.3.10 Common Aluminum Extrusion Defects

Detection of extrusion defects followed by analysis and preventive measures are three important factors to be considered in every extrusion plant to maintain the highest quality of the extrusion. To deal with those factors, it is necessary to have a complete understanding

FIGURE 4.27
Example of harder alloy light and medium gage extrusions. (From Universal Alloy Corporation.)

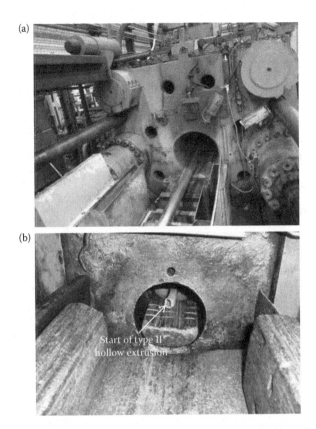

FIGURE 4.28
Aluminum hollow extrusion production, (a) type I aerospace tube extrusion production (From ALUnna Tubes, Germany.) and (b) type II hollow aerospace profile (From Universal Alloy Corporation, USA.).

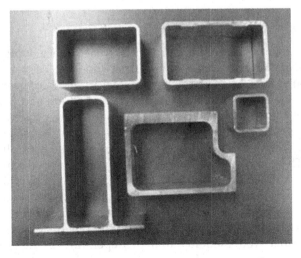

FIGURE 4.29
Example of type II harder alloy extrusions. (From Universal Alloy Corporation.)

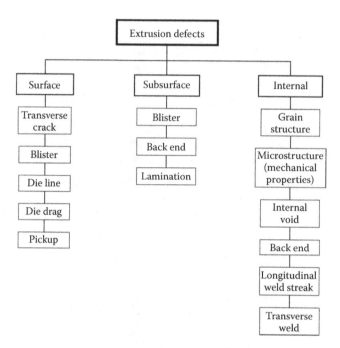

FIGURE 4.30
Types of aluminum extrusion defects.

of the mechanics and extrusion variables and their effects on extrusion. Any extrusion defect can be related to any of the extrusion variables or a combination of variables discussed in Section 4.3.1. The common extrusion defects can be categorized into three main types: (1) surface, (2) subsurface, and (3) internal as shown in Figure 4.30. Each type has its several individual defects related to the type of the defects. Examples of several surface-related defects including transverse crack (called speed crack), blister, die line, die drag, and die pick up have been shown in Reference 2.

4.3.11 Economics of Aluminum Extrusion

The cost of manufacturing aluminum extrusions for aerospace applications [8] is influenced by a number of factors associated with the various process steps including billet casting with special need, die design and manufacturing, billet preheating, controlled extrusion press parameters, solution heat treatment, stretching and profile correction, as well as extended and more complex precipitation hardening cycle with added quality-control requirements. The additional process steps include two separate complex heat treatment processes, as well as post-heat treat processes like controlled stretching and profile correction. These account for the basic differences in producing extrusions from high strength aluminum like 2xxx and 7xxx series alloys compared to the more common 6xxx alloys [2]. In general, the extrudability ratings of 2xxx and 7xxx alloys are very low compared to the 6xxx series alloys. Special attention is being given while designing and manufacturing the dies and tooling. Besides the initial raw material (billet) cost, die and tooling costs could range from moderate to high, depending on the type (solid or hollow or semi-hollow) and the geometry of an extrusion. The die cost is spread over the total number of billets extruded or the total quantity of extrusions through the particular die.

Also, die and set-up costs decrease as the quantity produced from the die increases. The die cost generally goes higher when the size of the extrusion gets larger. Heavy profile aerospace extrusions come from the larger capacity presses with larger dies. One of the cost measuring factors is the rate of production, and percentage recovery from an extrusion die, which is a function of many factors. These include die geometry and the design parameters and tolerances applied to the geometry. Extra fabrication costs being considered are due to many operations involved other than die and tooling cost. These include making use of additional equipment, energy, and manpower involved to produce the extrusions for use by the aerospace industry.

Cost inputs of producing extrusions for aerospace application are shown in the block diagram (Figure 4.31). Cost inputs consist mainly of energy required to run each operation, and the labor required to get the work done at each operation. Figure 4.31 also shows the billet as input to the extrusion press, and output as the net extrusion with final temper condition as required by the customer. There are loss materials between the gross weights of extrusion from the press to the net weight of extrusion for sale with a certain percentage recovery. Losses are mainly incurred during preheat treatment and post-heat treatment cycles. Percentage recovery is the driving factor for pricing the extrusion.

Major variables affecting cost of extrusions are explained using the sketches shown in Figure 4.32:

- Extrusion geometry
- Billet alloy
- Press size and order quantity
- Length of extrusion and final temper

Extrusion product shape and sizes are defined by their geometric configuration. Based on the geometry, the factors including cross-sectional area of the shape, perimeter of the shape, and the circumscribing circle diameter are commonly used in defining the shapes and sizes of extrusion. Shape factor/form factor were introduced [5] as a measure of the degree of difficulty regardless of classification into different types of sections. These factors help designing and manufacturing die and tooling. Tight tolerances and any critical tongue in the die will increase the cost of the die. Based on the degree of difficulty of producing the extrusion shape, press selection comes into consideration to determine the type of press (direct or indirect), and also the press capacity (large, small, or medium tonnage). Finally, productivity/recovery of the extrusion will be the determining cost factor of producing the extrusion.

As mentioned earlier, the extrudability ratings of billet alloys for aerospace extrusions are very low and as a result, lower extrusion speed is generally used to extrude with longer press time, which increases the price of extrusion. Additional price considerations come from the final recovery (%) of the extrusion from the particular billet alloy and the shape of extrusion.

The price of extrusion is very much dependent on the press size needed and the order quantity from the customer. The larger the press used, the higher the cost of energy to run the press and the cost of die and tooling. Running a small quantity will cost more due to shorter machine time with the same amount of machine setup time for die and tooling change. Press efficiency plays a certain role in controlling the price. Energy efficient modern presses with many closed-loop control systems can optimize the production process to increase the rate of production without changing the quality of extrusion.

The price of extrusion is also dependent on the required length and the final temper needed by the customer. It is required to determine the billet size to calculate the number

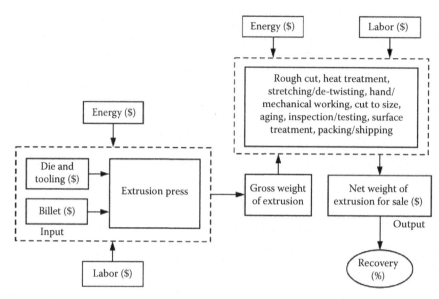

FIGURE 4.31
Energy and labor inputs at various stages for aerospace extrusion.

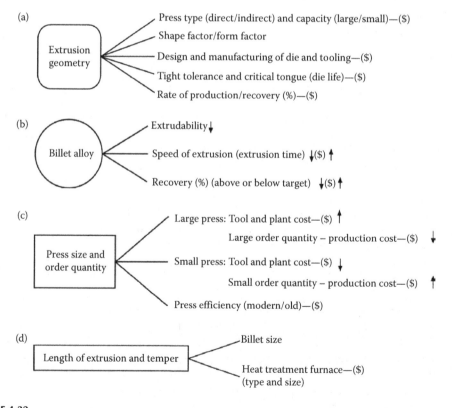

FIGURE 4.32
Major variables affecting cost of extrusion, (a) extrusion geometry, (b) billet alloy, (c) press size and order quantity, and (d) length of extrusion and temper.

of pieces of rough cut extrusions per billet prior to the heat treat process. Length of the extrusion in rough cut is also dependent on the type and size of the heat treatment furnace available or needed.

4.3.12 Extrusion of Titanium Alloys

Titanium extrusions are widely used almost exclusively for military aircraft. Titanium and its alloys have a good resistance to corrosion and a favorable strength-to-density ratio. Due to this advantage, titanium extrusions have drawn considerable demand to manufacture various components of commercial aircraft especially, since carbon fiber composites have been considered as structural materials. Due to galvanic reaction between aluminum and carbon fiber composite materials, the usage of titanium alloy products at the interface with the composite materials in the aircraft is growing rapidly. The fundamentals of extrusion technology of titanium alloys are very similar to those of aluminum extrusion technology with some exceptions. This is true mainly on the extrusion billet temperature, which is the elevated extrusion temperature in the range of 1292–1904°F (700–1040°C) for titanium. Due to the elevated billet temperature, a few more adjustments are associated with titanium extrusion, including billet temperature, die design, lubrication, die and tooling temperature, and extrusion speed and post-extrusion treatments as shown in Table 4.5.

Most commercial titanium alloys contain aluminum to increase the strength and vanadium, chromium, manganese, molybdenum, or copper to depress the α/β transformation temperature. The β-phase temperature range is preferred for extrusion. Billet temperature is a very critical factor in titanium extrusion. A visible surface oxide scale and oxygen-enriched α-stabilized layer (alpha case) is produced at high temperature above 1292°F (700°C). Alpha case layer is brittle in nature. To avoid formation of oxide and alpha case layers, it is recommended to go for rapid billet heating using induction heating or a furnace with a protective atmosphere. Table 4.6 shows some typical billet temperatures of different titanium alloys.

Similar to aluminum extrusion, a round cylindrical billet is the input material for titanium extrusion also. But processing steps of making titanium extrusion billet are different than that of aluminum billet preparation. Figure 4.33 shows the functional block diagram showing the major process steps of titanium billet making for extrusion. The processing of titanium billet involves three major steps: reduction of titanium ore into a porous form called "sponge;" melting of sponge with master alloys to form titanium ingot; and primary

TABLE 4.5

Difference in Fundamental Parameters between Aluminum and Titanium Extrusion

Extrusion Parameter/Post-Processing	Harder Aluminum Alloys	Titanium Alloys
Billet temperature	572–860°F (300–460°C)	1292–1904°F (700–1040°C)
Container temperature	Hot	Hot
Die and tooling temperature	Hot	Ambient
Die design	Flat face entry	Conical entry
Die lubrication	Graphite	Glass
Extrusion speed	Slow	High
Press tooling	Hot stem/dummy pad	Cold stem/dummy pad
Stretching/straightening	Ambient	Hot
Heat treatment/stress relieving	Solution treatment and aging	Annealing

TABLE 4.6

Billet Temperature for Titanium Alloy Extrusion

Material	Billet Temperature°F (°C)
Ti99.5	1292–1652 (700–900)
(α + β)—*Alloys*	
6Al4V	1652–1832 (900–1000)
7Al4Mo	1562–1742 (850–950)
6Al6V2Sn	1472–1904 (800–1040)
6Al4Zr2Sn2Mo	1472–1742 (800–950)
α—*Alloys*	
1Al2.5Sn	1544–1904 (840–1040)
8Al1Mo1V	1706–1850 (930–1010)

Source: Adapted from Laue, K. and Stenger, H., *Extrusion*, ASM International, Materials Park, Ohio, 1981.

Note: Container temperature is kept around 1004°F (540°C); maximum press speed in between 98 and 1148 ft/min (30–350 m/min) and the maximum extrusion ratio is kept 100:1.

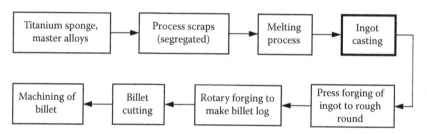

FIGURE 4.33
Major process steps of billet making.

fabrication to convert coarse structured ingot into billet. Melting is done by using different technologies including plasma, electron beam, and vacuum arc melting for the continuous casting method of producing titanium ingots. Cast ingots are processed through press forging for making rough round (Figure 4.34) shapes, before sending these to the rotary forge for making extrusion billet logs (Figure 4.35). Logs are cut into the respective billet length as needed for the extrusion geometry to be made in the extrusion press. Each billet is machined separately to remove the rough oxygen-rich outer surface of the billet and also to put a smooth finish with a radius at the entry side of the billet in the die to reduce friction at the die–billet interface.

Since flow stress of titanium alloys increases very rapidly with increasing strain rate, the extrudability of titanium alloys drops when extruded at high-speed to avoid excessive cooling of the billet in the container. As a result, titanium alloys are always extruded with a lubricant and conical dies with a large inlet radius (Figure 4.36) to enhance the extrudability of titanium alloys. Lubrication is done in both ways: (1) lubrication of the outer periphery of the billet to lubricate billet–container interface and (2) lubrication at the front of the die to lubricate die bearing-extruding metal interface. In 1950, Ugine Séjournet of France invented a process that uses glass as a lubricant for extruding steel [5]. The lubrication process called "Séjournet" process is used for titanium extrusion where the molten glass lubricates the high-temperature billets to retain heat and particularly provide

FIGURE 4.34
Press forging an ingot to rough round to feed to the rotary forge. (From ATI Wah Chang.)

titanium-shields against atmospheric contact during extrusion. The heated billet rolls across a glass frit table (Figure 4.37) on its way to the press, melting the glass and forming a thin layer of lubrication film 0.02–0.03 inch (0.5–0.75 mm), on the billet surface in order to separate it from container walls and allow it to act as a lubricant before it is pushed into the container (Figure 4.38) before extrusion.

Since, in direct extrusion, the friction between titanium billet and the container is severe, it is detrimental to produce inhomogeneous metal flow in the container before leaving the die opening, which can result in extrusion defects if a flat type of die is used. Conical entry dies (Figure 4.36a) are recommended to have more uniform flow through the die.

The following standard lubricated extrusion practices are generally followed:

- Machined billet to achieve good surface quality
- Induction heating of billet under a protective gas

FIGURE 4.35
Rotary forging making extrusion billet logs. (From ATI Wah Chang.)

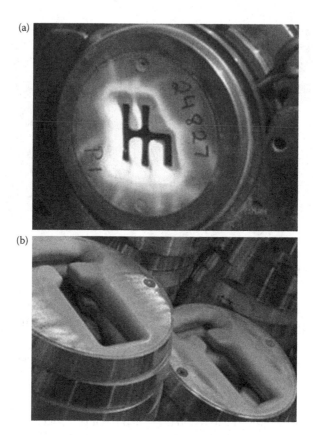

FIGURE 4.36
Titanium extrusion dies. (a) Conical entry die and (b) larger inlet radius. (From Plymouth.)

- Conical dies with large radius at the entry (Figure 4.36b)
- Billets of tube extrusion are pierced outside the press to avoid dropping of billet temperature

The schematic representations of direct extrusion of titanium alloys are shown in Figure 4.39. A round ring or cake of compressed glass powder attached to the die face (Figure 4.39a) is inserted into the container before the billet enters, sandwiching the cake between the billet and the die face. The glass powder cake melts as the billet is pushed through the die, distributing additional molten glass over the length of the extrusion (Figure 4.39b). Certain thickness of molten glass coating deposits on the billet front as it passes through the die, preventing metal-to-metal contact with the bearing surfaces. The extrusion will have a 0.01 inch (0.25 mm) thick layer of glass, which can be easily removed once it is cooled.

Figure 4.40 shows red hot titanium heavy extrusion for aerospace structural applications as it exits from the press. Die life is a very critical factor for titanium extrusion since the dies are always exposed at the extreme operating conditions, which are influenced mainly billet temperature. The shape of extrusion dies and the appropriate die material are very important in titanium and its alloy extrusions. The die has to satisfy two important factors:

1. Mechanical properties at high temperatures
2. Good resistance to wear

FIGURE 4.37
Titanium billet rolled down through glass powder before loaded into the container. (From RTI International.)

To enhance the wear resistance of a titanium extrusion die, ceramic–aluminum oxide coatings are applied to the working surfaces of the die. The coatings are applied perpendicular to the wearing surface to improve the bonding with the original die steel (H13 material).

Figure 4.41 shows the major process steps of titanium extrusion starting from the machined billet to the finished extrusion ready for fabrication of aircraft components. The first step after extrusion is the sand blasting to remove the glass from the extrusion, prior to hot stretch straightening.

Extrusions are stretch straightened using a special stretcher as shown in Figure 4.42. The stretch-straightening machines have headstocks that apply tension, reducing the cross section of the extrusion by about 1%. These headstocks can also rotate 360°, both clockwise and counterclockwise to make the extrusion straight and free from any angular twist. Stretching machine capacities generally fall between 125 and 300 t. Titanium, however,

FIGURE 4.38
Hot billet with glass lubricant entering into the press container. (From Plymouth.)

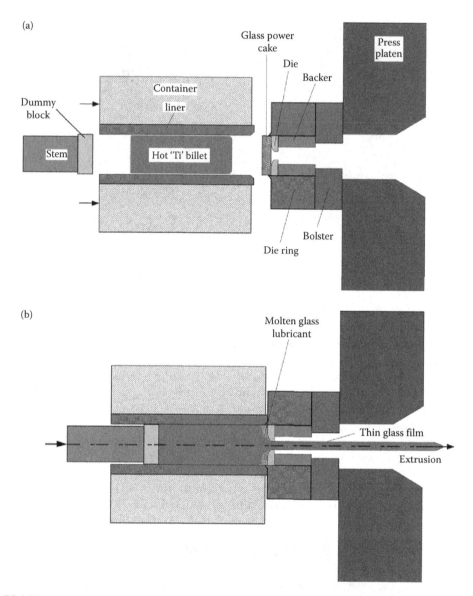

FIGURE 4.39
Schematic representation of direct extrusion of titanium alloys, (a) billet loaded into the container, glass cake attached to the die and (b) extrusion started.

cannot be cold worked, and it is resistance-heated to 1300°F (704°C) prior to stretching. Heating titanium extrusion in an exposed atmosphere forms a layer of alpha case over the entire surface of the extrusion. After hot stretch straightening, the extrusions are cut to the length required by the customer with the standard rejection of front and rear parts of the extrusion, which are clamped into the stretcher grips. Stretched extrusions are stress relieved by a subsequent annealing process to obtain the material properties required for the extrusion. Annealed extrusions are run through a pickling process to remove the alpha cases from the entire surface of the extrusion. The material removal from the pickling process falls normally within 0.005 inch (0.127 mm).

FIGURE 4.40
Heavy titanium aerospace extrusion from the press. (From RTI International.)

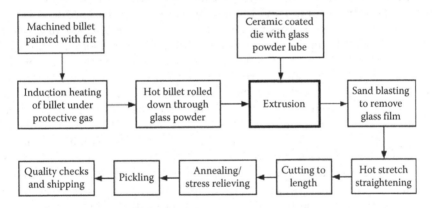

FIGURE 4.41
Major process steps of titanium extrusion.

FIGURE 4.42
Hot stretch straightening of titanium extrusion. (From Plymouth.)

4.3.13 Extrusion of Titanium Round Hollow

Seamless, type I round hollow extrusion presses and manufacturing processes for titanium alloys follow very much the same principles as explained in Sections 4.3.5 and 4.3.8 for aluminum hollow round-shaped extrusion. Since the seamless extruded tube will be used as an input raw stock for the tube reduction process, very special care needs to be taken during the extrusion process to obtain proper tube dimension and surface finish, and also uniform fine grain structure by controlling the following variables:

- Extrusion ratio
- Billet temperature
- Extrusion speed
- Lubrication

Maximum billet size and extruded tube sizes are dependent on the size of the press used by the manufacturer. The extruded tube size is also dependent on the final tube size to be produced in the tube reduction process. Extrusion ratio is generally kept between 5 and 10 and the billet preheat temperatures are kept between 1250 and 1800°F (675–980°C) depending on the extrusion ratio and the size of the press. Extrusion speed is varied by varying the ram speed of the press ranging from 1 to 20 inch (25–500 mm) per minute. Slower speeds provide better temperature control resulting in more uniform structure along the length of extrusion. Modern isothermal extrusion technology can produce fine uniform grain structure. Like solid extrusion, glass powder lubrication technique is used to lubricate the entire drilled billet and the die and tooling, since titanium has a tendency to gall and stick to the die surface to produce a poor surface finish. Figure 4.43 shows the production of a seamless titanium tube extrusion as it emerges from the press.

Surface preparation of the extruded tube is very important for the tube reduction process. Both inner and outer surfaces should be free from any flaws. As a part of the preparation of tubes for the tube reduction process, the outer surface requires grit blast to remove any lubricant film and any residual oxides. The tubes are next pickled in nitric-hydro-fluoric acid solution to prepare surface conditioning required for the cold reduction process. A titanium tube cold reduction process called the pilgering process will be discussed in Section 4.7.

FIGURE 4.43
Extrusion press extruding a hollow tube. (From ATI Wah Chang.)

4.4 Forging

Forging is the oldest metal working process, having its origin with the primitive black-smith of Biblical times. Forging is a heavy deformation process where heated metal ingot flows plastically by hammering or pressing between dies under high compressive force to result in desired shape. The basic principle of forging is shown in Figure 4.2. The process helps to refine the physical properties of the metal along with grain structure of the metal. Using proper die design, the metal's grain can be oriented in the direction of stress on the metal structure applied during actual application. Metal flow and grain structure are controlled in the forging process to have good strength and toughness. The advantages of the forging process include mainly:

- Parts have high strength to weight ratio—useful in the design of the aircraft structures
- It offers low cost of manufacturing

The forging machines are capable of making parts ranging from a bolt to a large landing gear beam of an aircraft, or even larger. Most forging works are carried out at elevated temperature (hot forging); some are also carried out at the room temperature (cold forging). Two major classes of forging equipment used for production are:

1. Drop hammer
2. Forging press

A forging hammer applies impact force on the material surface, whereas a forging press applies more slowly a compressive stress on the metal surface. Mainly, there are two types of forging processes used in the industry for manufacturing small-to-large metal parts of various shapes and geometry:

1. Open die forging
2. Closed die forging

Forging presses are designed and built to handle both open and closed die forging processes. Open die forging presses are designed to deal with upsetting ingots, slabs, and preforming shapes for closed die forging. Forging presses are mainly of the mechanical and hydraulic types.

4.4.1 Open Die Forging

In open die forging, a solid workpiece is placed between two flat dies and compressed by a vertically applied load by either a drop hammer or a forging press. Open die forging is also called upsetting or flat die forging. An example of a cylindrical solid billet being upset between two flat dies is shown in Figure 4.44. Figure 4.44a shows the workpiece ready for upsetting between two flat dies. Upsetting causes uniform deformation of the billet without friction (Figure 4.44b) whereas Figure 4.44c shows the barreling of the billet or ingot due to frictional force, τ_f, at the die–billet/ingot interface. Barreling could be controlled by using proper lubrication to reduce frictional stress at the interface. Shapes of simple

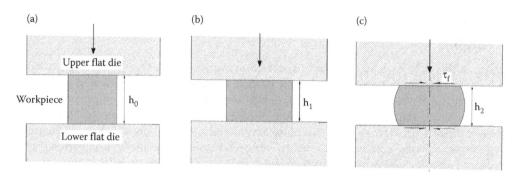

FIGURE 4.44
Schematic representations of open die forging/upsetting, (a) ready for upsetting, (b) ideal flow, and (c) friction effect.

geometry can be made by the open die process. In open die forging, a hammer comes down and deforms the workpiece, which is placed on a stationary anvil.

The flow pattern of local deformation can be visualized [9] by using the grid pattern applied to the test specimens as shown in Figure 4.45a. As deformation proceeds, the material adjacent to the dies remains stationary as shown in area I in Figure 4.45b. Material near the outer surface of the cylinder not in contact with the die is deformed since the center material moves radially outward as shown in area III in Figure 4.45b. The bulk of the deformation is concentrated in the remaining center part of the specimen as shown in area II of Figure 4.45b.

One of the basic forging operations is the upsetting of metal stock in between parallel flat dies [9]. The deformation is limited and is symmetrical around a vertical axis. Dies need to be made relatively narrow if preferred elongation is desired as shown in Figure 4.46a. When compressing a rectangular or round billet in between the flat dies, frictional forces at the billet and die interface play a role to control the metal flow along and transverse to the axial direction of the billet. The frictional force in the axial direction is smaller than that in the transverse direction. As a result, the metal flows more in the axial direction as shown in Figure 4.46b and c. Since the width of the die provides the frictional resistance, there will be some spread in the transverse direction. The ratio of elongation-to-spread depends on the width of the die. But too much of width reduction may cause cutting instead of elongating as shown in Figure 4.46c.

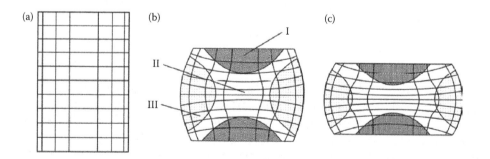

FIGURE 4.45
Schematic diagram of inhomogeneous deformation. (a) Grid pattern on test specimen, (b) 1st stage of deformation, and (c) final stage of deformation.

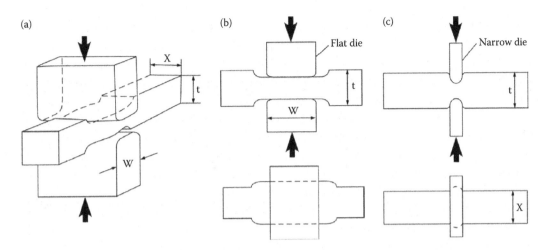

FIGURE 4.46
Compression between flat and narrow dies, (a), (b), and (c).

Forging billets are preshaped in open die forging usually by edging and fullering to move the metal in the right places for the next forging. The pre-shaped billet is then placed in the blocking die cavity for rough forging. The maximum change in the shape usually occurs in the rough forging step. The process concludes with the finishing die, where a workpiece is forged to the final shape within dimensional tolerances. For added economic advantage in the process, usually the blocking cavity and the final shaped cavity are made into the same die block. Fullering and edging geometries are often made on the edges of the die block to make the process more economical. Fullering and edging processes are explained [9] in Figure 4.47, based on the direction of metal flow in the dies with convex- and concave-shaped geometries, respectively. In the fullering process, the horizontal component of the applied forging force (except from the centerline) tends to move the metal away from the center of the die (Figure 4.47a). However in edging (Figure 4.47b), the horizontal force component tends to move the metal toward the center of the die increasing the thickness above the original bar thickness.

4.4.2 Closed Die Forging

Closed die forging is the process in which the metal deformation is confined within a shaped cavity in the dies to achieve a desired shape and geometry. Depending on the

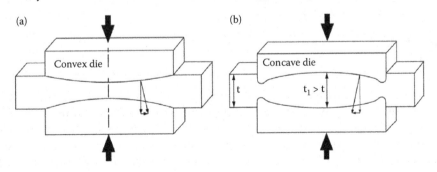

FIGURE 4.47
Schematic representations of forging process, (a) convex die (fullering) and (b) concave die (edging or rolling).

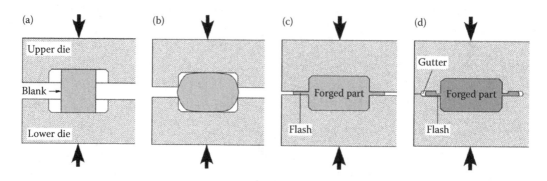

FIGURE 4.48
Schematic representations of closed die forging, (a) upsetting, (b) filling, (c) forged part with flash, and (d) forged part with flash in gutter.

shape and complexity of the part, dimensional tolerances, economics, and performance of the part, the dies are designed and fabricated to produce two-piece conventional top-and-bottom type dies. Hydraulic and mechanical presses are used to produce conventional and precision forgings in a variety of aluminum and titanium alloys for aircraft structural components. A schematic representation of the closed die forging process is shown in Figure 4.48 using a simple rectangular-shaped cavity in the dies. When two dies are brought together, the forging blank undergoes plastic deformation until the blank expands gradually to fill the die cavities eventually touching the side wall of the die impressions to complete the forging. At the end of the forging process, a small amount of metal begins to flow outside of the die impressions. This material is known as "flash."

In closed die forging, machined matching die blocks are used to produce forged parts to close dimensional tolerances. Since the forging dies are made from huge steel blocks, and a lot of machining processes are also involved, the dies are expensive. Large production runs are required to meet the cost of the dies. Forging blocks are prepared with additional material to allow the blocks to fill the cavity of the dies. Since it is difficult to keep the right amount of metal in the correct places during fullering and edging, it is necessary to use a slight excess of metal. During final closed die forging, when the upper and lower dies come close together, the excess material flows out as flash (Figure 4.48c). In order to prevent the formation of a very wide flash, a ridge, known as a flash gutter (Figure 4.48d), is usually provided in the die as part of a proper design. A couple of examples of flash gutter design [9] are explained in Figure 4.49. Figure 4.49a shows that conventional forging dies are designed with a cavity avoiding a place for the flash to flow. There are two important dimensions of the die in the flash region:

1. Flash thickness
2. Land width

When the land width increases or flash thickness decreases, more constraints are provided to the metal being forged. By changing the design to have minimum flash formation in forging, there will be a requirement for a higher forging load that imposes higher stresses on the forging dies. Figure 4.49b shows the design of tapered saddle in the die. Tapered saddle die design provides more constraint to metal flow than that of parallel saddle design. This design also minimizes metal loss due to flash. Tapered saddle design is generally used when the metal savings could justify the use of forging equipment with

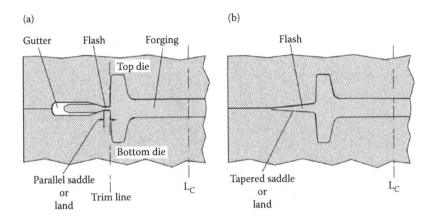

FIGURE 4.49
Flash gutter designs, (a) conventional and (b) taper saddle.

higher capacity. The choice of flash cavity design in the die may be a matter of economics of forging.

The final step of closed die forging is the removal of flash using a trimming die. Closed die forgings can be accomplished by using either drop forming hammer or a forging press. In press forging, each sequence is normally performed by a single application of pressure, whereas in drop forging, multiple blows are used for each step.

Closed die forging, also called impression-die forging, has been further improved in recent years through increased automation, which includes induction heating, mechanical feeding, positioning and manipulation, and the direct heat treatment of parts after forging. One variation of impression-die forging is called flashless forging, or true closed die forging. In this type of forging, the die cavities are completely closed, which keeps the workpiece from forming flash. The major advantage of this process is that less metal is lost to flash. Flash can account for 20%–45% of the starting material. The disadvantages of this process included additional cost due to a more complex die design, the need for better lubrication, and better workpiece placement.

4.4.3 Roll Forging

The roll forging process follows the principle of a rolling process where the billet is plastically deformed by passing it between the rolls. The roll forging process is used mainly to obtain a varying cross section along the longitudinal direction of the billet. During deformation between the rolls, the billet is subjected to high compressive stresses from the squeezing action of the roll geometry. Roll forging is carried out generally on two-high rolling mills as shown schematically [9] in Figure 4.50. Rolls are rotating in counter-rotational directions, and feeding the billet continuously in the same direction. However, there is some basic difference between conventional rolling and the roll forging process. The forging dies are embedded to the rolls and compress the billet metal causing it to conform to the die geometry as it passes through the rolls. Roll forging can be made continuous with the use of multiple rollers and dies. This is also a type of draw forging, since the billet is slowly drawn out of the forging die. Roll forging may be used as a main operation to make the part, or as a preform forging operation for the next closed die forging. Special designed dies are used for roll forging machines.

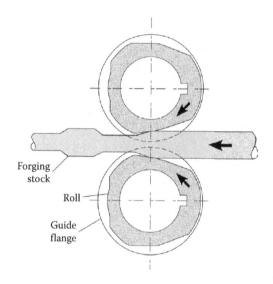

FIGURE 4.50
Roll forging principle.

4.4.4 Forging Variables

The major variables that influence the forging operations are as follows:

- Shape and size factor
- Temperature
- Material flow strength and forgeability
- Friction and lubrication
- Die temperature

Forgeability is defined as the tolerance of metal for deformation without failure, regardless of forging pressure requirements. Factors influencing forgeability are given below:

- Pure metal
- Alloy composition and purity
- Number of phases
- Grain size
- Temperature
- Strain rate
- Flow stress

4.4.5 Aluminum Forging

Closed die forging of aluminum alloys has a very wide application in manufacturing structural components of an aircraft. Aluminum forging starts with the cast billet or ingot (Figure 4.51) cut to the right size (Figure 4.52). Figure 4.53 shows various shaped aluminum hand forging stocks sometimes used as pre-forging stocks.

FIGURE 4.51
Cast aluminum billet stock waiting for forging. (From Weber Metals.)

FIGURE 4.52
Sawing cast aluminum billet to size before forging. (From Weber Metals.)

Figure 4.54 shows an example of an aluminum window frame forging for aerospace applications. Figure 4.55 shows the steps of window frame forging that starts with a pre-form hand forging, on to the die close, and finished with a final part shape with flash.

4.4.6 Titanium Forging

Applications of titanium forgings have grown a lot since the aircraft manufacturers started designing and manufacturing aircraft with carbon fiber composite materials. The fundamental difference between the aluminum and titanium forging process is the forging

FIGURE 4.53
Various shaped aluminum hand forgings. (From Weber Metals.)

FIGURE 4.54
Aluminum window frame forging of an aircraft. (From Weber Metals.)

temperature, as seen also in the extrusion of titanium alloys. Figures 4.56 through 4.58 show a few examples of titanium forging.

4.4.7 Common Defects in Forging

Common defects in a forging process are illustrated as explained in the book [1]. Incomplete forging penetration is detected in the macro-etching of a forging cross section when the deformation during forging is limited to the surface material instead of the entire thickness of the forging stock. Due to insufficient forging load or load limitation of the forging equipment, the interior metal structure of the forging stock is not breaking down properly. This type of defect is very common for large cross sections that are highly recommended to be made using a forging press instead of a forging hammer.

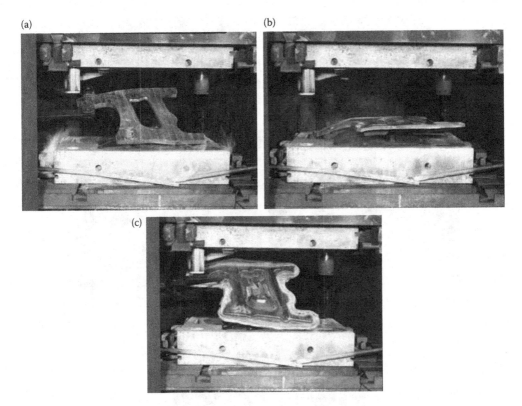

FIGURE 4.55
Forging steps of aluminum window frames of an aircraft. (a) Cut to shape of a preform hand forging ready for closed die forging, (b) forging is complete and removed from the bottom die, and (c) forging showing with flash. (From Weber Metals.)

FIGURE 4.56
Hand forging of titanium in 1650 T forging press. (From Weber Metals.)

FIGURE 4.57
Steps of forging of titanium for structural aircraft components. (a) Preheating the preform forging stock, (b) loading the preheated preform forging stock to the press die, (c) positioning the hot preformed part on the bottom die, (d) burning of the lubricant just before forging, and (e) removing the part from the bottom die. (From Weber Metals.)

FIGURE 4.58
Titanium roll forging. (From Allvac.)

There are several kinds of surface cracking defects in forging. One kind of surface cracking can occur due to excessive working of the surface layer at too low or too high a temperature at the die and working metal interface. Another kind of cracking occurs at the flash of closed die forgings. Cracking at the flash generally penetrates into the body of the forging when the flash is trimmed off. Flash cracking could be avoided by increasing the flash thickness or relocating the flash to a less critical region of the forging.

Cracking may also be produced in forging due to formation of secondary tensile stress. In the case of upsetting of cylindrical or round stock, circumferential tensile stresses are developed that cause internal cracks in the part. To minimize the bulging effect that causes the circumferential tensile stress during upsetting, the usual practice is to use concave faced dies. Internal cracking tendency is less in the case of closed die forging because the compressive stresses are developed from the reaction of the forging metal with the die wall.

Cold shut or fold is another kind of surface defect in closed die forging. Cold shut normally occurs when two surfaces of metal fold against each other without joining completely. A cold shut may also occur when some leftover material from the flash or fin of a previous forging is pressed into the metal surface during subsequent forging operations from the same die.

A certain degree of directionality appears in the microstructure where the second phases and inclusions are oriented parallel to the direction of greatest deformation during forging. When the microstructure is viewed in low magnification, it appears as flow lines or fiber structure. This kind of fiber structure is not generally considered a forging defect since it is characteristic of almost all forgings. From the mechanical properties point of view, the fiber structure resulted in lower tensile and fatigue properties in the transverse direction. Sometimes it is very important to keep the amount of deformation between a 50% and 70% reduction in cross section to achieve a proper balance of tensile properties in both longitudinal and transverse directions.

Examples of a few common external surface defects in aluminum forging are shown in this section. Figure 4.59a shows an instance where metal flows past the base of a rib resulting in rupture from the base structure. Figure 4.59b shows the micrograph at the section AA from Figure 4.59a. It shows an enlarged grain section through the rib and web of the forging, with a rupture of the grain structure shown by the arrow. Figure 4.59c shows an example of a metal cavity like metal sucked-in on one side to fill a projection (rib or boss) on the opposite side. Figure 4.59d shows an example of twist in a part. This is caused by a forging turning in the die during trimming. This may also occur from quenching during heat treatment. Figure 4.59e shows a couple of defects; arrow 1 indicates rough surface finish in the direction of metal flow, which may be caused by insufficient flow of lubrication. Arrow 2 indicates that the metal may tend to separate due to metal flowing too fast during forging. The end grain defect shown in Figure 4.59f occurred when the end of the forging bar stock remained in the forging periphery. The end grain is shown by the bright areas indicated by the arrows. Chemical etching will cause end grain pitting. This pitting on the surface needs to be machined otherwise it will reduce the fatigue performance of the forging. Figure 4.59g shows a sharp ragged edge on the forging caused by metal squeezed into the cracks in the die impression. The cracks in the die impression may be caused by using the die set for a certain length of time, and it is an indication that the die has reached its maximum life. Figure 4.59h shows mainly the trim torn surface. The trim die has cut into the forging during the trimming operation causing the surface to tear away. Figure 4.59i indicates a cold shut or fold-type defect appearing as a crack on the forging surface. It is the result of metal folding or flowing back on itself. Figure 4.59j shows the enlarged grain section through the cold shut or fold area of the previous Figure 4.59i. There is a clear indication of

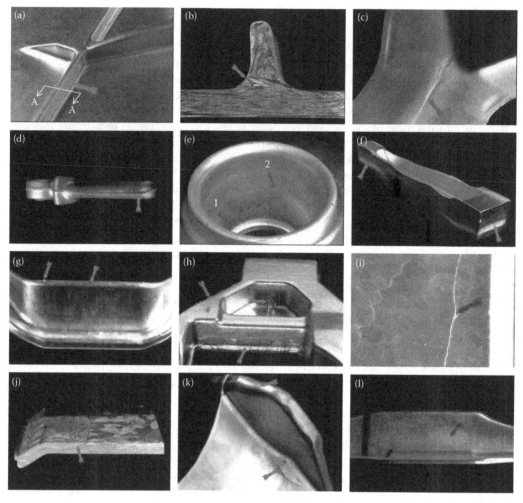

FIGURE 4.59
Example of surface defects on aluminum forgings, (a) flow thru rupture, (b) section AA of flow thru rupture, (c) suck-in, (d) twist, (e) metal separation, (f) end grain, (g) sharp ragged edge, (h) trimming effect, (i) cold shut or fold, (j) enlarged grain, (k) non-fill, and (l) pitting corrosion.

two separate layers of metal grains flowing in two opposite directions. Figure 4.59k shows an example of non-filling into the die impression. This occurs simply when the cavity or impression is not completely filled with metal as indicated by the top arrow. Bottom arrow indicates a metal shearing effect on the web of the forging. Figure 4.59l shows pitting corrosion, localized corrosion in small pits, or craters in the metal surface.

4.4.8 Residual Stresses in Forging

Since forging is carried out well within the hot working temperature region, the residual stresses produced in forgings due to inhomogeneous deformation are generally quite small. However, during heat treatment, some amount of residual stress and warping of long part may occur during quenching operation.

Cooling of large forgings from the hot working temperature is very critical. Improper cooling may subject to the formation of small cracks at the center of the forging cross section. In order to avoid developing high thermal or transformation residual stresses, large forgings are cooled in a controlled temperature from the working temperature to bring the forging to a safe temperature within certain period of time.

4.5 Rolling

High volume of aircraft metal is produced by rolling than by any other deformation process. The largest quantity is rolled flat products produced by hot rolling process followed by cold rolling. Rolling is a plastic deformation process when a block of ingot is rolled between the rolls as shown in Figure 4.60. This process is widely used metal forming process to produce high production flat sheet, plates, and slabs from different metals like aluminum, titanium, steel, and other metals for aerospace applications.

The major variables to control the rolling process are:

- Roll diameter
- Deformation resistance of metal
- Friction between the rolls and the metal
- Presence of front and back tension

Figure 4.61 shows the major rolling parameters including the geometry of the rolls and forces involved in deforming the metal by the rolling process [10]. A workpiece (sheet or plate) of thickness t_0 enters the roll gap with a velocity, V_0, and finally exists the roll gap with final thickness t_1 and a final velocity V_1. In order to maintain constant thickness in a pass, the exit velocity must be greater than the entrance velocity to maintain the volume constancy relationship.

$$t_0 V_0 = t_n V_n = t_1 V_1 \tag{4.17}$$

Therefore, the velocity of rolled sheet or plate product steadily increases from entrance to exit. There is one point along the surface of contact between the roll and the sheet or plate metal where the surface velocity of the roll, V, equals to the velocity of the sheet, V_n.

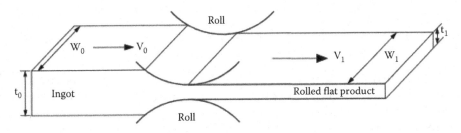

FIGURE 4.60
Schematic representation of rolling process.

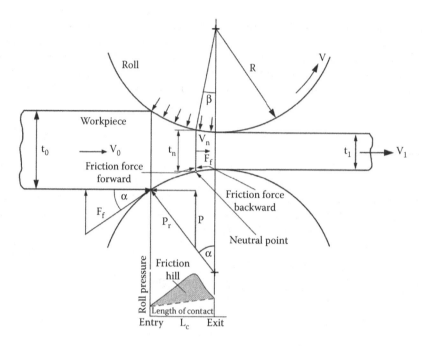

FIGURE 4.61
Rolling parameters in rolling of flat product.

This point is called the neutral point or is also called the no-slip point. There are two forces, tangential F_t and radial P_r acting at any point along the surface of contact. From the entrance to the neutral point, the product moves slower than the roll surface and the friction force acts as to draw the metal into the roll gap. On the exit side of the neutral point, the rolled product moves faster than the roll surface. The direction of friction force is then reversed as shown in Figure 4.61 to oppose the delivery of the sheet from the roll gap.

The vertical component of Pr is the rolling load, P, the force with which the rolls press against the metal. This force is also equal to the force exerted by the metal trying to force the rolls apart. This force is also called the roll separating force. There is another important term called specific roll pressure, the rolling load per unit contact area. The contact area between the metal and the rolls is equal to the multiplication of the width of the sheet, W_1, and the projected length of arc of contact, L_c.

The specific roll pressure, p, is given by

$$p = \frac{P}{W_1 L_c} \tag{4.18}$$

The roll pressure distribution along the arc of contact is also shown in Figure 4.61.

In a conventional hot or cold rolling process, the main objective is to reduce the thickness of the metal. Generally, a slight increase in width also occurs, so that the decrease in thickness results primarily in an increase in length of the product. In a flat rolling process, there is no loss of metal during transformation from initial slab type ingot to the breakdown flat shape. From the volume constancy relation, volume will remain constant; the metal displaced by reduction in thickness will go mostly to an increase in length of the flat product.

In the rolling of a sheet or plate, the thickness is reduced from t_0 to t_1 by rolls of radius R. The absolute reduction in single roll pass is given by

$$\Delta t = t_0 - t_1 \tag{4.19}$$

The thickness reduction in percentage is given by

$$R = \left(\frac{t_0 - t_1}{t_0} \right) \cdot 100\% \tag{4.20}$$

Reduction can be expressed in terms of natural strain

$$\varepsilon = \ln\left(\frac{t_0}{t_1} \right) \tag{4.21}$$

4.5.1 Friction Model in Rolling

To begin the rolling operation, the workpiece (slab, bloom, or billet) must be drawn into the roll gap due to friction between the workpiece and the roll interface surface as shown in Figure 4.61. This is also called the initial roll bite. Rolls may refuse to bite if the rolls are very smooth or well lubricated. Therefore, rolling will proceed when

$$F_f \, \mathrm{Cos}\,\alpha > P_r \, \mathrm{Sin}\,\alpha \tag{4.22}$$

According to the deformation in rolling process, the workpiece is subjected to high compressive stresses to squeeze the metal between the rolls to create surface shear stresses as a result of the friction between the rolls and the metal.

The friction between the roll and the exiting metal surface plays a very important role in the rolling process. Friction force not only pulls the metal into the rolls, but also affects the magnitude and distribution of the roll pressure. From the distribution of roll pressure along the contact length, L_c, of the roll, it is found that the larger the frictional forces, the greater must be the rolling load. The pressure builds up steeply and becomes a maximum at the neutral point where the frictional direction changes. Since the roll pressure distribution is very much dependent on the friction, the pressure distribution shown in Figure 4.61 is commonly called the "friction hill." High friction in rolling also produces greater lateral spread and edge cracking, whereas in cold rolling a very low friction value is obtained due to lubrication and a polished roll surface finish may also lead to difficulties in feeding the metal into the rolls.

The friction force acting between the entrance and the neutral point advances the sheet or plate between the rolls, while the friction force acting between the neutral point and the exit opposes the rolling action. The difference between the friction on the entrance side and the friction on the exit side provides the necessary power for rolling. The position of the neutral point is automatically determined by the power required to deform the strip and to overcome friction losses. The larger the thickness reduction, the further the neutral point moves toward the exit. The effect of front and back tension in rolling process basically moves the neutral point with respect to the roll entrance and the roll exit causing the change of roll pressure distribution.

4.5.2 Deformation in Rolling

The metal deformation in a rolling process [1] can be considered to be two-dimensional (2D). In sheet rolling, it is considered with good approximation that the reduction in thickness is transferred into an increase in length with a slight increase in the width. Thus, the plane strain model holds good for the mathematical analysis of rolling. The amount of lateral spread could be significant in the rolling of bars and shapes as compared to the lower importance in sheet or strip rolling. The spread may be a function of

- Diameter and the condition of the rolls
- Flow properties of the metal
- Reduction

In comparison with extrusion or forging processes, the deformation in rolling is relatively uniform. Figure 4.62 shows the typical grid deformation produced by the rolling of a flat bar. It is shown that the surface layers are both compressed as well as sheared.

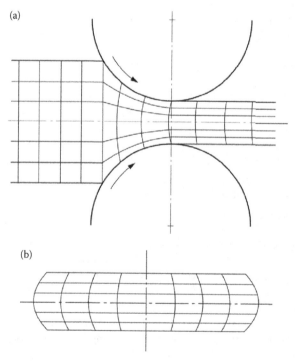

FIGURE 4.62
Deformation in rolling, (a) side view and (b) end view.

4.5.3 Classification of Rolling Mills

Rolling mills can be conventionally classified [1] with four basic configurations with respect to the number and arrangement of the rolls as shown in Figure 4.63. The simplest and most common type of rolling mill is the two-high mill (Figure 4.63a). Equal-sized rolls are rotated only in one direction. The metal stock is returned to the entrance, or rear, of the rolls for further reduction by hand carrying or by means of a platform, which can be raised to pass the work above the rolls. In a two-high reversing mill (Figure 4.63b), the workpiece can be passed back and forth through the rolls by reversing their direction of rotation. In a three-high rolling mill (Figure 4.63c), the upper and lower rolls are power driven and the middle roll is frictionally driven.

In a rolling mill, an appreciable amount of power savings could be achieved by using smaller diameter rolls supported by the larger diameter back up rolls. The simplest mill of this type is the four-high mill (Figure 4.63d). Very thin gage sheet metal with close dimensional tolerances can be produced with smaller diameter work rolls. The cluster mill as shown in Figure 4.63e where each of the work rolls is supported by two backing rolls is a typical mill of this kind.

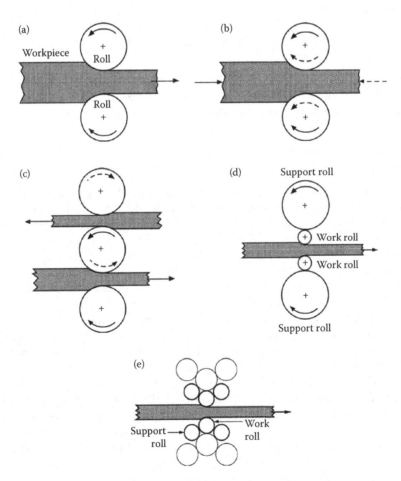

FIGURE 4.63
Different rolling mill configurations, (a) two-high, (b) two-high reversing, (c) three-high, (d) four-high, and (e) cluster.

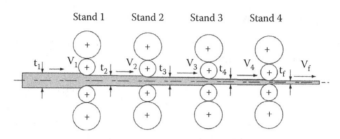

FIGURE 4.64
Schematic of continuous rolling mill.

In commercial practice, for high production, a series of four-high mills are installed one after another in tandem. Each set of rolls is called a stand (Figure 4.64). During the continuous rolling process, the metal exits at different velocities from each stand in the mill. Since there is no loss of material during rolling process, the volume constancy relation will hold good at each stand to produce a uniform flow of metal without any interruption in between the stands.

$$V_1 t_1 = V_2 t_2 = V_3 t_3 = V_4 t_4 = V_f t_f \tag{4.23}$$

Accordingly, the linear speed of each set of rolls in each stand is optimized using the volume constancy relation so that each successive stand takes the flowing metal at a speed equal to the exit speed of the preceding stand.

4.5.4 Aluminum Sheet and Plate Rolling

In aluminum rolling, the first step is the melting process and alloy production accomplished by melting primary aluminum ingots with scrap, along with alloying elements, and some additional elements to resist corrosion. The molten metal chemistry is checked to ensure controlled alloy chemistry. Molten metal at the right temperature is processed through the filtration process before it cast into rolling ingot. The filtration process removes any impurities as small as 20 μm in size. In the industry, producers such as Kaiser Aluminum have a capacity of 100,000–160,000 pound melting furnaces. Rolling ingots up to 30 inches thick, 68 inches wide, 237 inches long, and weighing as much as 20,000 pounds are used to roll sheet products. Figure 4.65 shows the major process steps used to produce aluminum sheet products. Prior to the first rolling process, the surface of the rolling ingot is scalped to remove the oxide layers generated during casting. After scalping, the ingots are put into large furnace called a "soaking pit" where the ingots are heated to a temperature as high as 1100°F. Precise temperature control is required to assure uniform metallurgical properties in the finished product. The next step is hot rolling where the hot soaked ingot is brought into the roll bites of the first hot rolling mill. Figure 4.66 shows the operator-controlled hot aluminum rolling mill. The numbers of hot rolling mills are determined depending on the final thickness desired. Computerized statistical process control closely monitors the metal as it is processed through the system, with a constant feedback of data to improve the process performance to obtain uniform gage thickness of rolled products. After hot rolling, annealing (heat treat to soften the metal) is needed for some metal to enable a further reduction in thickness. Partial annealing is also done to release internal stresses, which developed during rolling. Cold rolling is the next step, which is performed

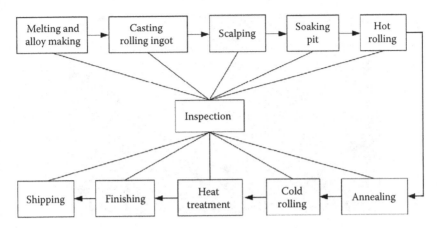

FIGURE 4.65
Process steps of rolling of aluminum flat sheet product.

FIGURE 4.66
Operator-controlled aluminum hot rolling mill. (From Kaiser Aluminum.)

to achieve further thinner gages metal with tight tolerances and improved surface finish. After cold rolling, the metal is heat treated, stretched to maximize flatness, and relieve internal stresses, stenciled, slit, or sheared to various widths, lengths depending on the customer requirements. Different kinds of inspection methods at each stage of the operation are conducted to ensure the highest quality of metal by verifying chemical, physical, surface texture, and finally mechanical properties and formability of the metal.

Sheet products are also supplied in the right size coil form as required by the aerospace customers to meet their specific application needs for manufacturing fuselage stringers and frames. Coils are made after rewinding (Figure 4.67) from the main rolled coil to different sizes (Figure 4.68) as required by the customer.

The differentiation between plate and sheet is determined by the thickness of the product. In general, plate has a thickness greater than ¼ inch (6 mm), although there are exceptions to this limit, depending upon the width. Sheet and strip refer to rolled products, which generally have a thickness less than ¼ inch (6 mm). In general, strip refers to the

FIGURE 4.67
Rewind after cold rolling. (From Kaiser Aluminum.)

rolled product with a width no greater than 12 inch (305 mm), while sheet refers to the product of greater width.

In aluminum plate rolling mills for aerospace applications, special care is given especially for heat treatment, handling, and final stretching operations as shown in Figures 4.69 through 4.71. This ensures the producer can achieve the final properties and temper of the plate providing the highest quality plate products for the user's application.

4.5.5 Titanium Plate Rolling

Rolling of titanium at elevated temperature permits greater thickness reduction due to the lower yield strength and change in friction coefficient between sheet or plate and the rolls. High temperature rolling introduces additional issues of surface contamination, and size and flatness variation of the rolled sheet or plates. There is an increase of power consumption with increasing roll diameter for a constant reduction and also higher friction conditions due to surface condition of rolls and sheet materials. Figure 4.72 shows manufacturing processes of titanium rolling mill for plate/sheet products for aerospace.

FIGURE 4.68
Cold rolled aluminum. (From Kaiser Aluminum.)

FIGURE 4.69
Exiting from horizontal heat treating furnace. (From Kaiser Aluminum.)

FIGURE 4.70
Handling system to move the heat treated plate. (From Kaiser Aluminum.)

FIGURE 4.71
Stretching of aluminum plate. (From Kaiser Aluminum.)

FIGURE 4.72
Rolling of titanium plate. (From VSMPO, Russia.)

4.5.6 Common Defects in Rolled Products

Defects in rolled products [1] can be introduced mainly during the ingot stage or during the rolling stage of production. Internal types of defects are normally derived from the ingots having some blowholes and longitudinal stringers of nonmetallic inclusions.

Since rolled products typically have a high surface to volume ratio, the surface condition of metal during each stage of operations is very important. In order to maintain the high standard of surface quality during a rolling process, the ingots have to be free from any surface contamination. Typical surface defects are caused by folded over fins, oxides being rolled into the surface, or scratches due to defective rolls or guides.

Thickness variation and inconsistent flatness are also common problems for rolled sheet, plates, and strips. Most of the variation of thickness along the rolling direction is caused by the rolling speed and strip tension. Variation of either thickness or hardness of input material entering the rolling mill may also cause the variation of final thickness of the product.

Variation of thickness along the width of the flat sheet products is generally caused by the deflection of rolls under rolling pressure. A set of parallel rolls will deflect under pressure and produce sheet having a thicker dimension at the center than the outer end of the product. To compensate for the amount of deflection of rolls at the center, some sort of camber or crowning is provided in the roll geometry, so that the working surface remains parallel at the rolling load. Since the crowning or camber in the rolls is dependent mainly on the applied load during rolling, a different set of rolls will be required to satisfy different load conditions for each product dimension.

4.6 Fundamentals of Tube Drawing

The principle of tube drawing is the same as wire or rod drawing, where the material is pulled through a die of smaller cross section than that of the original stock. Tubes are produced on a draw bench with different die combinations. Figure 4.73 shows a schematic

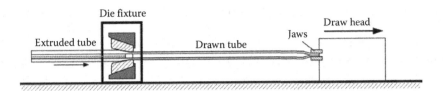

FIGURE 4.73
Schematic of a tube drawing operation.

representation of a tube drawing operation where tube stock is pulled through a die by the draw head using jaws to grip the free end of the tube.

In order to reduce the wall thickness and accurately control the inside diameter, the inside of the tube is supported by a mandrel while the tube passes through the die. The mandrel inserted into the tube stock could be fixed or floating as shown in Figure 4.74. In the case of a fixed mandrel, the mandrel is fastened to the end of a stationary rod attached to the tube stock end of the draw bench and is positioned so that the mandrel is located in the throat of the die where a constant gap is maintained between the mandrel outer diameter (OD) and die inner diameter (ID) at the die bearing area (Figure 4.74a). Floating mandrel drawing (Figure 4.74b) is advantageous because it provides a higher material yield than any of the other processes, and it provides long-length capability with a smooth ID surface. Die and mandrel design for floating mandrel drawing is more critical than for the fixed mandrel process. The die bearing must be long enough to permit the mandrel to seat in the tube ID. But too long at a bearing length also increases friction, which could cause additional problems in production. In addition to the die design, lubrication and tube cleanliness are also critical for successful floating mandrel drawing. Tube drawing

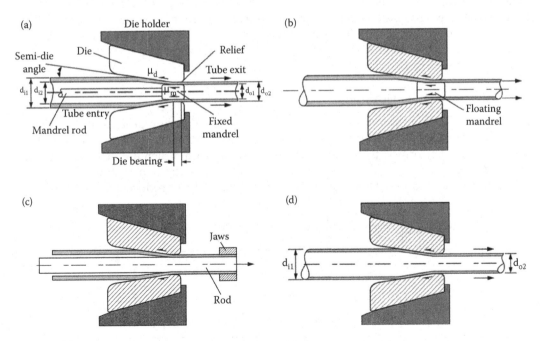

FIGURE 4.74
Schematic diagrams showing different tube drawing processes, (a) fixed mandrel, (b) floating mandrel, (c) moving rod, and (d) tube sinking.

also can be done with a moving mandrel (Figure 4.74c) by pulling a long rod through the die with the tube. Due to the difficulties in using long rods as mandrels, tube drawing with a rod is usually limited to the production of small-sized tubing.

Tube sinking (Figure 4.74d) is yet another tube processing technology, where there is no need of using any mandrel to support the inside of the tube during drawing through the draw die. Since the inside of the tube is not supported in tube sinking, the wall thickness of the tube will either increase or decrease depending on the die parameters. Generally, lower die angles tend to cause wall thickening, whereas higher die angles cause wall thinning. Tube sinking uses a long bearing length to achieve the correct size and optimal roundness, making this process suitable for a final sizing operation. On a commercial basis, tube sinking is used only for the production of small diameter tubes. Sinking is also used for the production of rectangular-shaped tubes. The standard procedure is (1) sinking the round tube to a rectangular tube and (2) further drawing with a rectangular die/mandrel.

The total force required for drawing through the die is dependent on the major variables in tube drawing:

1. Die angle
2. Reduction (%)
3. Flow stress of tube material
4. Die bearing length
5. Die friction

Die friction is a function of die material, the lubrication, and the drawing speed. In the case of a wire or rod drawing process, friction is acting only at the die and the drawing material interface, whereas in the case of tube drawing process, friction is acting both on the die side as well as the mandrel or plug side. The cold tube drawing process provides a better surface finish, close dimensional tolerances, thinner walls or smaller diameter, irregular shapes, and can finally increase the mechanical properties.

4.6.1 Aluminum Tube Drawing

Seamless round aluminum tubes are produced by a tube drawing process using a seamless extrusion tube. Manufacturing of seamless hollow extrusion is also a special extrusion operation performed by using an extrusion press with a piercing or fixed mandrel attached to the extrusion main ram as explained in the extrusion section. Figure 4.75 shows the major process steps that are normally followed to produce aluminum drawn

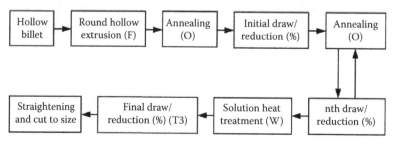

FIGURE 4.75
Major manufacturing steps of aluminum tube drawing process.

FIGURE 4.76
Seamless aluminum tube manufacturing processes. (a) Tube drawing principle, (b) production die and mandrel/plug, (c) production of drawn tube, and (d) drawn aluminum tubes on the table. (From ALUnna Tubes, Germany.)

tube. The process starts with the seamless extrusion from aluminum billets from the extrusion press. Tubes are cut to size and annealed to make the material softer from the as-extruded "F" temper before being fed into the first draw reduction step, normally with a fixed mandrel as shown in the schematic diagram (Figure 4.76a). Also shown is the real production setup of a draw bench with draw die, mandrel, or plug and mandrel rod as shown in Figure 4.76b and c. Figure 4.76d shows some drawn tubes that are kept on the storage conveyer table for the next operation. Due to work hardening of tube material during an individual cold draw run, tubes are subsequently annealed for the next draw. The resulting final diameter and the wall thickness parameters are very close to the customer requirements. There could be more than one draw step prior to each subsequent annealing, based on the alloy and the amount of reduction to be given. Drawn aluminum tubes are normally supplied in "T3" temper. To make the tubes in "T3" temper, all the tubes received from the nth draw are sent for solution heat treatment to condition the tubes to the "W" temper, which is an unstable stage. In the next step, tubes are sent for final draw with very minor reduction to obtain the final dimensional tolerances, and in the process the tubes are becoming a stable condition at a "T3" temper. A final draw is given sometimes from larger outer diameter to the final smaller diameter without changing the wall thickness by the tube sinking process. This process is used for making tubes of different outer diameter with same wall thickness. The final step involves stretch straightening and a cut to size operation. Tubes of square or rectangular shapes can be produced by drawing large round tubes through the shaped dies.

4.6.2 Common Defects in Aluminum Tube Drawing

A few examples of tube drawing process-related defects are discussed in this section. Surface-related defects in tube drawing are very common and those are clearly visible with the naked eyes. Die lines like "V"-type grooves on the tube outside surface are associated with the interaction between the die surface and the drawing tube outer surface with the lubricant at the interface. This could be due to any foreign particles trapped into the surface or local breakdown of lubricant at the interface. Such lines normally extend over the entire length of the drawn tube. In extreme cases, a band of transverse crack-type surface features may also be observed as shown in Figure 4.77a. A similar example of line marks on the ID of the tube associated with the mandrel OD at the tube ID and the mandrel interface with the presence of lubricant as shown in Figure 4.77b. Saw chips or other foreign particles can cause defects by creating some kind of indentation marks on either of the tube surfaces (Figure 4.77c). Indentation marks are dependent on the size and geometry of the foreign particles trapped during the pulling of tube material either at the die or mandrel interfaces. Figure 4.77d shows a trapped saw chip embedded on the tube OD. Particles remaining on the tube surfaces are difficult to assess especially on the ID.

FIGURE 4.77
Surface-related common defects in drawn tube. (a) Broken surface, (b) longitudinal line on ID, (c) indentation on tube OD, and (d) entrapped saw chip on OD. (From ALUnna Tubes, Germany.)

4.7 Fundamentals of Pilgering

Conventional tube drawing processing as shown in aluminum tube manufacturing is not used for the titanium tube reduction process. Cold pilgering [11,12] is the most commonly used technique applied mainly for the titanium tube reduction process. This technique can produce very large reductions before an annealing process is required to soften the material for the next reduction. It is possible to have about 60%–70% reductions in between annealing steps. The cold pilgering process is illustrated in the schematic diagram (Figure 4.78) showing the vertical section through the tube reducer at the start and end of a reduction stroke. The tube is advanced a small increment and rotated about 60° while in this position. Rolls and die move laterally and rotate as shown to reduce the tube in small increments. The tube is pulled over a tapered stationary mandrel by using two grooved rolls or dies which roll back and forth in a constant cycle. The rolls with tapered grooves are mounted in a saddle, which is moved back and forth by a crank arrangement. Gears are mounted on the roll shafts and these are engaged with the racks mounted on the machine frame so that, as the saddle moves, the rolls rotate in precise relation to each other with the stationary mandrel. The incoming hollow tube is rotated and advanced by a small increment at the beginning of each stroke of the crank. The tube is reduced in a series of small increments that combine to achieve the large overall reduction in size. The shape of the tapered roll groove and the tapered mandrel control the reduction. Depending on the final size, it may take five or six reductions to complete the work. Very small tube size less than ¼ inches (6.35 mm) OD is generally cold drawn from tube-reduced tubing. The feed increment of the input tube side is an indication of the overall efficiency of the cold pilgering mill. The output performance of the pilgering mill depends on

- The tube material
- Dimensional tolerance
- Tool design
- Tool quality
- Lubrication

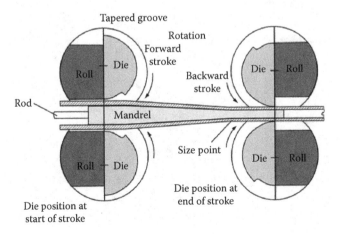

FIGURE 4.78
Schematic representation of tube reduction process. (From Sandvik Special Metals, Kennewick, Washington, USA.)

4.7.1 Seamless Titanium Tube

The titanium seamless extrusion process has attained a higher priority now for the fabrication of thin-wall round tubes for aerospace applications. Titanium 3Al-2.5V seamless tubes are generally used for various control systems in the aircraft, including high-pressure hydraulic systems to move various flight control surface components of the wing and empennage, and also landing gear actuators and engine systems.

Figure 4.79 shows the tube reduction process used in production of titanium tubing for aerospace applications. The titanium tube manufacturing processes involves many steps, starting from raw material ingot to the final packaging for shipping. The major process steps are shown in the process flow chart in Figure 4.80. As explained previously in the titanium extrusion section, the round-shaped titanium cast ingots are forged in a series of reductions to a round bar of smaller diameter. Reheating temperature is controlled to a certain limit to establish optimum grain size and phase distribution. The oxygen-enriched surface layer formed during forging at an elevated temperature is removed by grinding and or machining operations to convert the forging ingots to round-shaped hollow billets for seamless hollow extrusion.

The tubes received after extrusion may need initial annealing to enable the start of cold reduction. Annealing in a vacuum heating furnace is done to make sure the tubes are free from any undesired oxides and oxygen-enriched surfaces, and to bring the tubes into a softer stage for cold reduction. After the first reduction, the tubes are dimensionally checked. The steps shown within the dotted line range are repeated until the tubes reach their final dimensions.

In tube reduction by the cold pilgering process, it is possible to reduce wall thickness as well as diameter of the tube simultaneously as shown schematically in Figure 4.81. Q factor is defined as the ratio of wall thickness reduction to diameter reduction. The tube has a diameter d_1 and wall thickness t_1 before pilgering, and is changed to diameter d_2 and wall thickness t_2 after tube reduction, which will determine the Q factor as

$$Q = \frac{t_1 - t_2}{d_1 - d_2} \tag{4.24}$$

FIGURE 4.79
Photograph showing the titanium tube reduction process in production. (From Sandvik Special Metals, Kennewick, Washington, USA.)

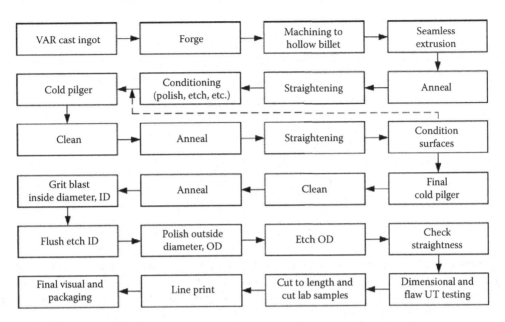

FIGURE 4.80
Titanium tube manufacturing process steps.

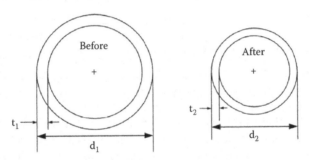

FIGURE 4.81
Tube reduction.

The value of Q can be adjusted in the pilgering process by changing the taper angle of the mandrel. The Q factor is a high number when the wall thickness reduction is much more than the diameter reduction.

4.8 Seamless Steel Tube

Most seamless steel tubes are first rotary pierced [1] to make the raw hollow stock before using the pilgering or other processes for further reduction. Figure 4.82 shows the schematic representation of rotary piercing process of making steel seamless tubing. In production, heated cylindrical billet is placed between rotating rolls in the horizontal plane inclined at an angle. Since the work rolls are set at an angle to each other, the billet is given a forward helical motion toward the piercing-pointed mandrel or plug. The roll pressure

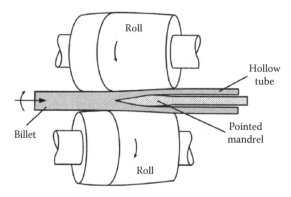

FIGURE 4.82
Schematic of rotary piercing.

acting around the periphery of the billet creates an opening at the center of the billet right in front of the pointed mandrel and continues opening further to form the inside diameter of the tube. Piercing is the most severe metal forming operation for tube production, and is limited to certain high-quality steels.

4.9 Inspection and Testing of Tubes

The utilization of tubing in aerospace is very critical when designing the hydraulic systems of various aircraft models. To assure the highest quality requirements are met, the manufacturers are required to provide inspection and testing methods to meet and assure the tube specifications. Tests and inspection methods must meet industry standards, and must be precise and accurate. Test samples used for destructive testing to qualify the material are taken from each lot of material processed in production. The following tests are generally designed to qualify the parts:

1. Material chemistry
2. Tensile testing (strength and percentage elongation)
3. Bend test
4. Flare test
5. Flattening test
6. Pressure test
7. Microstructure (grain size)

References

1. Dieter, G.E., *Mechanical Metallurgy*, International Student Edition, McGraw-Hill, Kogakusha, Ltd., Tokyo, 1961.

2. Saha, P.K., *Aluminum Extrusion Technology*, ASM International, Materials Park, Ohio, 2000.

3. Saha, P.K., Thermodynamics and tribology in aluminum extrusion, *Wear*, 218, 179–190, 1998.

4. Saha, P.K., Quality issues of hollow extrusions for aerospace applications, In *Proceeding of the 9th International Aluminum Extrusion Technology Seminar*, Orlando, FL, May 2008.

5. Laue, K. and Stenger, H., *Extrusion*, ASM International, Materials Park, Ohio, 1981.

6. *Aluminum Extrusion Alloys*, Kaiser Aluminum and Chemical Sales, Inc., Oakland, California, 1964.

7. Davis, J.R. (ed.), *Aluminum and Aluminum Alloys*, ASM International, Materials Park, Ohio, 1993.

8. Saha, P.K., Economics of aluminum extrusion for aerospace applications, In *Proceedings of the Tenth International Aluminum Extrusion Technology Seminar*, Miami, FL, May 2012.

9. Jenson, J.E, *Forging Industry Handbook*, Forging Industry Association, Cleveland, Ohio, 1970.

10. Schey, J.A., *Tribology in Metalworking*, ASM International, Metals Park, Ohio, 1984.

11. Forney, C.E. and Meredith, S.E., *Ti-3Al-2.5V Seamless Tubing Engineering Guide*, Sandvik Special Metals Corporation, Kennewick, Washington, 1990.

12. Stapleton, G., *Cold Pilger Technology*, 1st ed., Glen Stapleton, Sheridan, IN, 1996.

5

Introduction to Composite Materials for Aerospace

5.1 Introduction

Composite materials are engineered materials produced from two or more material components with significantly varying physical and chemical properties and that remain separate and distinct on a macroscopic scale within the finished product. Manufacturing composite structures are much more complex compared to the manufacturing of metal structures. To make a composite structure, the input raw material, in the form of fibers, tapes, or fabrics are laid out in a mold under heat and pressure. The resin as a matrix material flows under heat and pressure and is solidified when the heat is removed. It can be formed into various shapes according to the geometry of the mold. One of the very useful features of composite materials is that the material can be sandwiched with different layers with the fibers layered in different directions, which allows the engineers to design the parts to satisfy strength and other mechanical requirements.

Application of composite materials in military aircraft has been established a long time ago. After researching for quite some time, composite technology has come back recently to commercial aircraft manufacturing in a much larger scale, due to the following advantages:

- Lighter weight
- Optimum strength and stiffness
- Improved fatigue strength
- Resistance to corrosion
- Reduced assembly cost (less fasteners)
- Less detailed parts and fasteners
- High strength to weight ratio
- Higher modulus
- Improved performance, greater payloads
- Improved fuel efficiency, longer range of flight

In spite of the many advantages, composite materials have some disadvantages that need to be considered before selecting the right composite materials for the right part:

- High raw material costs
- High fabrication costs

- Less strength in the out-of-plane fiber direction
- Weak in impact
- Delaminating or ply separations
- Repair difficulties as compared to metal structure
- Adverse effects from both temperature and moisture

Modern commercial aircrafts, including the Boeing 787 and Airbus A350, utilize a significant amount of composites to substitute for heavier metal parts to make the aircraft more fuel-efficient. Applications of composite materials are expanding in both commercial and military aircraft design and manufacturing. As an example, Boeing has increased usage by a very significant amount of composite materials from 1% (747) to 50% (new 787) of the structural weight of the individual aircraft. Similarly, F-22 uses composites for at least one-third of the structural weight. Experts are predicting that future military aircraft will be more than two-thirds composite materials. Military aircrafts use substantially greater percentages of composites than commercial aircraft, primarily due to different ways of carrying out maintenance of the aircraft. Generally, military aircrafts are constantly being maintained, more so than commercial aircraft, which need to improve cost effectiveness requiring less maintenance. If the composites are damaged in service, they require immediate repair as compared to metals such as aluminum, which can survive longer before the material fails. This chapter introduces the fundamentals of the input fiber–matrix composite raw materials used for manufacturing various structural and nonstructural components for aerospace applications.

5.2 Components in Composite Material

The major components of composite materials are fibers and matrix, with their interface zone of reinforcement of fibers within the matrix as shown schematically in Figure 5.1. Fibers are generally strong and stiff, whereas the matrix is weaker and less stiff. The matrix material surrounds and supports the reinforcement materials by maintaining their relative positions. The reinforcement fibers control the mechanical and physical properties of the composite material. A wide variety of matrix and reinforcement fiber materials makes variation of mechanical and physical properties of the final composite materials

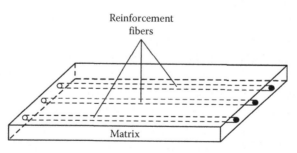

FIGURE 5.1
Material components in composite material.

that allow the designers to choose an optimum combination to produce aircraft parts. The most commercially used composite is polymer-matrix composite, which has glass or carbon fibers reinforced in matrices based on thermosetting polymers, such as epoxy or polyester resins. Sometimes, thermoplastic polymers may be preferred due to its flexibility in production. There are also two more classes of composite materials in which matrix is metal or a ceramic.

5.3 Fibers

The major role of the fibers in a composite material is to provide strength and stiffness. In general, the fibers are brittle in nature with linear stress–strain behavior having low percentage of elongation. Fibers are manufactured and sold in various forms mainly as fibers, filament, strand, tow, roving, yarn, tape, and woven cloth. Fiber is a general term for a material that has a long axis, many times greater than its diameter. The term aspect ratio, which is defined as the ratio of fiber length to its diameter, is generally used to specify the fiber. The aspect ratio is generally greater than 100 for most fibers. Filament is the single fiber having an extreme length ratio of the fibrous material. The filament unit is formed by a single hole in the spinning process. A strand is generally associated with glass fiber, which is untwisted primary filament bundle, which can be either a continuous filament or staple fiber. Tow is the term used for carbon and graphite fibers, similar to a strand for glass fibers, to describe the number of untwisted filaments produced at one time. Tow size is usually expressed as Xk (3 k tow contains 3000 filaments). Roving is a number of strands or tows collected into a parallel bundle without twisting. Roving can be chopped and used for short fiber segments for sheet molding compounds (SMC), bulk molding compounds, or injection molding operations. Yarn is an assembly of one or more strands twisted together. Tape is a composite form in which a large number of parallel filaments (tows) are reinforced together with an organic matrix material (epoxy). Woven cloth is another composite product made from yarn or tows woven in various patterns to provide reinforcement in both 0° and 90° directions.

The following factors are mainly considered for selecting the right type of fibers for different applications:

- Tensile strength and modulus/stiffness
- Compression strength and modulus/stiffness
- Density
- Coefficient of thermal expansion
- Impact strength
- Environmental resistance
- Cost

Based on the above factors, each type of fiber can be compared with any other type. Table 5.1 compares advantages and disadvantages of different type of fibers used for making composite materials [1].

TABLE 5.1

Comparison between Typical High-Strength Fibers

Type of Fiber	Advantages	Disadvantages
E-, S-glass	High strength Low cost	Low stiffness Short fatigue life High temperature sensitivity
Aramid (Kevlar)	High tensile strength Low density	Low compressive strength High moisture absorption
Boron	High stiffness High compressive strength	High cost
Carbon (AS4, T300, IM7)	High strength High stiffness	Moderately high cost
Graphite (GY-70, pitch)	Very high stiffness	Low strength High cost
Ceramic (silicon carbide, alumina)	High stiffness High use temperature	Low strength High cost

5.3.1 Glass

Glass fibers are extensively used in commercial composite applications due to their low cost, high tensile strength, high impact resistance, and good chemical resistance. The three most common glass fibers are E-glass, S-2 glass, and Quartz, respectively. Glass is an amorphous material consisting of silica (SiO_2) with other oxide components to provide specific compositions and properties. The major processing steps of manufacturing fiberglass fibers are shown in Figure 5.2. A batch mixture of silica and other material such as glass is melted, and the molten mixture is extruded through a small bushing hole in the form of filaments. Glass filaments are processed in bundles called strands. A strand is a collection of continuous glass filaments. The strands are then collected into a bundle, which is called roving, after going through a few processes, including traversing, winding, and curing. Roving refers to a bundle of untwisted strands, packaged like thread on a large spool.

The diameter of the strand is controlled by the size of bushing hole, extrusion speed, temperature, viscosity of the molten mixture, and the cooling rate. A number of individual strands are usually incorporated into a roving to provide a convenient form for further processing. Rovings are wound onto individual spools. Fiber glass rovings can either be

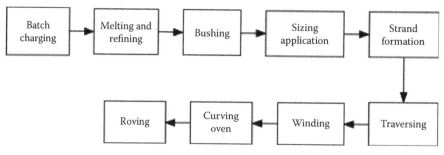

FIGURE 5.2
Manufacturing process steps for fiberglass fibers.

used to directly create a product or can be wound into fabrics for the common manufacturing techniques including pultrusion, gun roving, filament winding, and molding. The fabrics include:

- Chopped matt
- Unidirectional
- Woven

If the glass fiber is to be used for weaving, it is usually twisted into a yarn to provide extra strength during the weaving process. Since glass fibers are amorphous, they are considered as isotropic. The fiber diameters are in the range 10–20 μm (0.0004–0.0008 inch).

5.3.2 Kevlar (Aramid)

Kevlar or aramid fibers are actually organic fibers with strength and stiffness in between glass and carbon fibers. Kevlar fibers are manufactured by dissolving the polymer (aromatic polyamide) in sulfuric acid and extruding through small holes in a rotating device. The typical diameter of Kevlar fiber is 12 μm (0.0005 inch). Kevlar fibers are sensitive to moisture absorption. Due to their high molecular orientation, they are very anisotropic in both mechanically and thermally. Kevlar fibers are available in tows or yarns of various weights for further conversion to woven cloth or chopped fiber mat.

5.3.3 Boron

Boron and other ceramic fibers including silicon carbide, SiC, and aluminum oxide, Al_2O_3, are characterized by high strength, high stiffness, and high temperature application. Those fibers are not commonly reinforced with polymeric matrices rather used with metal or ceramic matrices for high temperature applications.

5.3.4 Carbon and Graphite

In high-performance composite structures, carbon and graphite fibers are the most prevalent fibers. Carbon fiber diameters are generally in the range of 7–8 μm (0.0003 inch). The raw material used to make carbon fiber is called the precursor. About 90% of the carbon fibers produced are made from polyacrylonitrile (PAN). The remaining 10% are made from rayon or petroleum pitch. All of these materials are organic polymers, characterized by long strings of molecules bound together by carbon atoms. The exact composition of each precursor varies from one manufacturer to another and is generally considered a trade secret. The method of production is essentially the same for any type of precursor material. The major process steps of manufacturing carbon fibers are shown in Figure 5.3. The precursor material is drawn or spun into a thin filament, which is then heated slowly in air to stabilize it, preventing it from melting at the high temperatures. The stabilized fibers are heated at high temperature in an inert atmosphere to drive out noncarbon constituents of the precursor material. This process is called the pyrolysis process, also known as carbonization, which changes the fibers from a bundle of polymer chains into a bundle of ribbons of linked hexagonal graphite plates, oriented randomly through the fiber. PAN-type precursor is stretched during the stabilization process and heated to 200°C (400°F) in atmospheric air.

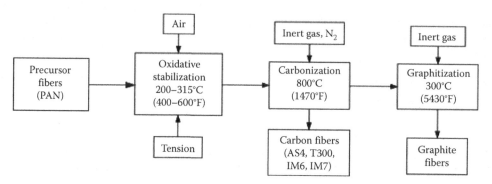

FIGURE 5.3
Major process steps of carbon fiber manufacturing.

Graphite fibers that are a subset of carbon fibers are produced by further processing at very high temperatures up to 3000°C (5400°F). This process is called graphitization, which results in enhanced crystallinity and produces ultra-high stiffness graphite fibers with increased thermal conductivity in the axial direction.

An example of a carbon fiber manufacturing process line in the industry is shown in Figure 5.4. The fibers in the bobbins are loaded into the creel that feeds the PAN fiber through a series of specialized ovens during the most time-consuming stage of production, oxidation. Figure 5.5 shows various stages during the oxidation process of the carbon fibers in the manufacturing process line. In the oxidation process, oven airflow plays a critical role in controlling process temperatures and preventing exothermic reactions. Airflow designs may be single flow, which are parallel or perpendicular to the tow band or multipath flow. Figure 5.5c shows center-to-end parallel airflow technology applied to a manufacturing oxidation oven.

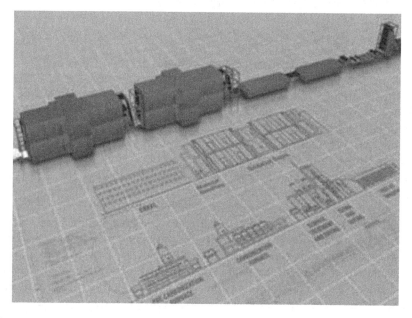

FIGURE 5.4
Carbon fiber manufacturing industry process line. (From Despatch Industries.)

FIGURE 5.5
Oxidation process of carbon fibers. (a) PAN fibers split into 80 k tows in the first passes through oxidation oven. (The fibers start out white and turn to gold and then black), (b) PAN fibers in a 320 k tow band after making several passes through a pilot-scale oxidation oven, (c) center-to-ends airflow technology, (d) PAN passing through the end-slots of a large-scale oxidation oven, and (e) fibers are manipulated to work out any knots as the fibers begin running through ovens. (From Despatch Industries.)

5.3.5 Properties of Fibers

Like the physical and mechanical properties of wrought aluminum alloys and other metals, the properties of fibers are very important to know while designing the aircraft components, and also in the application of that component to the airframe structure. Table 5.2 [1] shows properties of typical fibers used to manufacture various composite materials.

Figure 5.6 [2] shows schematically the stress–strain curves of types I and II carbon fibers. Based on final heat treatment temperature, carbon fibers are generally classified into type I (high heat treatment fibers heat treated above 2000°C (3632°F), associated with high modulus type fiber) and type II (intermediate heat treatment fibers heat treated around or above

TABLE 5.2

Properties of Typical Fibers

Type	Diameter μm (10⁻³ inch)	Density g/cm³ (lb/inch³)	Modulus GPa (Msi)	Tensile Strength MPa (ksi)
Glass				
E-glass	8–14 (0.30–0.55)	2.54 (0.092)	73 (10.5)	3450 (500)
S-glass	10 (0.40)	2.49 (0.090)	86 (12.4)	4500 (650)
Carbon				
AS4	7 (0.28)	1.81 (0.065)	235 (34)	3700 (535)
T300		1.76 (0.063)	230 (33)	3100 (450)
T-400H		1.80 (0.065)	250 (36)	4500 (650)
IM-6	4 (0.16)	1.80 (0.065)	255 (37)	4500 (650)
IM-7	4 (0.16)	1.80 (0.065)	290 (42)	5170 (750)
Graphite				
T-50		1.67 (0.060)	390 (57)	2070 (300)
GY-70		1.86 (0.067)	520 (75)	1725 (250)
P100 S	10 (0.4)	2.02 (0.073)	720 (105)	1725 (250)
Boron	140 (5.6)	2.50 (0.090)	395 (57)	3450 (500)
Aramid				
Kevlar 49	12 (0.47)	1.45 (0.052)	131 (19)	3800 (550)
Kevlar 149	12 (0.47)	1.45 (0.052)	186 (27)	3400 (490)
Silicon carbide				
SCS-2	140 (5.6)	3.10 (0.112)	400 (58)	4140 (600)
Nicalon	15 (0.6)	2.60 (0.094)	172 (25)	2070 (300)
Alumina				
FP-2	–	3.70 (0.133)	380 (55)	1725 (250)
Nextel 610	–	3.75 (0.135)	370 (54)	1900 (275)
Saphikon		3.80 (0.137)	380 (55)	3100 (450)
Sapphire whiskers		3.96 (0.143)	410 (60)	21,000 (3000)
Silica	–	2.19 (0.079)	73 (10.5)	5800 (840)
Tungsten	–	19.3 (0.696)	410 (60)	4140 (600)

1500°C (2732°F), associated with high strength type fibers). The vertical arrows indicate complete failure. Figure 5.6 shows that the high modulus fibers have a much lower strain to failure at 0.50% strain compared with the high strength fibers (1.0%).

5.4 Fabrics

Woven fabrics are another form of constructing composite materials from the primary fibers. Straight yarns are often used in the form of a woven fabric or textile. Woven products are usually constructed in an orthogonal (0° and 90°) form, which consists of two sets of interlaced yarns. Figure 5.7 shows the schematic representation of the weaving process [4]. Fabrics are made by interlacing two orthogonal sets of yarns (warp and fill).

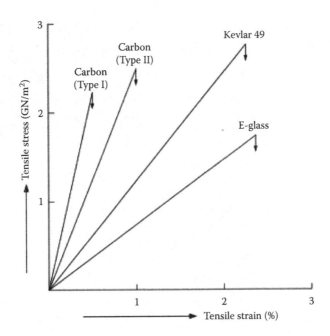

FIGURE 5.6
Stress–strain plots of fibers, vertical arrow indicates complete failure.

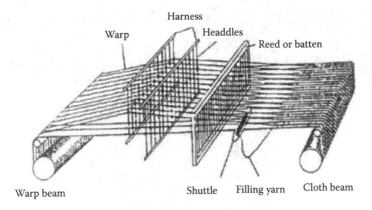

FIGURE 5.7
Schematic of basic weave process.

The longitudinal direction of the fabric is called "warp" and the fabrics in the transverse direction are called fill. This process produces woven cloth or fabric by having a uniform gap between warp yarns and insertion of the fill yarns.

Figure 5.8 shows few examples of common fabrics from three different fabrics used in the composite manufacturing industry.

Fabrics are generally classified by six different weave patterns of interlacing of warp and filling yarns including plain, satin, twill, basket, leno, and mock leno. The weave style can be varied according to crimp and drapeability. Low crimp gives better mechanical performance because straighter fibers carry greater loads; a drapeable fabric is easier to lay up over complex forms. Figure 5.9 shows an example of mostly used plain, satin, and twill type woven fabrics.

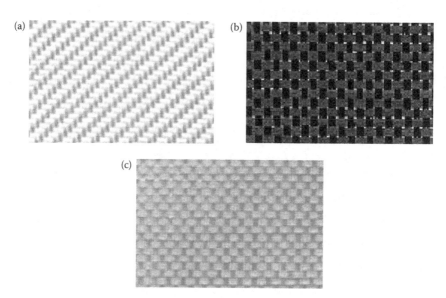

FIGURE 5.8
Example of common cloth/fabric. (a) Glass fabric, (b) carbon fabric, and (c) Kevlar. (From US Composites.)

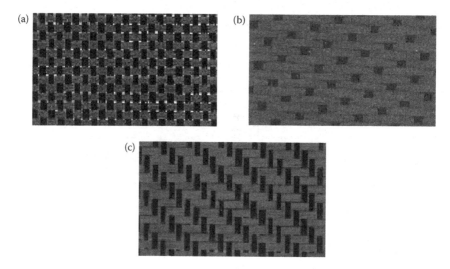

FIGURE 5.9
Example of carbon fabric weave patterns. (a) Plain, (b) satin, and (c) twill. (From US Composites.)

In the plain weave, each yarn is interlaced over every other yarn in the other direction. Plain weave has low drapeability and high crimp. In satin weaves, each yarn is interlaced over every fourth, fifth, and so on in the other direction. Satin weave provides good drapeability and low crimp. In twill woven fabric, each yarn is interlaced over every third yarn in the other direction. Twill weave provides average drapeability and average crimp.

Most weaving fabrics have the same material in both warp and fill directions, with a similar number of fibers. But special fabrics can be made using warp of one material with fill of a different material, which are called hybrid weaves. Frequently used hybrids are made with glass warp and carbon fills. The types of material used in the hybrid are

FIGURE 5.10
Example of carbon/Kevlar hybrid weave (carbon 0°, Kevlar 90°). (From US Composites.)

required to obtain specific properties of the composites such as mixing carbon with Kevlar (Figure 5.10) to take advantage of the toughness of Kevlar with carbon fibers.

Since each pattern has its own characteristics based on the interlacing per unit area, the selection of pattern is variable according to ease in handling during impregnation, as well as difficulty to form in complex contours, conformability (drapeability) of the fabrics over a complex surface, surface finish, and finally to achieve the strength and stiffness required of the composite. Woven fabrics are frequently the choice for fabricating complex geometries, but it is difficult to maintain the directionality of the yarns on compound contours and other complex shapes. The behavior of fabric reinforcement in the composite is measured by the fabric crimp, which is the measure of the yarn waviness. The crimp fraction decreases and the drapeability of the fabric increases as on transition from plain to twill and multi-harness satin weaves.

5.5 Matrices

The matrix is the bonding material for fiber reinforcement within the composite material and determines the quality of the surface. The main function of the matrix is to

- Hold fibers in their respective position
- Protect the fibers from abrasion
- Transfer load between fibers
- Provide inter-laminar shear strength

There are a variety of matrix materials used to provide the highest advantages, along with customer need and satisfaction. Matrix materials can be polymers, metals, ceramics, or carbon. The cost of each matrix varies with temperature resistance. There are several broad categories, each with numerous variations. The most common are known as

1. Polymeric matrix—epoxy
2. Polyimide

3. Polyamide

4. Polyester

5. Vinyl ester

6. Phenolic

7. Thermoplastic

8. Polypropylene

9. Polyether ether ketone (PEEK)

Polymer matrices are the most widely used for composites in commercial and high-performance aerospace applications. The most widely used polymers are thermoset resins. Polymers (poly means many and mer means unit or molecule) exist in at least three major forms shown in Figure 5.11 [3]: (1) linear, (2) branched, and (3) cross-linked. A linear polymer is merely a chain of mers. A branched polymer consists of a primary chain of mers with other chains that are attached in three dimensions just like tree branches. Cross-linked polymer has a large number of three-dimensional (3D) highly interconnected chains. Linear polymers have the least strength and stiffness, whereas cross-linked polymers have the most strength and stiffness because of their inherently stiffer and stronger internal structure.

The selection of the right matrix for the right fiber is very important to assure proper bonding and curing to get the desired strength and stiffness required to build a structural part. The matrix selection criterion is as follows:

- Resistance to heat, chemicals, and moisture
- High strain resistance to failure
- Cure at as low temperature as possible
- A long pot or out-time life
- Nontoxic

The physical and mechanical properties of the composite materials are very much dependent on the structure and properties of the fiber–matrix interface as shown schematically in Figure 5.12. There is a large difference between the elastic properties of fibers and matrix

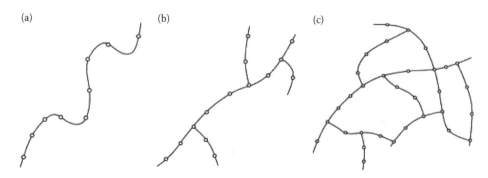

FIGURE 5.11
Polymer structure. (a) Linear, (b) branched, (c) cross-linked.

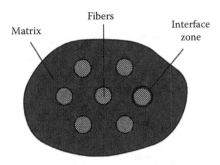

FIGURE 5.12
Schematic of fiber–matrix interface.

materials, therefore, the interface strength is very important to transfer stress acting on the matrix to the fiber. Bond strength at the interface is due to adhesion between fiber and matrix. Adhesion between fiber and matrix can be attributed to five main mechanisms:

1. Adsorption and wetting
2. Interdiffusion
3. Electrostatic attraction
4. Chemical bonding
5. Mechanical adhesion

Detailed explanations of each mechanism were discussed in *Engineering Mechanics of Composite Materials* [1].

5.5.1 Classification of Polymer Matrices

There are three main classes of structural polymer matrices:

- Rubbers (cross-linked)
- Thermosets (cross-linked)
- Thermoplastics (branched)

Matrices for polymeric composites can be compared in either thermosets or thermoplastics. Thermoset resins usually consist of a mixture of a resin and a compatible curing agent to form a liquid of low viscosity, which cures due to exothermic or externally applied heat. Thermoset materials are usually liquid or malleable prior to curing and designed to be molded into their final form, or used as adhesives.

Figure 5.13 [3] shows schematically the stress–strain plots of a general purpose grade polyester resin tested in tension and compression. The full lines are the experimental stress–strain plots of a polyester resin tested in uniaxial tension and compression, respectively. The dotted line is predicted by assuming the ratio of yield strength in compression and yield strength in tension equals to 1.3. In tension, fracture occurs at 63 MN/m^2 at a strain of 2% and there is no sign of yielding before fracture. Large-scale yielding occurs in compression with upper yield strength of 122 MN/m^2.

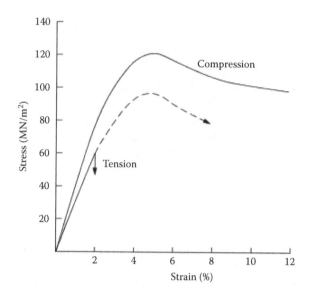

FIGURE 5.13
Stress–strain plot of a general purpose polyester resin.

TABLE 5.3

Comparison between Thermosets and Thermoplastics Resins

Thermosets	Thermoplastics
• Mostly used	• Mostly suited for automated production
• Does not flow under heat and pressure	• Softens under heat and pressure, so it flows readily
• Amorphous	• May be crystalline
• Applied to reinforcement as low viscosity liquids or varnishes	• High viscosity even in the melt. Reinforcement by dry or melt compounding
• Emits volatiles on curing (not epoxies or polyesters)	• No emissions of volatiles during molding

Thermoplastics start with fully reacted high viscosity materials that do not cross-link on heating. Heating at a high enough temperature, thermoplastics either soften or melt and they can be reprocessed a number of times. Table 5.3 shows the comparison between thermoset and thermoplastic resins.

5.6 Classification of Composites

Composites are classified into different categories [1] according to the type of fibers and fabrics and their placements within the matrix as shown in Figure 5.14. The composites with the particulate fillers consist of varying sizes and shapes, which are randomly dispersed within the matrix. Due to the randomness of particle distribution, particulate composites can be considered as quasi-homogeneous and quasi-isotropic on a macroscale, which is larger than the particle size and spacing. Particulate composites may consist of

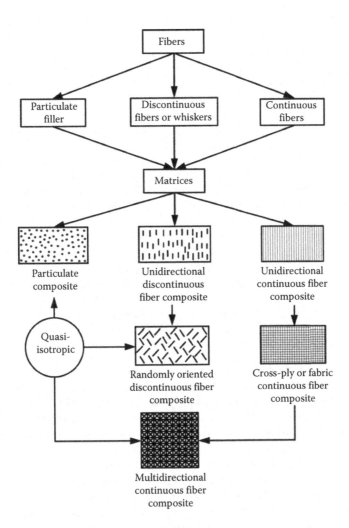

FIGURE 5.14
Classification of composites.

nonmetallic particles in a nonmetallic matrix; metallic particles in nonmetallic matrices; metallic particles in metallic matrices; or nonmetallic particles in metallic matrices.

In discontinuous or short-fiber composites, the fibers are in the form of short fibers, nanotubes (1 nm in diameter and 1000 nm in length approximately), or whiskers. Although the fibers are short, the lengths of the fibers are long compared to their diameters (high aspect ratio). Short fibers can all be oriented unidirectional or in a random orientation pattern. In general, composite materials tend to be markedly anisotropic, discontinuous fibers can be regarded as a quasi-isotropic composite.

In continuous-fiber composites, the long continuous fibers are reinforced. Continuous fibers could be placed all parallel (unidirectional) to each other, could be right angle to each other (cross-ply or woven fabric type), or can be oriented along several directions (multidirectional continuous-fiber composite). With a certain number of fiber directions and distribution of fibers, multidirectional continuous-fiber composites can be regarded as a quasi-isotropic material.

TABLE 5.4

Type of Fiber Composite Materials

Matrix Type	Fiber	Matrix
Polymer	E-glass	Epoxy
	S-glass	Phenolic
	Carbon (graphite)	Polyimide
	Aramid (Kevlar)	Bismalemide
	Boron	Polyester
		Thermoplastics (PEEK, polysulfone, etc.)
Metal	Boron	Aluminum
	Borsic	Magnesium
	Carbon (graphite)	Titanium
	Silicon carbide	Copper
	Alumina	
Ceramic	Silicon carbide	Silicon carbide
	Alumina	Alumina
	Silicon nitride	Glass-ceramic
		Silicon nitride
Carbon	Carbon	Carbon

Fiber-reinforced composite materials can be classified into broad categories based on the type matrix used for reinforcement as shown in Table 5.4 [1]. Polymer-matrix composites include thermoset (epoxy, polyimide, and polyester) or thermoplastic (PEEK, polysulfone) resins reinforced with glass, carbon (graphite) aramid (Kevlar), or boron fibers. Polymer-matrix composites are primarily used for relatively low temperature applications. Metal matrix composites consist of metals or alloys of aluminum, magnesium, titanium, and copper reinforced with fibers of boron, carbon (graphite), or ceramic fibers. Their maximum temperature use is limited by the softening or melting temperature of the matrix metal. Ceramic composites consist of ceramic matrices (silicon carbide, aluminum oxide, glass-ceramic, and silicon nitride) reinforced with ceramic fibers. Ceramic-matrix composites are best suited for high-temperature applications. Carbon/carbon composites consist of carbon or graphite matrix reinforced with graphite yarn or fabric. They have unique properties of relatively high stiffness and moderate or low strength at high temperatures coupled with low thermal expansion and low density.

5.7 Lamina

The basic terminology of fiber-reinforced composite materials will be introduced as lamina and laminates in this section. Composites can be described as layers or plies of high strength fibers embedded in a matrix of plastic resin. The components are layered generally in laminate form. A lamina is a planed (sometimes curved) layer of unidirectional fibers or woven fibers or fabrics in a matrix. A unidirectional fiber reinforced in the matrix is also referred to as unidirectional lamina. The lamina is an orthotropic material with principal material axes in the direction of the fiber (longitudinal), perpendicular to the fiber direction (transverse), and normal to the plane of the lamina as shown in Figure 5.15a.

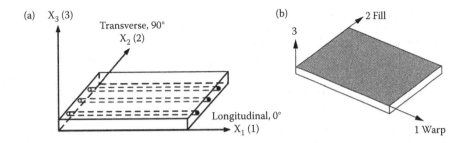

FIGURE 5.15
Two types of lamina. (a) Unidirectional fibers and (b) woven fibers.

The principal axes are designated as 1, 2, and 3, respectively. For woven type fiber/fabric composite, the warp is in the principal direction 1 and fill is along the principal direction 2 as shown in Figure 5.15b [1]. The fibers are the principal reinforcing or load-carrying agents, whereas matrix is supporting and protecting the fibers in place within the lamina.

5.7.1 Geometrical Properties of Lamina

Major properties of the fiber-reinforced composite materials are strongly dependent on microstructural parameters including fiber diameter, fiber length, fiber length distribution, volume fraction of fibers, and the alignment and packing arrangement of the fibers in the matrix. The effect of each parameter varies from one property to another. It is required to characterize these parameters for effective processing of composite materials, efficient design, and manufacture of components. Figure 5.16 shows a unidirectional lamina where all the fibers are aligned parallel to each other. High-performance aircraft components usually consist of layers of lamina stacked up in a predetermined arrangement to achieve optimum properties and performance.

Fiber placement within the matrix could be any regular or irregular geometrical order. In an ideal situation, the fiber placement can be considered to be arranged on a square or hexagonal lattice as shown in Figure 5.17 with each fiber having a circular cross-section and the same diameter. In practice, glass fibers are closely considered to be circular

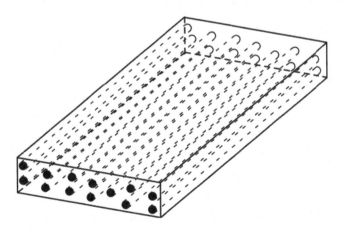

FIGURE 5.16
Unidirectional lamina with fiber placing.

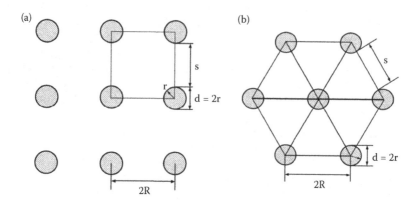

FIGURE 5.17
Geometrical aspects of fiber placing. (a) Square packing and (b) hexagonal packing.

cross-section with smooth surface finish, whereas the carbon fibers are considered as rough circular cross-section with very irregular surfaces. In practice, there is also some variation of fiber diameter for all types of fiber materials that are dependent on their manufacturing procedures.

Composite lamina is characterized by the following geometrical properties:

$$\text{Fiber volume/weight ratio} = \frac{\text{Volume/Weight of fibers}}{\text{Volume/Weight of composites}}$$

$$\text{Matrix volume/weight ratio} = \frac{\text{Volume/Weight of matrix}}{\text{Volume/Weight of composites}}$$

$$\text{Void volume ratio} = \frac{\text{Volume of voids}}{\text{Volume of composites}}$$

Assuming the fibers are round in cross-section, the volume of fiber is calculated from a simple geometry, the fiber volume ratio and the spacing between the fibers in a square pack are given by

Fiber volume ratio:

$$Vf = \frac{\pi}{4}\left(\frac{r}{R}\right)^2 \tag{5.1}$$

Fiber spacing:

$$S = 2\left[\sqrt{\frac{\pi}{4Vf}} - 1\right]r \tag{5.2}$$

where, r is the radius of fiber and 2R is the center distance between two fibers. For maximum fiber concentration, s will be equal to zero, 2R = 2r, fiber volume ratio would be 0.785.

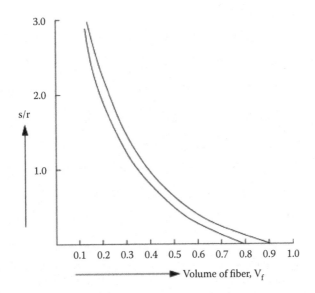

FIGURE 5.18
Relation between spacing and fiber volume. (From Hull, D., *An Introduction to Composite Materials*, Cambridge University Press, UK, 1981.)

Similarly, the fiber volume ratio and the spacing between the fibers in a hexagonal pack are given by

$$Vf = \frac{\pi}{2\sqrt{3}} \left(\frac{r}{R} \right)^2 \tag{5.3}$$

$$S = 2 \left[\sqrt{\frac{\pi}{2\sqrt{3}\,Vf}} - 1 \right] r \tag{5.4}$$

For maximum fiber concentration, fiber volume ratio for hexagonal pack will be 0.907, as shown in Figure 5.18.

5.8 Laminate

A laminate is a stack of lamina with various orientations of principal material directions in the lamina as shown in Figure 5.19. The layers of each lamina are bonded together by the same matrix material that is used in individual lamina. Laminates can be composed of plates of different materials or layers of fiber-reinforced lamina. The major purpose of lamination is to get the directional dependency of strength and stiffness of a composite material to satisfy structural needs. Any number of layers of a multilayer laminate could be oriented in any combination of directions to have the maximum strength and stiffness

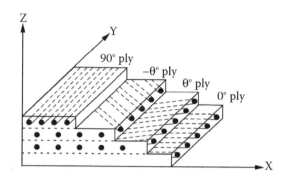

FIGURE 5.19
Laminate.

of the final laminates. Since the orientation of the principal material axes varies from layer/ply to layer/ply, it is more convenient to analyze laminates using common (X, Y, and Z) coordinate axes. The orientation of a given ply is considered by the angle between the reference X axis and the major principal material axis (fiber orientation or warp direction) of the ply, measured in a counterclockwise direction of the X–Y plane. Composite laminates are designated in a manner indicating

- The number of plies
- Type
- Orientation
- Stacking sequence

Few examples of laminate designations are shown in Table 5.5 [1].

TABLE 5.5

Laminate Designation

Laminates	Designation
Unidirectional six ply	$[0/0/0/0/0/0/] = [0_6]$
Cross-ply symmetric	$[0/90/90/0] = [0/90]_s$
	$[0/90/0/90/90/0/90/0] = [0/90]_{2s}$
	$[0/90/0] = [0/\overline{90}]_s$
Angle-ply symmetric	$[+45/-45/-45/+45] = [\pm45]_s$
	$[30/-30/30/-30/-30/30/-30/30] = [\pm30]_{2s}$
Angle-ply asymmetric	$[30/-30/30/-30/30/-30/30/-30] = [\pm30]_4$
Multidirectional	$[0/45/-45/-45/45/0] = [0/\pm45]_s$
	$[0/0/45/-45/0/0/0/0/-45/45/0/0] = [02/\pm45/0_2]_s$
	$[0/15/-15/15/-15/0] = [0/\pm15/\pm15/0]_T = [0/(\pm15)_2/0]_T$
Hybrid	$[0^K/0^K/45^C/-45^C/90^G/-45^C/45^C/0^K/0^K]_T = [0_2^K/\pm45^C/\overline{90}^G]_s$

Note: s = symmetric sequence, T = total number of plies, – (over bar) = laminate is symmetric about the mid-plane of the ply, number subscript = multiple of plies or group of plies.

5.9 Type of Composites

Due to high demand of composite materials in the aerospace industry, continuous developments in composite materials are taking place. There are a wide range and many forms of polymeric, metallic, and ceramic materials being used both as fibers and as matrix materials, with the aim of improving strength, toughness, stiffness, resistance to high temperatures, and reliability in service, particularly under severe adverse environmental conditions.

5.9.1 Polymer

Polymer composites are classified according to the type of matrix and fiber materials used prior to bonding. A very wide variety of properties can be achieved by choosing the type, length, and direction of reinforced fibers (glass, carbon or graphite, aramid, and boron) within the matrix (thermosetting or thermoplastic). Where necessary, the composite can have different properties in different directions. When the possibility of changing the matrix material is included, it can be seen that polymer composites form a vast family of engineering materials.

5.9.2 Hybrid

Hybrids are material forms that make use of two or more fiber types. Lay ups of different materials in the hybrid composite are shown as an example in Figure 5.20. The major types are

1. Inter-ply hybrid
2. Intra-ply hybrid
3. Super hybrid

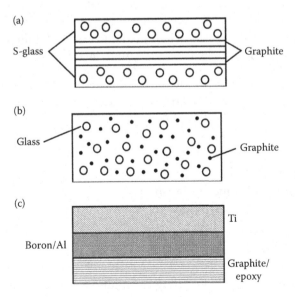

FIGURE 5.20
Hybrid composites. (a) Inter-ply, (b) intra-ply, and (c) super hybrid.

Common hybrids generally include glass/carbon, glass/aramid, and Kevlar/carbon fibers. The major advantage of hybrids is the utilization of properties and features of each reinforcement type.

5.9.3 Metal Matrix

The advantage of a metal matrix over polymer matrix is its higher resistance to elevated temperatures and higher ductility and toughness. The limitations of metal matrix composites are higher density and greater difficulty in processing. The matrix materials in these composites are usually aluminum, aluminum–lithium, magnesium, and titanium, although other metals are also being investigated. The fiber materials typically are graphite, aluminum oxide, silicon carbide, and boron, with beryllium and tungsten as other possibilities. Because of their high specific stiffness, light weight, and higher thermal conductivity, boron fibers in an aluminum matrix have, for example, been used for structural tubular supports in the space shuttle orbiter. Other applications include stabilizers for aircraft and helicopters. An important consideration is the proper bonding of fibers to the metal matrix.

5.9.4 Ceramic Matrix

Fibers used in ceramic-matrix composites are classified according to their diameter and aspect ratio. There are three general categories:

1. Whiskers
2. Monofilaments
3. Textile multifilament fibers

Reinforcements in the form of particulates and platelets are also used. Ceramic-matrix selection is usually controlled by thermal stability and processing considerations. Melting point of the matrix is the first indication of high temperature stability.

Fiberglass is the best known of the composite materials. The fibers are woven into fabric or tape and impregnated with epoxy. The greatest benefits of fiberglass are high strength-to-weight ratio and low cost. The disadvantage of slow-cured fiberglass composite structures is their propensity to creep or distort at high ambient temperature.

There are many desirable properties of monolithic ceramic materials:

- High moduli
- High compressive strengths
- High temperature withstanding capability
- High hardness and wear resistance
- Low thermal conductivity
- Low chemical inertness

Due to its high temperature withstanding capability, ceramics have become very popular for extremely high-temperature applications. But ceramics are limited to certain structural applications due to their very low fracture toughness properties.

5.10 Manufacturing Technology of Input Raw Materials for Composite Fabrication

Reinforcement of fibers in the matrix can be combined in the form commonly called prepreg, which is manufactured by impregnating fibers with a controlled volume of resin. Prepreg is the input raw material for fabrication of particular composite parts made by stacking and shaping layers of prepreg materials. Prepreg tapes or fabrics are used for many different types of layup operations including hand, automated tape, and fiber placement, and in some filament winding. Unidirectional tape where the fibers are placed in parallel in the longitudinal direction is commonly used. Prepregs are also made with woven fibers offering reinforcement in two dimensions. Prepregs made by impregnating fiber performs and braids provide 3D reinforcement. A prepreg tape with any combination of layers of parallel or woven fibers pre-impregnated with resin is partially cured to a certain degree. Prepregs are made to meet certain specifications including fiber volume ratio, ply thickness, and degree of partial cure. The manufacturing process of prepregs is illustrated in Figure 5.21. The process consists of

- Fiber control
- Fiber collimation
- Resin impregnation
- Tape production

Prepregs are made in roll forms of varying widths supported on backing paper. The resin content is usually 32%–42% by weight. The tapes are characterized by self-adhesion, drapeability, or ability to conform to shapes, shelf life, out time, and gel time. The prepreg must be stored refrigerated at approximately –18°C (0°F) before final use.

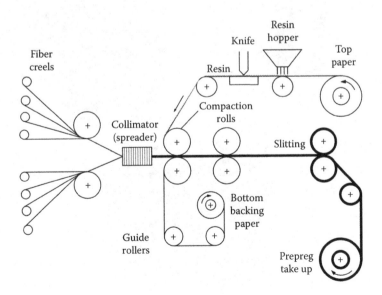

FIGURE 5.21
Schematic of prepreg process.

5.10.1 Type of Prepregs

The prepregs are generally rolls of uncured materials in the following forms: tow, tape, and woven fabric. Tows are the "threads" that make up the cloth or tape. Tape refers to weaved cloth, typically in assorted inch widths as shown in Figure 5.22:

- Unidirectional tow (unidirectional fiber tow impregnated with resin)
- Unidirectional tape (large number of unidirectional tows are impregnated with resin)
- Woven fabric (fabric is impregnated with resin)

Prepregs are generally manufactured in two different forms based on the type of matrix/resins used to prepare the prepregs, which includes

1. Thermoset
2. Thermoplastic

Thermoplastic prepregs are composite reinforcements that are pre-impregnated with thermoplastic resin. Thermoplastic prepregs can be provided in unidirectional tape, or in fabrics that are woven or stitched. The primary difference between thermoset and thermoplastic prepreg is that thermoplastic prepregs are stable at room temperature, and, generally, do not have a shelf life. More commonly used in prepreg composite manufacturing is thermoset prepregs. The primary resin matrix used is epoxy. With a thermoset prepreg, the thermosetting resin starts as a liquid and fully impregnates the fiber reinforcement. Excess resin is precisely removed from the reinforcement. Meanwhile, the epoxy

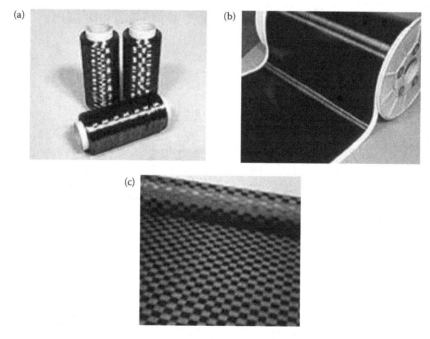

FIGURE 5.22
Example of prepregs. (a) Tow, (b) tape, and (c) woven fabric. (From Mitsubishi Rayon Company, Ltd.)

resin undergoes partial curing, changing the state of the resin from a liquid to a solid. In this stage, the resin is partially cured, and usually tacky. When the resin is brought up to an elevated temperature, it often returns briefly to a liquid state prior to hardening completely. Once cured, the thermoset resin that was in partial curing stage is being converted to fully cross-linked stage. Since the epoxy is in a partial curing stage, thermoset prepreg is required to be stored either refrigerated or frozen prior to use.

The great advantage of using prepregs is their ease of use in one ready form instead of using fibers/fabrics, epoxy resin, and the hardener for the epoxy resin separately and also dealing with resins in a liquid state could be messy. Thermoset prepregs are generally supplied with a backing film on both sides of the fabric to protect it during transit and preparation. The prepreg is then cut to the desired size and shape, the backing is peeled off, and the prepreg is then laid into the mold or tool geometry. Both heat and pressure are applied for the specified amount of time to cure.

5.11 Introduction to Manufacturing Processes of Composite Materials

There are many manufacturing processes of composite materials for fabrication of aircraft components. Although this chapter does not cover manufacturing processes, Chapter 18 has been created to discuss major manufacturing processes of composite materials for aerospace. An overview of the primary manufacturing processes is given in Table 5.6 [2]. Processes 1–5 are considered to be open-mold processes, whereas processes 6–11 are considered to be closed mold processes.

5.12 Physical Definition of Materials

Table 5.7 shows the physical definition of materials used in the fundamental of mechanics of composite materials. The behavior of isotropic, orthotropic, and anisotropic materials under loadings of normal stress and shear stress is shown in Figure 5.23. The mechanics of composite materials [1–3] are a separate subject matter being taught in the graduate level study for various discipline of engineering.

Under uniaxial tensile loading along nonprincipal material directions, an isotropic material undergoes an axial deformation having strains, ε_x, in the loading direction and transverse deformation strain, ε_y, with no shear deformation (Figure 5.23a).

$$\varepsilon_x = \frac{\sigma_x}{E} \tag{5.5}$$

$$\varepsilon_y = -\nu \cdot \varepsilon_x = -\frac{\nu \sigma_x}{E} \tag{5.6}$$

$$\gamma_{xy} = 0 \tag{5.7}$$

where ε_x, ε_y, and γ_{xy} are axial, transverse, and shear strains, respectively, σ_x is the axial stress, E is Young's modulus, and ν is Poisson's ratio.

TABLE 5.6

Primary Manufacturing Processes of Composite Materials

Process	Outline
1. Hand lay-up	Chopped-strand mats, woven roving, and other fabrics made from fibers are placed on the mold and impregnated with resin by painting and rolling. Layers are built up until design thickness is reached. Molding cures without heat or pressure
2. Spray-up	Chopped rovings and resin are sprayed simultaneously into a prepared mold and rolled before the resin cures
3. Vacuum bag, pressure bag, autoclave	The prepreg sheets are stacked on the mold surfaces in predetermined orientations, covered with a flexible bag, and consolidated using a vacuum or pressure bag in an autoclave at the required curing temperature
4. Filament winding	Continuous rovings or strands of fibers are fed over rollers and guides through bath of resin and then wound, using a computer-controlled machine onto a mandrel at predetermined angles. The resin is partially or completely cured before removing the component from a mandrel type mold
5. Centrifugal casting	Mixtures of the fibers and resin are introduced into a rotating mold and allowed to cure *in situ*
6. Hot press molding, compression molding	Heated match dies or tools are loaded with raw material, sheet molding compound (SMC), dough molding compound cloth, or unidirectional prepreg pressed to the shape of the cavity and cured
7. Injection molding, transfer molding	Molten or plasticized polymer mixed with short fibers is injected usually at high pressure into the cavity of a split mold and allowed to cure
8. Pultrusion	A continuous feed of fibers in pre-selected orientations and then impregnated with resin and pulled through a heated die to give the shape of the final section
9. Cold press molding	A low pressure and low temperature process in which fibers are impregnated with resin and then pressed between matched dies. Heat is generated during the cure
10. Resin injection	Fibers in cloth form are placed in the tool, which is then closed. The resin is injected at low pressure into the cavity and flows through the fibers to fill the mold space
11. Reinforced reaction injection molding	A rapid curing resin system involving two components, which are mixed immediately before injection is used. Fibers are either placed in the closed mold before resin is injected or added as short chopped fibers in one of the resin components to form slurry before injection

TABLE 5.7

Common Physical Definitions

Physical Designation	Definitions
Homogeneity	Properties of material are the same at every point or the properties are independent of location
Hetrogeneity/inhomogeneity	Properties vary from point to point or are dependent on location
Isotropy	Properties remain same in all directions or are independent of the orientation of the reference axes
Anisotropy	Properties at a point vary with direction or are dependent on the orientation of the reference axes
Orthotropy	Properties dependent on orientation, but are symmetric about three mutually perpendicular planes

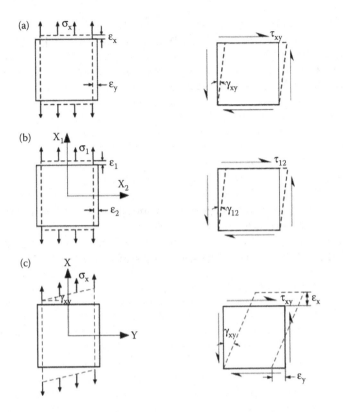

FIGURE 5.23
Response under uniaxial normal stress and pure shear stress on various types of materials. (a) Isotropic material loaded along nonprincipal directions, (b) orthotropic material loaded along principal material directions, and (c) anisotropic or orthotropic material loaded along nonprincipal directions.

Under pure shear loading, τ_{xy}, the material undergoes pure shear deformation. The shear strain and the normal strains are as follows:

$$\gamma_{xy} = \frac{\tau_{xy}}{G} = \frac{2\tau_{xy}(1+\nu)}{E} \tag{5.8}$$

$$\varepsilon_x = \varepsilon_y = 0 \tag{5.9}$$

where τ_{xy} = shear stress and G = shear modulus.

An orthotropic material loaded in uniaxial tension along one of its principal material axes ($X1$ and $X2$) undergoes deformation (Figure 5.23b) similar to those of an isotropic material as given by

$$\varepsilon_1 = \frac{\sigma_1}{E_1} \tag{5.10}$$

$$\varepsilon_2 = -\nu_{12} \cdot \varepsilon_1 = -\frac{\nu_{12}\sigma_1}{E_1} \tag{5.11}$$

$$\gamma_{12} = 0 \tag{5.12}$$

where ε_1, ε_2, and γ_{12} are axial, transverse, and shear strains, respectively, σ_1 is the axial stress, and v_{12} is Poisson's ratio associated with loading in the 1 principal direction and strain in the 2 principal direction.

Under pure shear loading, τ_{12}, the material undergoes pure shear deformation along the principal material axes. The strains are given by

$$\gamma_{12} = \frac{\tau_{12}}{G_{12}} \tag{5.13}$$

$$\varepsilon_1 = \varepsilon_2 = 0 \tag{5.14}$$

The shear modulus G_{12} = shear modulus is an independent material constant and is not directly related to Young's moduli and Poisson's ratio.

Anisotropic material, under uniaxial tension, undergoes axial, transverse, and shear deformation (Figure 5.23c) and the corresponding strains are given by

$$\varepsilon_x = \frac{\sigma_x}{E_x} \tag{5.15}$$

$$\varepsilon_y = -v_{xy} \cdot \varepsilon_x = -\frac{v_{xy}\sigma_x}{E_x} \tag{5.16}$$

$$\gamma_{xy} = \eta_{xs} \cdot \varepsilon_x = \eta_{xs} \cdot \frac{\sigma_x}{E_x} \tag{5.17}$$

where ε_x, ε_y, and γ_{xy} are axial, transverse, and shear strains, respectively, σ_x is the axial normal stress, E_x is Young's modulus in the x direction, v_{xy} is Poisson's ratio associated with loading in the x direction and strain in the y direction, and η_{xs} is the shear coupling coefficient.

Under pure shear loading, the material undergoes both shear and normal deformations. The shear and normal strains are given by

$$\gamma_{xy} = \frac{\tau_{xy}}{G_{xy}} \tag{5.18}$$

$$\varepsilon_x = \eta_{sx} \cdot \frac{\tau_{xy}}{G_{xy}} \tag{5.19}$$

$$\varepsilon_y = \eta_{sy} \cdot \frac{\tau_{xy}}{G_{xy}} \tag{5.20}$$

where G_{xy} is the shear modulus referred to the x and y axes and η_{sx} and η_{sy} are the shear coupling coefficients.

The physical properties of composite materials are generally not isotropic in nature. The structures are rather typically orthotropic. Unlike the stiffness of an isotropic metal, the

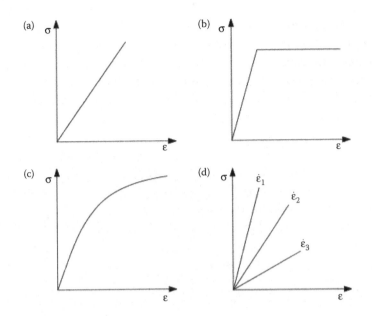

FIGURE 5.24
Various stress–strain relation. (a) Linear elastic, (b) elastic-perfectly plastic, (c) elastic–plastic, and (d) viscoelastic.

stiffness of a composite panel depends upon the directional orientation of the applied forces and or the applied moments. Panel stiffness is dependent on the design of the panel:

1. Fiber reinforcement and matrix
2. Method of build
3. Thermoset versus thermoplastic
4. Type of weave
5. Orientation of fiber axis to the primary force

The stress–strain relation of fibers used in different types of composites is classified in four different classes as shown in Figure 5.24 [3]. Fibers generally exhibit linear elastic behavior, although reinforcing steel in concrete is more nearly elastic-perfectly plastic. Fiber-reinforced composite materials such as boron- and graphite-epoxy are usually treated as linear elastic materials because the essentially linear elastic fibers provide the majority of strength and stiffness.

5.13 Comparison between Composites and Conventional Metals

A brief comparison between the fiber-reinforced composite materials and the conventional metal has been made (Tables 5.8 through 5.14) based on the following aspects [1]:

1. Micro-mechanics
2. Macro-mechanics

3. Material characterization
4. Design and optimization
5. Fabrication technology
6. Maintenance, serviceability, and durability
7. Cost effectiveness

TABLE 5.8

Micro-Mechanics

Materials	Comparison (Advantages/Limitations)
Composites	• High strength, stiffness of fibers • High fracture toughness • Capability of interfacial stress transfer (it does not allow the cracks propagate in longitudinal direction) • High scatter in fiber strength • Low interface bond • High localized stresses (strain) concentration
Conventional metals	• Local irregularities (or microstructure) control brittle or ductile behavior

TABLE 5.9

Macro-Mechanics

Materials	Comparison (Advantages/Limitations)
Composites	• Anisotropy and heterogeneity enable tailoring material according to stress fields • Ability to control and predict mechanical performance • Analyses are complex • Elaborate computer programs
Conventional metals	• Analyses are more straightforward

TABLE 5.10

Material Characterization

Materials	Comparison (Advantages/Limitations)
Composites	• Prediction of macro-properties from properties of constituents • Complex experimental program
Conventional metals	• Only two elastic parameters (modulus of elasticity, Poisson's ratio) • Two strength parameters (tensile, compression)

TABLE 5.11

Design and Optimization

Materials	Comparison (Advantages/Limitations)
Composites	• Ability to design material and structure in one continuous process • Enough degrees of freedom of optimization • Material selection and design procedure are more complex because of many options • Shortage of analytical tools for optimization
Conventional metals	• Few geometrical parameters to vary • Optimization is limited

TABLE 5.12

Fabrication Technology

Materials	Comparison (Advantages/Limitations)
Composites	• Building of materials and structure profiles • Simple tooling • One-step assembly—fewer parts • Hand labor skill is needed • Standardization and automation are limited • Quality control is more complex
Conventional metals	• Material and structural fabrication are separate operations • Need for sophisticated multi-joint assembly

TABLE 5.13

Maintenance, Serviceability, and Durability

Materials	Comparison (Advantages/Limitations)
Composites	• Long-term performance in hostile environments • Enhanced fatigue life • Easy maintenance and repair • Sensitivity to hydrothermal effect • Difficult to detect internal damage in service • Need for protective coatings
Conventional metals	• Sensitivity to corrosion • Easy propagation of cracks leading to catastrophic failure

TABLE 5.14

Cost Effectiveness

Materials	Comparison (Advantages/Limitations)
Composites	• Cost savings due to weight reduction • Lower tooling cost • Lower cost due to lower number of parts • Lower assembly cost • Raw materials are expensive • High cost of primary materials (skins, cores)
Conventional metals	• Raw materials are inexpensive • High cost of machining, tooling, and assembly

References

1. Daniel, I.M. and Ishai, O., *Engineering Mechanics of Composite Materials*, 2nd Edition, Oxford University Press, UK, 2006.
2. Hull, D., *An Introduction to Composite Materials*, Cambridge University Press, UK, 1981.
3. Jones, R.M., *Mechanics of Composite Materials*, 2nd Edition, Taylor & Francis, Inc., UK, 1999.
4. Campbell, F., *Manufacturing Processes for Advanced Composites*, Elsevier Advanced Technology, UK, 2004.

6

Structural and Operating System Components of an Aircraft

6.1 Introduction

An aircraft is an integration of many large, medium, and small metal and nonmetal structural and nonstructural components. Within each large, medium, or small component, there are many components and subcomponents. Figure 6.1 shows major sectional breakdown including its structural and operating system components.

There are three major structural components including fuselage, wing, and stabilizers that determine the external shape of an aircraft, as outlined in Figure 6.2. In addition to these, there are two other undercarriage structural components including keel beam, which is considered to be the backbone of the fuselage structure and the landing gear and which support the aircraft on the ground during taxiing, takeoff, landing, and braking. It absorbs shocks of landing and taxiing. These components serve no function during flight. Due to the weight of the forward and aft sections of the aircraft, a large bending moment acts on the center section of the fuselage. To withstand this bending moment, a strong keel beam structure is provided at the undercarriage of the fuselage sections.

The power plant or engine supplied by the engine manufacturers is built with some external components called nacelles, which mainly protect the engine and improve aerodynamic performance of the aircraft during takeoff, in-flight, and landing. As mentioned in Figure 6.2, another critical structure of the aircraft is the pylon/strut, which attaches the engine to the aircraft. Engines can be rear mounted, wing mounted, or a combination of both. Most modern aircrafts designed by both Airbus and Boeing have the engines mounted on the wings.

The operating system components are mainly the secondary structures attached to the wing and stabilizers of an aircraft. This chapter introduces an overview of major structural and operating system components of an aircraft [1], along with major elements of each major component and their corresponding input metal shapes, before introducing any manufacturing processes to fabricate any metal and composite material component of an aircraft.

6.2 Fuselage

The body (fuselage) is a pressurized semi-monocoque structure. Monocoque (pronounced mon-a-coke) means "single shell" in French. In the semi-monocoque design, the structure is reinforced with circumferential frames and longitudinal stringers. The challenge of fuselage design is to make a structure strong enough to withstand various levels of

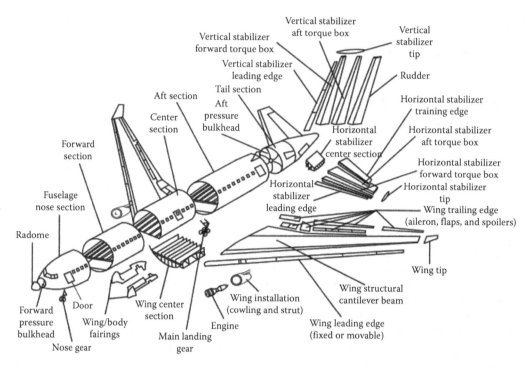

FIGURE 6.1
Schematic representation of structural and operating system components of an aircraft.

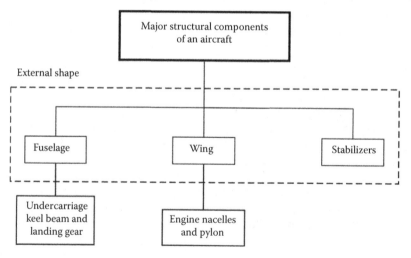

FIGURE 6.2
Major structural components of an aircraft.

stresses including tension, compression, bending, shear, and torsion. The major structural components of the fuselage section are shown in the schematic diagram in Figure 6.3. Figure 6.4 shows a close view of stringer and frame attached to the skin. Directly under and supporting the skin are the stringers. Stringers run longitudinally along the length of the fuselage to provide resistance in both torsion and bending stress. A frame is the

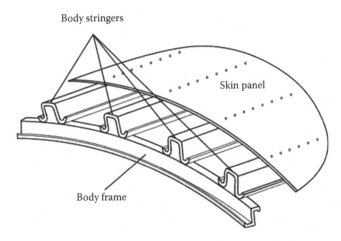

FIGURE 6.3
Major components of fuselage structure.

FIGURE 6.4
Example of stringer and frame attached to skin. (From The Boeing Company.)

circumferential structural member in the body structure that supports the stringers in the longitudinal direction and the body skin. In this structure, shear and bending loads in the skin are transmitted to stringers and frames. The basic raw materials used to make stringer, skin, and frames are shown in Figure 6.5. The skin panels for metal airplanes are generally made from flat aluminum aircraft alloy sheet including a combination of bare and clad materials. The stringers are made either from sheet metal or from extrusion shapes. Frames are generally made from sheet products, and sometimes include a combination of both sheet and extrusion products.

Figure 6.6 shows examples of upper lobe and lower lobe fuselage sections showing the three major components skin, stringer, and frames integrated with the fasteners.

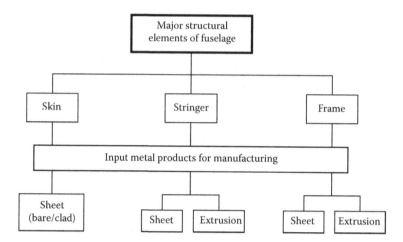

FIGURE 6.5
Major structural components of fuselage and input raw materials for manufacturing.

Figure 6.7 shows the schematic view of the external shape of the fuselage structure with several sections of the fuselage and its major structural elements connected at different sections of the fuselage. Pressure bulk heads attached at the forward and aft ends of the fuselage structure form a very large pressure vessel.

In actual aircraft manufacturing, the entire length of the fuselage structure is divided and identified by section numbers. In this example (Figure 6.7), the section numbers are designated by letters A–G for further explanation. Section A contains the following items:

- Radome
- Flight deck
- Forward pressure bulkhead
- Forward equipment center
- Nose gear wheel well
- Main equipment center
- Passenger entry door
- Forward cargo door
- Forward part of the forward cargo compartment

Section B contains the aft part of the forward cargo compartment. Section C (upper lobe) and section D (lower lobe) are combined as one complete section as a center section of the fuselage structure. The center section contains the following major items:

- Wing center section
- Keel beam
- Main landing gear wheel wells

Section E has two major items:

- Aft cargo door
- Aft cargo compartment

FIGURE 6.6
Example of fuselage section. (a) Upper lobe and (b) lower lobe. (From The Boeing Company.)

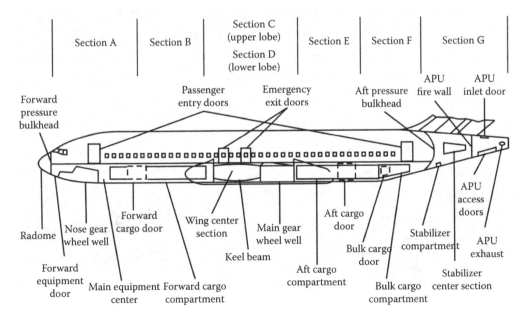

FIGURE 6.7
Schematic of fuselage section components of an aircraft.

Similarly, section F has two major items:

- Bulk cargo door
- Bulk cargo compartment

Section G has the following items:

- Aft pressure bulkhead
- Stabilizer compartment
- Stabilizer center section
- APU firewall
- APU inlet and exhaust
- APU compartment

Figure 6.8 shows an example of different fuselage sections being assembled in the production line. Figure 6.8 also shows the center section of the fuselage where the wing is attached.

Internal structural elements of the fuselage have the following major components:

- Passenger floor beam and supporting post
- Cargo floor beam and post
- Passenger cabin:
  - Ceiling structure
  - Stow bin structure
- Pressure bulkheads (forward and aft)

FIGURE 6.8
Fuselage section components in the production assembly line. (From The Boeing Company.)

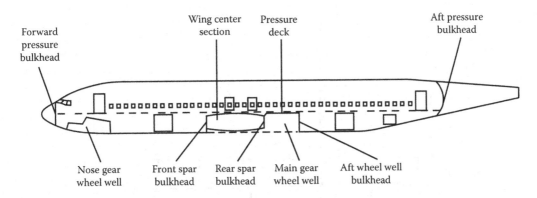

FIGURE 6.9
Pressure and other bulkhead locations within the fuselage.

Maintaining the pressurized environment is a critical factor to the fuselage design and the functions of skin, stringer, and frame of the semi-monocoque structure are all essential to meet this design need. A pressure bulkhead either in the form of flat discs or curved bowls is a heavy structural member in the fuselage used to seal the airplane in order to maintain cabin pressure at a certain altitude. Figure 6.9 shows not only the pressure bulkheads but also other bulkheads located at front spar, rear spar of wing box, and aft wheel well at the lower compartments of the fuselage to protect the center fuel tank and the main landing gear in case of a crash. Figure 6.9 also shows fuselage interfaces with center wing box, nose wheel well, and main gear wheel well.

Due to the weight of the forward and aft part of the aircraft, large bending moments occur on the center section. To carry these bending moments, a strong keel beam, the backbone of the aircraft is provided. The keel beam is generally placed along the center line below the center wing box and continuing through the main landing gear bay. At the forward end, it attaches to a bulkhead frame, goes under the wing center section fuel tank, through the main landing gear bays, and attaches to another bulkhead at the rear of the wheel wells. A bulkhead is a structural partition, usually located in the fuselage, which normally runs perpendicular to the keel beam (Figure 6.10). Metal keel beams are made with a combination of sheet metal, extrusion, rolled product, and forging components.

Figures 6.11 and 6.12 show few examples of internal structural components of an aircraft including aft pressure bulkhead, passenger floor beam, and floor beam with stanchion post attached to each frame location. Figure 6.13 shows an example of overhead structure to attach the stow bean.

6.2.1 Doors

The doors are an integral part of the fuselage mainly used for loading and unloading of passengers and cargo, also included are some other access and service doors located at various fuselage locations of the aircraft as shown in Figure 6.14. An example of a passenger and service entry doors is shown in Figure 6.15. Total number of passenger entry doors is dependent on the airplane model. Figure 6.16 shows a few examples of cargo doors.

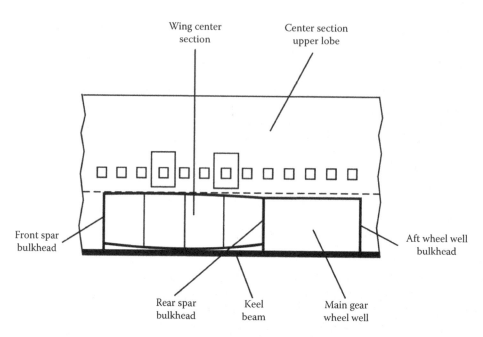

FIGURE 6.10
Schematic representation of keel beam.

FIGURE 6.11
Example of aft pressure bulkhead is attached to the fuselage tail section. (From The Boeing Company.)

Major structural and operational elements of aircraft door are:

- Skin
- Stringers
- Frames

- Fittings and hinges
- Actuators
- Torque tube

6.3 Wing

The airplane wing is an airfoil-shaped structure symmetrically attached on each side of the fuselage of an aircraft whose main function is to provide lift and aerodynamic

FIGURE 6.12
Structural elements of floors. (a) Passenger floor beam and (b) floor beam with stanchion post. (From The Boeing Company.)

(a)

(b)

FIGURE 6.13
Example of internal structural elements in cabin ceiling and stow bin area. (a) Cabin ceiling and (b) stow bin. (From The Boeing Company.)

flight control. The wings attached to the fuselage center section store fuel and also contain fuel system components. The wings also have additional structural attachments including engine, landing gear, and leading edge and trailing edge flight control surfaces. The major elements of the wing primary structure including the input raw materials to manufacture each element including skin panel, spar, stringer, and rib of the wing structure are shown in Figure 6.17.

Wing structure is a tapered cantilevered box type beam structure looking from the inboard end to the outboard end supporting the entire airplane structure like the wing of

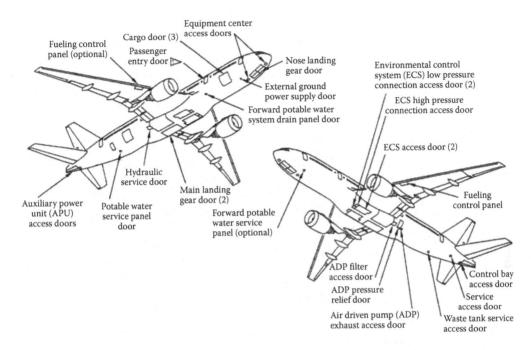

FIGURE 6.14
Various doors and their locations. (From The Boeing Company.)

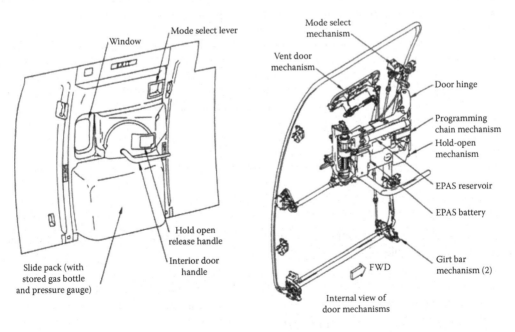

FIGURE 6.15
Passenger and service entry doors. (From The Boeing Company.)

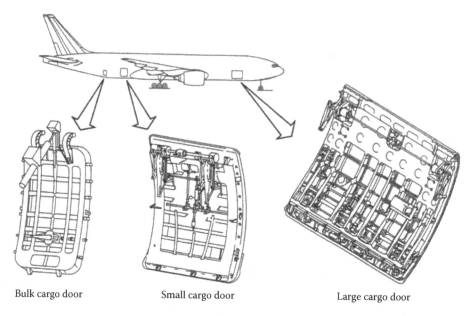

FIGURE 6.16
Cargo doors. (From The Boeing Company.)

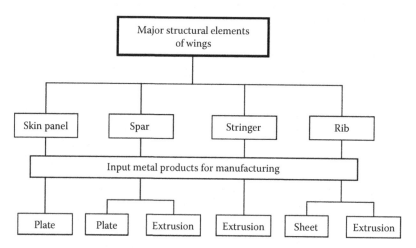

FIGURE 6.17
Major structural components of wing and input raw material for manufacturing.

a flying bird. A wing box has the following major components as shown in the schematic diagram as an example (Figure 6.18):

1. Front spar (A) that includes two front spar chords and front spar web
2. Rear spar (C) that includes two rear spar chords and rear spar web
3. Upper skin panels (B) with upper stringers
4. Lower skin panels (D) with lower stringers
5. Series of in-spar ribs

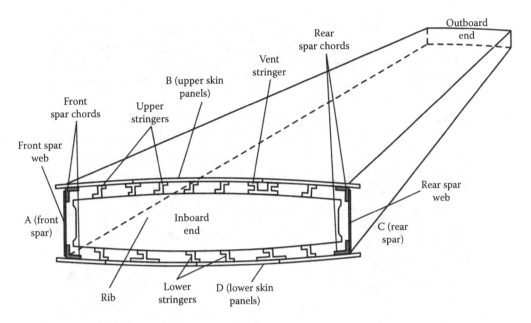

FIGURE 6.18
Schematic diagram showing an example of a metal wing box.

Figure 6.19 shows an example of a spar chord and web for installation to make a spar assembly for the wing. The spar looks like simple channel type beam. The spar of the wing is the most heavily loaded part of an aircraft structure, which carries much more stress at the root and less stress at the tip of the wing. Due to normal upward bending of the wing, a spar usually carries shear forces and bending moments. Aerodynamic forces not only bend the wing but also try to twist the wing. To prevent both bending and twist, a torsion box type structure is designed as shown in Figure 6.18.

The wing skin panels serve as part of the aerodynamic lifting surface and they are expected to meet certain aerodynamic smoothness criteria. In metal aircraft, wing skins are made of aluminum alloy plates machined in tapering thickness from inboard end to the outboard end. The upper and lower skin panels cover the entire tapered wing box beam from the inboard end to the outboard end. In service, the upper panels are generally subjected to compression and shear-type load, whereas the lower skin gets tension and shear-type load.

Stringers are the longitudinal members attached to the skin along the length of the wing as shown schematically in Figure 6.20. Stringers are generally extruded aluminum profiles machined to taper down from the inboard end to the outboard end. Stringers generally transfer skin loads to the wing ribs.

Series of ribs are placed and attached between front and rear spars along the length of the wing from inboard to the outboard direction. The spar ribs establish and maintain the aerodynamic shape of the wing box. The external air loads and internal fuel and pressure loads are applied to the panels, which are supported by the spar ribs and these loads are finally transferred to the front and rear spar, which are the ultimate load-carrying members of the wing. The ribs also carry and transfer crushing and shear loads from the skin panels due to wing bending, panel curvature. Ribs also serve as baffles against excessive fuel slosh. Figure 6.20 shows some elements of the rib structure including rib web, rib stiffener, and rib chord and how the rib structure is attached to the stringer and the skin.

(a)

(b)

FIGURE 6.19
Wing spar. (a) Wing spar components, one web and two spars and (b) assembled spar is ready for installation. (From The Boeing Company.)

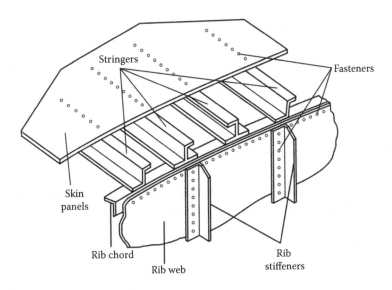

FIGURE 6.20
Schematic diagram of joining of stringer, skin, and rib of wing box.

6.3.1 Wing Configuration

Figure 6.21 shows the schematic representation of the wing of an aircraft including the wing control surfaces for the flight control. The wing control surfaces include the leading edge slats, trailing edge flaps, aileron, flaperon, and spoilers. Figure 6.22 shows the wing

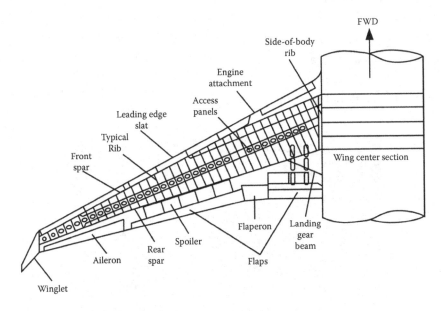

FIGURE 6.21
Major wing components.

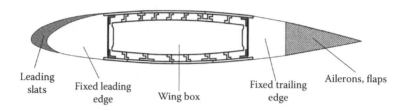

FIGURE 6.22
Schematic of wing section with leading and trailing edge.

section with leading and trailing edge. The leading edge slats attach to the front spar. The items attached to the rear spar and auxiliary structure are:

- Trailing edge flaps
- Aileron
- Flaperon
- Spoilers

Figure 6.23 shows an example of an assembled wing ready for installation to the fuselage center section.

The winglet is an aerodynamic fairing that is attached to the end of the wing. Winglet devices are usually intended to improve the efficiency of the fixed wing aircraft. Winglets generally provide potential performance benefits including:

- Improved takeoff performance
- Fuel savings
- Reduce engine maintenance costs
- Increased payload range
- Reduce noise and emissions
- Improved operational flexibility

FIGURE 6.23
Pair of wing assemblies, ready to join with fuselage. (From The Boeing Company.)

FIGURE 6.24
Examples of winglet devices for different airplane models. (a) 787, blended rake and (b) 737, blended extends up. (From The Boeing Company.)

Winglets add little extra length to the airplane's total wingspan. Wing tip structure has a similar concept to a very small wing-like structure, including spar and skin based on the design of the aircraft manufacturer. Figure 6.24 shows couple of examples of winglets.

6.4 Stabilizer

The aft portion of the aircraft, called also the empennage, usually consists of a group of stabilizers: horizontal and vertical (Figure 6.25). The horizontal stabilizer is similar to the wing in its basic structure, consisting of left, center, and right wing boxes. These contain:

- Torque box spars (front and rear spars)
- Ribs
- Stringers
- Skins

The vertical stabilizer has similar structural components like front spar and rear spar, ribs, stringers, and skins, which form a vertical structural beam. The leading edge and vertical tip are removable like horizontal stabilizer. The structure aft of the rear spar has a series of ribs that incorporate hinge bearings for the rudder and the tab. Exterior access panels and an interior crawlway are provided for inspection and maintenance purposes.

The stabilizer is installed or removed as a single unit as the wings are, with replaceable leading edge and stabilizer tips. Internal access is provided by an external access door and internal crawlways. The optional fuel tank consists of the stabilizer center section and outboard sections to the rib at stabilizer. An example of stabilizers being installed in an airplane is shown in Figure 6.26.

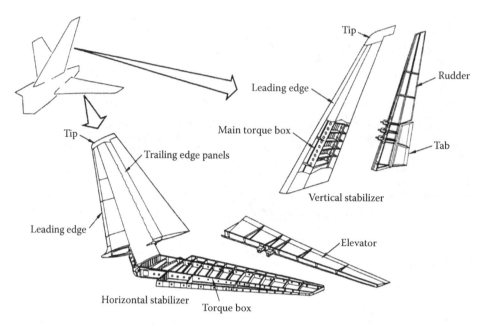

FIGURE 6.25
Schematic view of stabilizer structure. (From The Boeing Company.)

FIGURE 6.26
Installing stabilizers. (From The Boeing Company.)

In addition to the main structure of the stabilizers, three control surfaces are attached to the stabilizers to control the aircraft in three-dimensional air space. These are mainly:

1. Rudder
2. Tab
3. Elevator

6.5 Landing Gear

Most major passenger aircrafts have the tricycle landing gear system, which has two main landing gear under the wings and one nose landing gear. The number of wheels on each truck depends on the model of the aircraft, which includes size and shape of the aircraft.

The main landing gear is attached to the wing rear spar and the landing gear beam. It has the following structural elements as shown in an example (Figure 6.27a):

- Shock strut
- Trunnion (forward and aft)
- Side brace
- Drag brace
- Retract actuator
- Down lock actuator
- Lock link
- Truck
- Truck tilt actuator
- Springs (lock and strut)

The main landing gear strut includes an air–oil shock absorber. The drag brace and a side brace transmit loads from the strut to the airplane structure. The main landing gear shock strut, trunnion mounted between the wing spar and the landing gear support beam are hydraulically operated by the retract actuator. Two folding type side and drag braces absorb side and drag forces when the gear is down. Hydraulic actuators and bungee springs provide over-center locking of the drag and side braces through a jury strut and a lock link. The shock strut is serviced by an air- and oil-charging valve.

Like main landing gear, the nose landing gear (Figure 6.27b) strut includes an air–oil shock absorber. A folding drag brace transmits loads from the strut to the airplane structure. At full extension or retraction of the nose gear, the over-center mechanism of the lock link locks the drag brace. Figure 6.28 shows the major structural components of landing gears. Generally, forging is the input metal shaping process for manufacturing the structural components of the landing gear including the landing gear beam attached between the wing and the fuselage wheel well area.

The major components of landing gear doors are:

- Composite door structure
- Actuators
- Hook

The doors on each main landing gear wheel well close and open during gear retraction and extension. There are two door systems in the nose landing gear. The forward and aft doors are mechanically actuated during gear retraction. The forward doors are closed when the gear is either up or down and the aft doors are closed when the gear is up. The landing gear doors are made from lightweight composite materials.

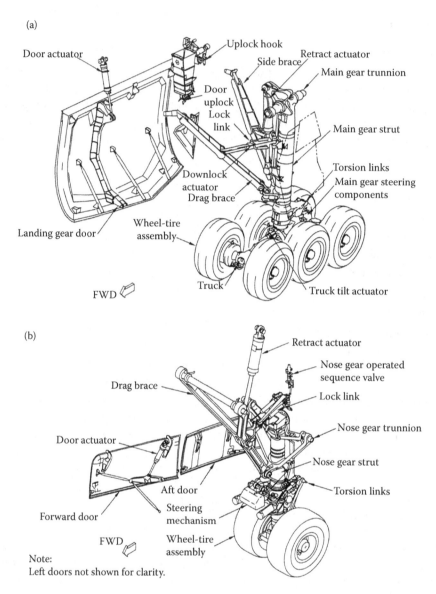

FIGURE 6.27
Schematic representations of landing gear major elements. (a) Main landing gear and (b) nose gear. (From The Boeing Company.)

The landing gear operates with center hydraulic pressure from the center system to retract and extend. During normal operation, the sequence valves control the gear and door movement. When the landing gear is fully retracted in flight, valves automatically remove hydraulic pressure from the landing gear. The landing gear control lever has two positions and electrically controls the landing gear selector valves for landing gear operations. A landing gear door opens and closes on each main gear wheel well during gear retraction and extension.

Similarly, in a normal operation, the nose landing gear uses center system hydraulic pressure to retract and extend. Sequence valves control forward door and landing gear

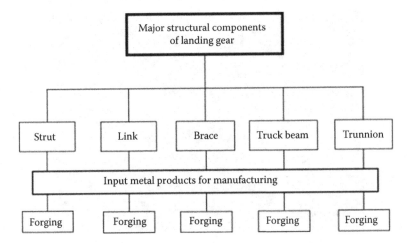

FIGURE 6.28
Major structural components of landing gear and input metal shapes for manufacturing.

movement. The forward doors of the nose gear wheel well operate hydraulically during gear retraction and extension. The aft doors operate by mechanical linkages that connect to the nose gear. The aft doors close only when the gear retracts.

6.6 Engine

The engine of an aircraft is generally manufactured by supplier specializing in manufacturing jet engines. The engines are built to satisfy the needs of the aircraft manufacturers, and finally satisfy the airlines customers' need. A few components associated to the engine and its mounting to the wing are:

- Power plant
- Engine nacelle
- Pylon
- Fairing
- Heat shield

The cowls shown in Figures 6.29 and 6.30 allow smooth airflow through and around the engine. Cowls also protect engine components from any external damage. Fixed and hinged cowls make the engine nacelle. The fixed cowls included the inlet cowl, nozzle, and plug as designed by the individual engine manufacturers, which are attached to engine flanges. Hinged cowls included the fan cowl and thrust reverser assembly. Those are hinged on the strut and latched on the bottom.

The pylon is an important structural element connecting the aircraft engine normally to the wing shown as an example in Figure 6.31. The main pylon is composed of a few substructures depending on the aircraft manufacturer's design. Pylon structure is made of

FIGURE 6.29
Power plants with fixed inlet cowl ready for installation. (From The Boeing Company.)

different metal components with different input raw materials mainly with titanium alloy plate forging and sheet products.

6.7 Major Operating System Components

Major operating system components are highlighted in Figure 6.32. This section is designed to help the beginner as well as aerospace practicing engineers to familiarize with various operating systems required to take the flight in the air and control it, and finally land the aircraft safely on the ground. Operating system of each individual component has its own design with few functional units to operate the component from the cockpit. Various kinds of materials are involved to manufacture each element and the final assembly of each unit of any operating system component.

6.7.1 Auxiliary Power Unit

Figure 6.33 shows schematic representation of an auxiliary power unit (APU). The APU is an external engine installed at the tail cone of an aircraft. APU supplies electrical and pneumatic power to the airplane, which allows independent ground operation. The APU is also available for use in flight.

6.7.2 Environmental Control System

Environmental system controls the following aspects of an aircraft to satisfy the comforts of a passenger on board:

- Air conditioning (temperature and moisture)
- Oxygen
- Pressurization and depressurization

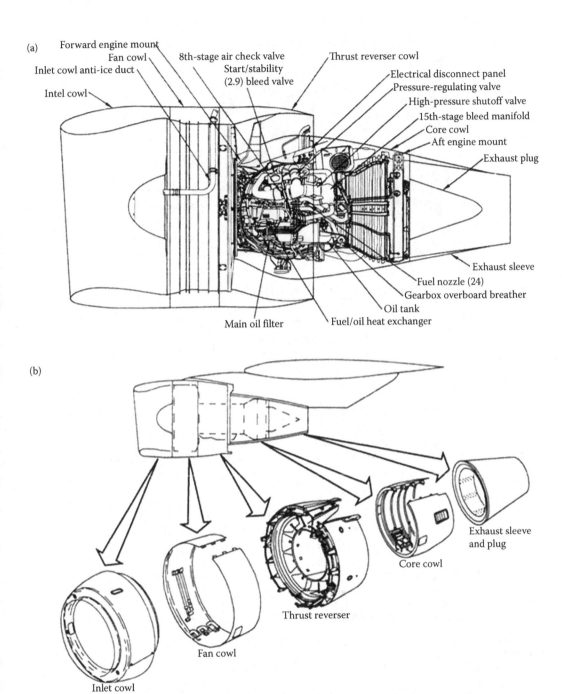

FIGURE 6.30
Schematic view of an engine. (a) Engine components and (b) engine cowling. (From The Boeing Company.)

FIGURE 6.31
Engines attached to the wing of an airplane. (From The Boeing Company.)

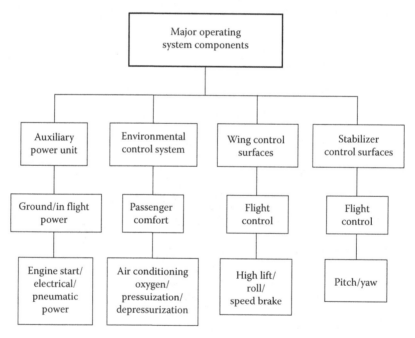

FIGURE 6.32
Major operating system components.

There are several applications of metal tubes and ducts that are associated with the manufacture and installation of the environmental control system in the aircraft as shown in Figure 6.34.

6.7.3 Wing and Stabilizer Control Surfaces

Figure 6.35 shows an example of flight control surfaces attached with the wing and the stabilizer of an aircraft.

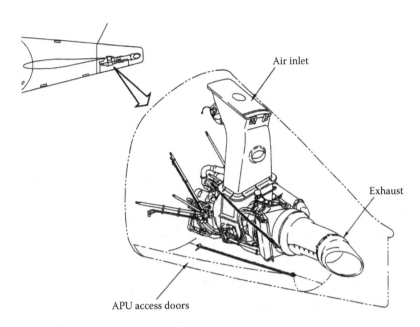

FIGURE 6.33
Auxiliary power unit. (From The Boeing Company.)

There are two types of flight control systems: (1) the primary flight control system and (2) the high lift control system. The primary flight control system is an electronic fly-by-wire system. It provides roll, pitch, and yaw control to the aircraft. For primary flight control, the control surfaces are as shown in Figure 6.36.

Both flight control systems are used to control the flight at takeoff, climb, cruise altitude, and also during landing operations by controlling the leading edge slats and the trailing edge flaps, ailerons, flaperons, and spoilers. The ailerons are the movable control surface attached to the trailing edge of the outboard end of the wing, whereas flaperons are attached to the trailing edge of the inboard end of the wing. Spoilers are attached to the trailing edge of the wing to spoil or disrupt the smooth flow of air over the wing to reduce the lift force of the wing. Left and right spoilers can be raised alternately for high-speed lateral control or can be raised together as speed brake during landing. Elevators are attached to the movable horizontal stabilizers, which control the pitch of the aircraft and finally rudder attached to the vertical stabilizer controls the yaw of the aircraft.

6.7.4 Wing High Lift Control Surfaces

Wing high lift control surfaces are shown in Figure 6.37. The high lift control system is also a fly-by-wire system to operate inboard and outboard trailing edge flaps, leading edge slats, and Krueger flaps for increased lift at lower speeds during takeoff and landing. Some of the airplane models have fixed leading edge with high lift trailing edge flaps. Slats and a combination of slats and flaps are the movable auxiliary airfoil attached to the leading edge of the wing of some of the aircraft models (Figure 6.38). The trailing edge of the wing is hinged with flight control flaps that can be lowered and extended within a definite curvature required during takeoff and landing of the aircraft (Figure 6.39). When lowered, the

(a)

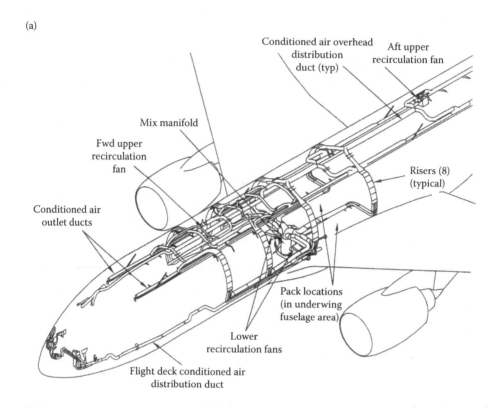

(b)

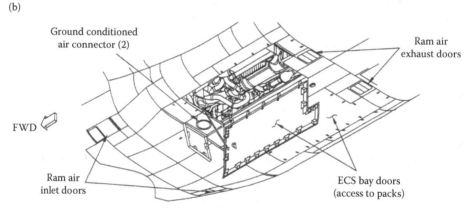

FIGURE 6.34
Environmental control system. (a) Fuselage main cabin area and (b) under wing fuselage area. (From The Boeing Company.)

flaps increase airplane lift by extending the surface area of the airfoil during takeoff. When the flaps are extended all the way down, it worked like air brake increasing air drag during landing. During taking off, the leading edge is kept open and it forms a slot and increases lift. A slot is an elongated passage through the wing to improve the airflow over the wing at high angles of attack.

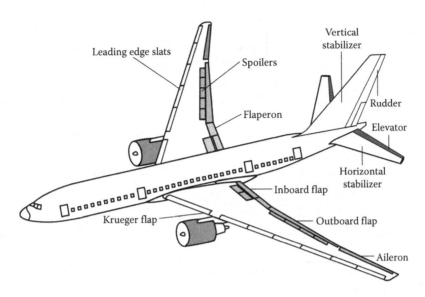

FIGURE 6.35
Example of flight control surfaces.

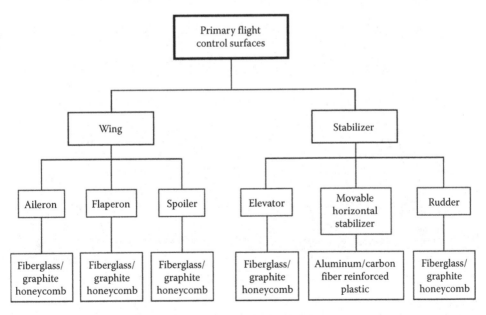

FIGURE 6.36
Primary flight control surface elements.

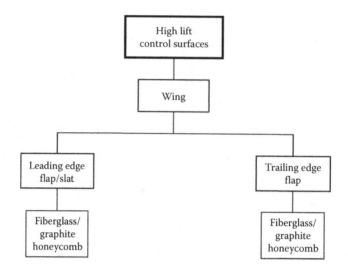

FIGURE 6.37
High lift control surfaces.

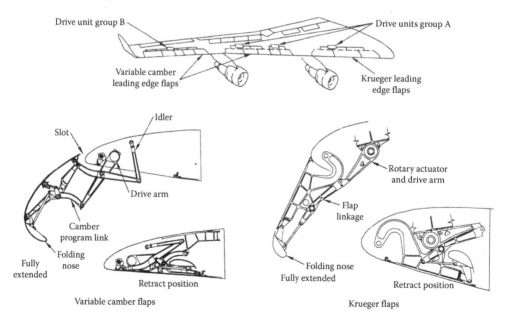

FIGURE 6.38
Example of leading edge slat control surface. (From The Boeing Company.)

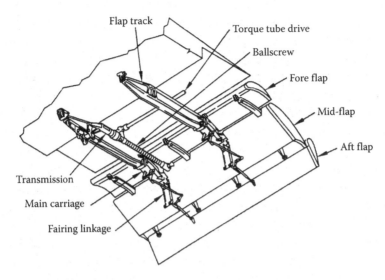

FIGURE 6.39
Example of trailing edge flap high lift control surfaces. (From The Boeing Company.)

Reference

1. www.boeing.com

7

Introduction to Manufacturing Processes of Metal Components of an Aircraft

7.1 Introduction

Chapter 6 discussed an overview of structural and operating system components of an aircraft. The type of input raw material and its associated product geometry required for manufacturing aircraft components varies according to the engineering design needs of the part for the particular model of aircraft. A variety of manufacturing technologies are applied to major input raw materials in the form of sheet metals, bars, plates, extrusions, and forgings. Manufacturing cost and quality of a component is directly related to the manufacturing technology involved. This is added to the research and development as needed to find the right technology for the right product component to result in optimum cost of manufacturing and highest quality as highlighted in Figure 7.1.

Major manufacturing processes as applied to aerospace component hardware will be covered in this book to provide a fundamental knowledge of aerospace manufacturing processes to the reader. This chapter introduces manufacturing processes used to produce various metal structure and operating system components of an aircraft. Manufacturing technology related to composite materials will be discussed in Chapter 18.

7.2 Major Manufacturing Processes in Aerospace

Figure 7.2 provides an overview of major manufacturing processes applied for both non-ferrous (mainly aluminum and titanium) and ferrous (alloy and stainless steel) metals used to produce various aircraft components.

7.2.1 Major Forming Processes of Various Metal Products

Metal-forming technology has wide application in making aircraft parts of various sizes and geometries from different input raw materials. Many different forming processes have been developed to produce parts from various shapes of metallic materials including sheet products, plates, extrusions, tubes, and more. Depending on the type of metal and the geometry of the part to be formed, it will determine proper selection of metal-forming processes to be utilized. The processes fall under either cold (ambient temperature) or hot (elevated temperature) forming. In general, forming aluminum alloy products falls under

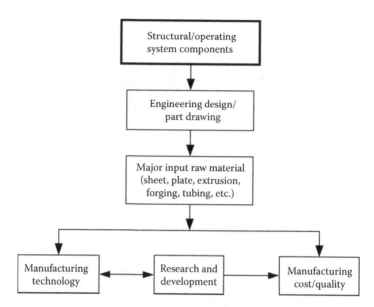

FIGURE 7.1
Aircraft components and cost relationship in manufacturing.

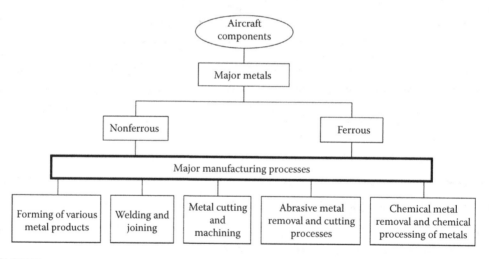

FIGURE 7.2
Major manufacturing processes of metal components of an aircraft.

cold forming, whereas titanium alloy products often falls under the hot forming category. Figure 7.3 shows the outline of forming of various metal products to manufacture three major structural components of the general aircraft configurations including fuselage, wing, and stabilizers of an aircraft.

Major metal forming processes required for forming flat sheet, plate, extrusion, and tube and duct for making various structural and operating system components of an aircraft are shown in Figure 7.4. All the forming processes will be discussed separately in Chapters 8 through 13.

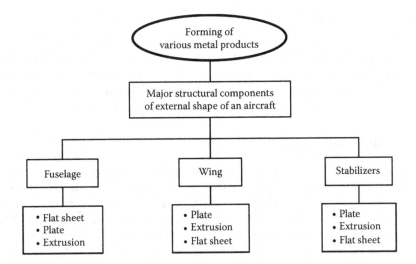

FIGURE 7.3
Forming of various metal products for major structural components of an aircraft.

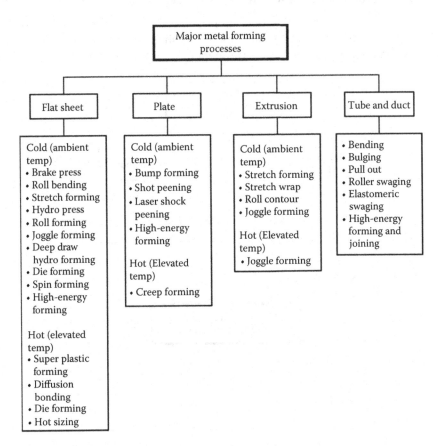

FIGURE 7.4
Major metal forming processes.

7.2.2 Welding and Joining

Welding and joining technologies are not commonly used to join metal parts or to make near-net-shape pre-machining stock for aerospace parts. Research is being continued to explore various welding technologies in the field of aerospace manufacturing and to qualify the process for part acceptance criteria. The major welding and joining technologies as shown in Figure 7.5 are being considered in the aerospace industries.

7.2.3 Metal Cutting and Machining

The demand of metal cutting and machining technology is growing heavily in the aerospace industry due to increase in application of special metals including titanium, inconel, and other high-strength metals. The application of machining is to remove metals from the starting metal stock in the form of extrusion, plate, bar, and forging to make variety of finish products for the aircraft using high-speed computerized numerical control (CNC) machining center. The three major metal cutting and machining processes shown in Figure 7.6 can cover machining of various small-, medium-, and very large-size aircraft parts of different materials including metals and composite materials.

7.2.4 Abrasive Metal Removal and Cutting Processes

Abrasive metal removal and cutting operations require a wide variety of abrasive products with different size, shape, and composition. The shape, size, and type of abrasives are selected based on the type of machine equipment needed for various process applications

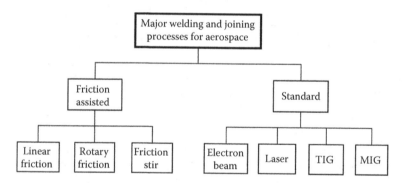

FIGURE 7.5
Major welding and joining processes.

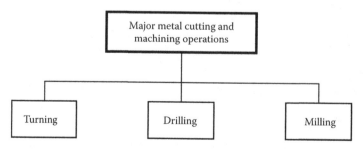

FIGURE 7.6
Major metal cutting and machining operations.

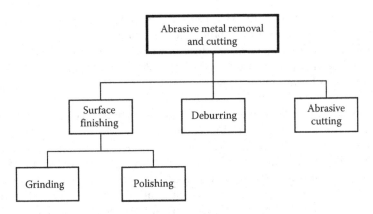

FIGURE 7.7
Abrasive metal removal and cutting processes.

as shown in Figure 7.7. Unlike other operations, abrasive tools and abrasive media selection are dependent on the following abrasive product specifications mainly:

- Type of abrasive
- Grit size
- Grade or hardness
- Structure
- Bonding

7.2.5 Chemical Metal Removal and Chemical Processing of Metals

Chemical metal removal process, which is commonly known as chemical milling process, is widely used in aerospace manufacturing. Although, chemical milling is an environmentally hazardous process it is widely used in the aerospace industry for generating weight reduction pockets in the inner side of the fuselage panels having uniform or compound radius of curvatures (ROCs) in the part geometry. Mechanical milling of same pocket geometry on the curved panel may require a very cost-effective special purpose machine.

A number of chemical processing technologies are used to clean and prepare both aluminum and hard metal parts for anodic/conversion coatings and plating to fulfill the engineering need of

- Corrosion protection
- Improve paint adhesion
- Engineering functions—hardness, wear resistance, abrasion resistance, and more
- Decoration—improved texture, color, and appearance

7.3 Introduction of Tribology in Manufacturing Processes

What does tribology mean and how does it influence manufacturing processes? It was not until 1966 that, at the instigation of the Peter Jost Committee in England, a new word, derived from the Greek word TRIBOS meaning rubbing, was adopted [1]. Its literal translation would be "the science of rubbing." Since then tribology has gained wide acceptance in the industry and it is defined in a recent edition dictionary as "the science and technology of interacting surfaces in relative motion and of related subjects and practices." Figure 7.8 shows the major subject matters of tribology which include (1) surface topography, (2) friction, (3) lubrication, and (4) wear. The subject matters are shown interrelated to each other considering the cause and effect of each subject to the other subject. As an example, it can be explained that higher surface roughness can cause high friction between the tooling and workpiece interface or due to high friction or lack of lubrication at the interface could cause bad surface finish, and finally lead to wear on the tooling surface or material transfer to the tooling or workpiece. This subject matter has an important role in metal manufacturing processes, since the tribological considerations are being used in determining process feasibility, economy of operation, and the quality of the manufactured product.

Major manufacturing processes included those such as metal forming, joining, cutting, and metal removal processes, which mostly entail contact between two objects: the tool or die and the workpiece. Before beginning a discussion on major manufacturing processes, the author would like to provide an overview of the role of tribology in manufacturing processes as shown in Figure 7.9.

In metal forming operations, friction control is a very important factor for a successful process. Frictional variability raises problems in many forming processes. For an example, the strain distributions in sheet metal forming are strongly influenced by friction. This can result in part failure due to tearing at a location where strain builds up or due to wrinkling. Tribology and thermodynamics and their relationship play a tremendous role in friction-assisted welding technologies. Frictional heat generated at the workpiece–tool interface brings both the joining objects into the plastic deformation phase within a certain depth at the interface. In metal cutting and machining, heat generation occurs due to friction at the chip–tool interface and the tool–workpiece interface. Heat is also generated due to plastic deformation in the shear zone. The life of a cutting tool is very much dependent on frictional heat control, which is the function of the cooling and lubricating system. Abrasive

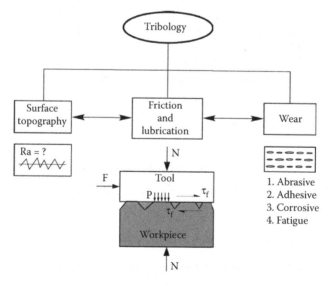

FIGURE 7.8
Subject matters of tribology.

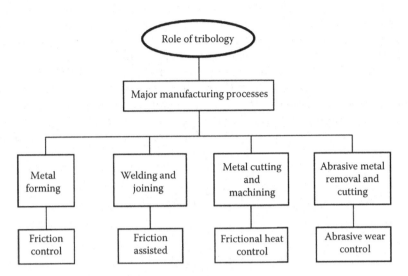

FIGURE 7.9
Role of tribology in manufacturing.

metal removal and cutting processes follow an abrasive control mechanism. An abrasive process is fundamentally a wear mechanism when two surfaces with a relative motion interact with each other. Wear is considered to be a harmful phenomenon for forming and cutting tools, but the wear mechanism in the abrasive process is rather beneficial to remove some thickness of material from the workpiece surface by mechanical action.

7.3.1 Surface Topography

All engineering surfaces are irregular on a microscopic scale. The irregularities can be considered to consist of series of peaks or ridges (usually called asperities) and valleys of different elevations, shapes, and spacing as shown in Figure 7.10. These features taken together constitute the texture of a surface. The irregularities on the surface are classified into roughness or waviness depending on their size. Tribological phenomena such as friction and wear depend mainly on the real area of contact between the surfaces depending upon the distributions, sizes, and shapes of the asperities. Measurements of these features are very important to study all kinds of surface contact phenomena for both qualitative and quantitative correlation between theoretical estimation and the practical evidence. Many methods are available for the measurement of the micro- or macro-level geometrical

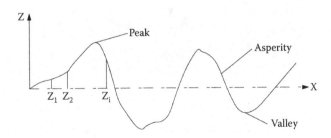

FIGURE 7.10
Surface topography.

features of surfaces, including optical methods using electron, interference, or reflection microscopy, and mechanical methods such as oblique sectioning and profilometry. Selection of measurement systems is very much dependent on the purpose and accuracy of the measurement. In the metal forming industry, the very common method of checking the surface is the stylus type profilometry method for obtaining the surface profiles. In manufacturing processes, either of the two parameters is usually used to define the texture of the surface. These parameters are the C.L.A. (center-line average), Ra, and R.M.S. (root mean square); Rq. C.L.A. and R.M.S. values are calculated by using the following expressions [1]:

$$C.L.A.\ (Ra) = \frac{1}{n} \sum_{i=1}^{n} |Z_i| \tag{7.1}$$

$$R.M.S.\ (Rq) = \left[\frac{1}{n} \sum_{i=1}^{n} (Z_i)^2 \right]^{1/2} \tag{7.2}$$

where n is the number of points on the center line at which the profile deviation Z_i is measured (Figure 7.10). The center line is taken as a line that divides the profile in such a way that the sums of the enclosed areas above and below it are equal.

7.3.2 Role of Friction

Friction in metal forming causes tangential forces that act at the interface between tool and workpiece and resists relative movement of the two surfaces at the asperity contacts (Figure 7.11). For most contact between tool and workpiece, only a small fraction of the surface area actually is in contact. In metal forming operations, use of a constant friction coefficient may be questionable, since the friction is dependent more on the real area of contact instead of apparent area of contact. Friction plays an important role in metal forming operations. The deformation process, design of tools, surface quality, selection of lubricants, and also the prediction of failures are largely dependent on understanding friction. Friction is not an independent parameter but is a function of many variables such as surface roughness, lubricant, surface chemistry, relative sliding velocity, and forming pressure. The frictional force at the tool–workpiece interface can be readily changed by lubrication and by changing the surface roughness of both workpiece and tooling.

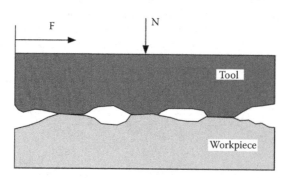

FIGURE 7.11
Surface interactions at the asperity contacts.

7.3.2.1 Adhesive Friction

Bowden and Tabor [2] used as a starting point the fact that when the tool and workpiece are loaded against each other, they make contact only at the tips of the asperities. Because the real area of contact is small, the pressure over the contacting asperities is assumed high enough to cause the softer workpiece to deform plastically. Plastic deformation at the contacts causes an increase in real area of contact that is just sufficient to support the applied load from the tool as shown in Figure 7.12.

Normal force is given by

$$N = A_c \cdot p_a \tag{7.3}$$

Friction force is given by

$$F = A_c \cdot \tau_a \tag{7.4}$$

Friction coefficient is given by

$$\mu = \frac{F}{N} = \frac{A_c \tau_a}{A_c p_a} \tag{7.5}$$

7.3.2.2 Plowing Friction

Bowden and Tabor also included plowing friction where the total friction force is caused by asperities of a harder tool surface plowing into soft workpiece by plastic deformation as shown in Figure 7.13. This is a major friction component during abrasion processes. Plowing is also important when the adhesion term is small as in the case of well-lubricated surfaces where the shear strength of the interfacial film is low.

In the case of sheet metal forming, the original roughness of the sheet and its roughening prior to die contact are important factors in determining lubricant entrapment. Tribology is of particular importance in sheet metal forming where the high ratio of surface to cross-sectional areas ensures that friction plays an important role in deciding forming forces and resulting strain distributions. Significant attention is required to develop tooling and lubrication systems for a new part manufacturing, which can avoid tearing and wrinkling

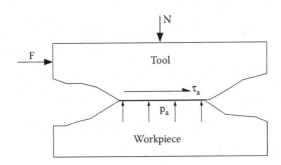

FIGURE 7.12
Adhesive friction.

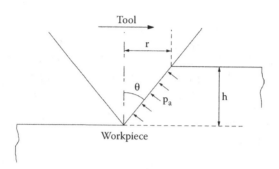

FIGURE 7.13
Plowing friction.

problems. Tooling wear and material transfer or pick up are often of great importance to productivity in high-volume sheet metal forming. Cleanliness and surface appearance are important in many sheet metal forming operations and are directly impacted by tribological considerations.

In most metal forming processes, low friction is beneficial since it reduces interface pressure, force and energy requirements, and heat generation. In some processes, like sheet metal forming and rolling operations, friction can be desirable and even essential to ensure proper gripping, contact, and plastic deformation during the operation. Therefore, a careful design of the lubrication system to achieve a suitable friction level is essential for a successful metal forming operation [3]. In stretching operations, the shape is developed by thinning the sheet while sliding over a die. Friction, together with the tensile properties of the sheet metal, governs the strain distribution, the attainable depth of stretching, and the location of fracture. Therefore, these features can be used for ranking lubricants.

Because of the low contact pressure in sheet metal forming, the mechanical and chemical properties of asperities control frictional shear forces at the interface. The surface topography of both sheet metal and tooling is very important in friction modeling. In the case of smooth tooling, the workpiece asperities are progressively flattened as tensile-induced deformation proceeds and the real area of contact increases [4]. In that case, the friction stress is proportional to the adhesion stress. On the other hand, the rough tooling asperities plough across the smooth workpiece, so that the friction contains both adhesion and plowing components. Hence, friction increases with local pressure, with increasing strain rate and with decreasing sliding velocity.

In general, friction could change according to the type of deformation occurring during two categories of forming processes:

1. Low pressure (sheet metal forming), where the growth of contact area controls the friction [5]:
   a. Friction increases with pressure
   b. Friction increases with strain
   c. Friction independent of sliding speed
2. High pressure (bulk processes), where the complete contact takes place:
   a. Friction independent of pressure
   b. Friction decreases with increasing strain rate (saturates)
   c. Friction increases with sliding speed (saturates)

7.3.3 Friction Models

There are three classical laws of friction (Amontons–Coulomb model) [6]. The three laws are as follows: (1) the friction is independent of the apparent area of contact between the contacting bodies, (2) the friction force is proportional to the normal load between the bodies, and (3) the kinetic friction force is nearly independent of the speed of sliding. The second law enables us to use a constant, known as the coefficient of friction. While the constant friction model is very useful in the analysis of many mechanical systems, it is of limited value in modeling friction at the tooling–workpiece interface in metal forming operations. The possibility of large fractional contact areas and the general plastic deformation of the workpiece surface in metal forming make the model of questionable value.

The classical theory of adhesive friction developed by Bowden and Tabor in 1953 [2] assumes that when metal surfaces are loaded against each other, they make contact only at the tips of the asperities and the friction is generated by adhesion at the tips. Their theory provides an explanation of two of the laws of friction, that is, that the friction is independent of the apparent area of contact and the friction force is proportional to the load.

Orowan (1941) [7] was among the first to point out the inconsistency in using a constant friction coefficient at high normal pressures. When analyzing the plate rolling process, he proposed a friction model in which the friction stress increases in proportion to pressure, with consequently constant friction coefficient, until a critical value of pressure is reached (Figure 7.14). Above the critical pressure, which is associated with the real area of contact becoming equal to the apparent area, the friction stress is constant. In other words, his model predicts constant friction coefficient at low normal pressure and constant friction stress at high normal pressure as shown in Figure 7.14.

Shaw et al. [8] have given a more precise explanation for this, saying that the ratio between real area of contact and apparent area increases with increasing pressure and approaches unity asymptotically, since very high pressures will be needed in the last phase of flattening. The connection between friction stress and normal pressure will be as shown in Figure 7.15 according to adhesion theory.

Wanheim and Bay [9] allow for a more gentle transition between the constant coefficient of friction at low pressures and the constant friction stress or friction factor at high pressure.

In some metal forming operations, such as stretch forming of sheet metal parts, the interface pressures are relatively low compared to those in bulk forming processes. Hence, the really unique feature of the workpiece–tooling interface in metal forming is that the workpiece is undergoing bulk plastic deformation and local workpiece deformations associated with asperity interactions are occurring on a plastic substrate.

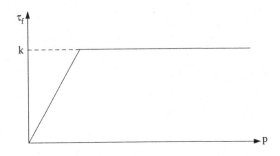

FIGURE 7.14
Orowan's friction model.

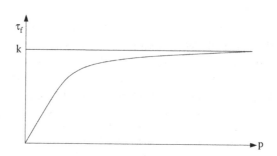

FIGURE 7.15
Shaw's model.

Wilson and Sheu [10] used upper-bound methods to investigate workpiece asperity flattening or surface indentation in the presence of bulk plane-strain plastic deformation. They found that the effective hardness is related to the ratio of bulk strain rate in the underlying material to the typical strain rate associated with the asperity flattening or surface indentation. As the ratio is increased, the effective hardness is reduced. A quantitative comparison of the predictions of the theory with experimental measurements of the flattening of model asperities in rolling showed excellent agreement with rolling experiments. Later, Saha and Wilson [11,12] verified Wilson and Sheu's [10] models of asperity flattening and its effect on variation of friction in sheet metal forming using an innovative sheet metal forming simulator as shown in Figure 7.16 [13]. The model is supported by the increase in friction with contact pressure and the angle of wrap. However, the predicted increase with strain in contact due to increased strain rate or decreased sliding speed was not observed. This may be due to roughening or some other competing mechanism.

In metal forming operations, depending on the tooling and workpiece material hardness and surface conditions (smooth or rough), the interaction between tool and workpiece at their interface could be categorized as

1. Smooth tool with rough workpiece
2. Rough tool with smooth workpiece

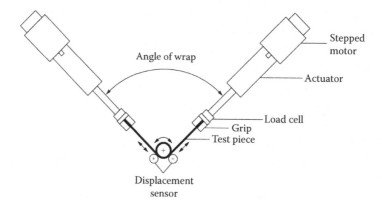

FIGURE 7.16
Schematic of sheet metal forming simulator. (Adapted from Saha, P.K. and Wilson, W.R.D., *Proceedings of the NAMRC XIX*, May, 1991.)

Considering these two tooling and workpiece conditions, mostly adhesive and plowing friction components are being considered to deal with metal forming processes.

7.3.4 Lubrication

Lubrication is vitally important in metal forming, metal cutting, and machining processes, and also abrasive metal removal processes. In most cases, reduction of friction has a beneficial effect by lowering the forces required for a given operation. This reduces the stresses imposed on tooling. Effective lubrication can improve product quality by eliminating surface defects, such as scoring or cracking, through the reduction of metal-to-metal contact and avoiding harmful residual stresses and internal defects through promoting more homogeneous deformation condition. Tooling life can be extended by the presence of a lubricant film, which will prevent wear and may act as a thermal insulating shield between the tooling and a hot workpiece.

A lubricant (solid, liquid, or grease type) is some material that is applied in between the mating surfaces to reduce friction and wear mechanisms. Lowering of friction and wear may be achieved by either physical or chemical means of the lubricant. In metal forming operations, it is required to consider the following factors for an effective lubrication system:

- Process geometry and conditions
- Tooling and workpiece topography
- Application method and removal

An effective lubrication system will tend to reduce friction or to provide a controlled level as appropriate. Effective lubrication can also improve product quality by reducing surface defects such as scoring, which are associated with metal-to-metal contact. Tooling life may also be extended by the presence of a lubricant film, which will reduce wear and act as an insulting shield between the tooling and workpiece. On the other hand, if the lubrication system is too effective it may engender rough product surfaces, cleaning problems, or corrosive wear of the tooling.

A realistic friction model must take account of the fundamental mechanisms of friction and lubrication at the tooling–workpiece interface. The situation is complicated by the fact that, under lubricated conditions, any of several different regimes of lubrication can occur at the tooling–workpiece interface. These regimes can be characterized by the thickness of the lubricant film related to the surface roughness and by the fraction of the interface load that is carried by the contact roughness peaks or asperities. Four lubrication regimes [3] have been described as shown in Figure 7.17:

1. Thick film ($h/R_c > 10$)
2. Thin film ($3 < h/R_c < 10$)
3. Mixed film ($h/R_c \leq 3$)
4. Boundary film ($h/R_c \leq 3$)

where h is the mean lubricant film thickness and R_c is the root mean square composite roughness of the surfaces. Composite roughness is the square root of the sum of the square of the individual surface roughness.

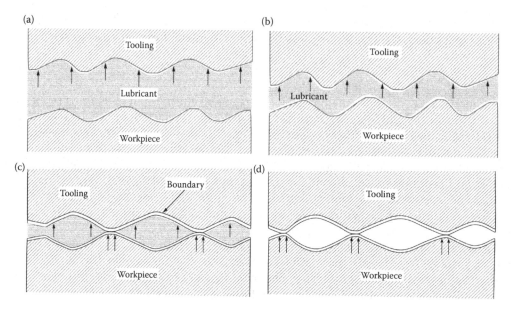

FIGURE 7.17
Lubrication regimes, (a) thick film, (b) thin film, (c) mixed film, and (d) boundary film.

In the thick film regime, the surfaces are separated by a continuous film of lubricant, which is much thicker than the roughness of either of the surfaces or the size of the lubricant molecules. Thus, surface roughness has little influence on lubrication and the lubricant can be modeled as a continuum between smooth surfaces. Lubrication is essentially a mechanical process and friction is decided by the shear resistance of the lubricant film under the conditions at the interface.

In the thin film regime, roughness influences lubricant flow but the film thickness is still generally much larger than the lubricant molecular size and "contacts" between individual asperities carry a negligible fraction of the interface load. Thus, the lubricant can be modeled as a continuum between rough surfaces. As in the thick film regime, lubrication is essentially mechanical but surface roughness modifies the formation and transport of lubricant films and the resulting friction.

In the mixed film lubrication regime, the interface loading is shared between the pressurized lubricant film in roughness valleys and "contact" at asperity peaks. In reality, if the lubricant is properly formulated with additives, actual metal-to-metal contact will not occur at asperity peaks because of the presence of tightly adhering boundary films formed as a result of chemical reaction or physical absorption of active lubricant additives at the interface. Even though these films are very thin (of the order of the lubricant molecular size), they prevent metal-to-metal contact, adhesion, and pick-up during asperity interactions.

In the boundary film regime, the interface load is carried by the thin film at the contacting asperity peaks. The mechanics of asperity deformation and the surface physics and chemistry of the workpiece and the tooling, and active components of the lubricant control the lubrication process. The Amontons–Coulombs [6] friction model (constant coefficient of friction) is applicable to the boundary lubrication of lightly loaded elastic contacts. The conditions of high loading and plastic deformation of the workpiece combine to make this model questionable in metal forming operations.

7.3.5 Wear

Wear occurs when two surfaces with a relative motion interact with each other. Wear in metal forming processes could be caused due to many reasons including:

- High contact pressure
- Lack of lubrication
- Surface condition and hardness of tool material
- Heat generation at the interface

Wear is a vast subject matter in tribology. Halling [1] has given a complete study of wear processes in detail. In order to study and gain a better understanding of wear, it is essential to recognize that several distinct and independent mechanisms are involved. Burwell [14] listed four wear mechanisms:

1. Adhesive
2. Abrasive
3. Corrosive
4. Surface fatigue

Schey [15] provided a useful review of the wear process in metal forming operations. Abrasive wear covers generally two types of situations. In both cases, wear occurs by the plowing-out of softer material of a given volume by the harder indenters of an abrasive surface. In the first instance, a rough, harder abrasive surface slides against a softer metal surface. In the second case, abrasion is caused by loose hard particles sliding between the rubbing abrasive and metal surfaces. Kalpakjian [16] also provided an excellent review of abrasive processes and finishing operations. A case study of a unidirectional abrasive metal removal process applied on aluminum flat sheet for aerospace parts in manufacturing processes was presented by Saha [17].

7.3.6 Surface Quality

The surface quality of formed metal products is defined by many factors. Two of the most important factors are:

1. Surface topography
2. Cleanliness

Both are closely related to the tribological (friction, lubrication, and tool wear) conditions used during their production. It is also important to note that conditions that lead to low friction and tool wear may not yield high product surface quality.

The surface topography of the product is often critical to both its appearance and functionality. Dull finish or rough surface appearance is often undesirable not only for the poor appearance, but also for the adverse effects of deep surface valleys and high peaks on fatigue, and corrosion type failure issues. However, very smooth surfaces are difficult to produce, and also the newly generated surface may not perform well in subsequent forming operations. The lubrication conditions used during forming operations have a profound effect on surface topography and appearance of the formed parts. The dry process produced a bright appearance, while the lubricated process produced a dull surface appearance.

The cleanliness of the surface is also strongly related to tribology. Chemically active lubricants may be very difficult to remove after forming. This can lead to problems with subsequent welding process, and also for painting. Wear debris from the tooling surface or product surface may also cause cleanliness problem.

References

1. Halling, J., *Principles of Tribology*, Macmillan Education Ltd, London, UK, 1989, pp. 1–39.
2. Bowden, F.P. and Tabor, D., *The Friction and Lubrication of Solids*, Part II, Oxford University Press, UK, 1964, pp. 91–95.
3. Wilson, W.R.D., Friction and lubrication in bulk metal forming processes, *Journal of Applied Metalworking*, VI, 1–19, 1979.
4. Wilson, W.R.D., Friction models for metal forming in the boundary lubrication regime, *Friction and Material Characterization*, ASME, 1988, pp. 13–23.
5. Saha, P.K., *Boundary Friction Measurements in Sheet Metal Forming*, PhD Dissertation, Northwestern University, Evanston, Illinois, 1993.
6. Amontons, G., *Hist. Acad. R. Soc.*, Paris, 1699.
7. Orowan, E., The calculation of roll pressures in hot and cold flat rolling, *Proceedings of the Institution of Mechanical Engineers*, 150, 140–167, June, 1943.
8. Shaw, M.C., Ber, A., and Mamen, P.A., Friction characteristics of sliding surfaces undergoing subsurface plastic flow, *Journal of Basic Engineering*, 82(2), 342–345, June, 1960.
9. Wanheim, T. and Bay, N., A model for friction in metal forming processes, *Annals of CIRP*, 27(1), 189–193.
10. Wilson, W.R.D. and Sheu, S., Real area of contact and boundary friction in metal forming, *International Journal of Mechanical Sciences*, 30(7), 475–489, 1988.
11. Saha, P.K. and Wilson, W.R.D., Friction in forming electro-galvanized steel sheet with tool steel and carbide coated tools, *Symposium on Contact Problems and Surface Interactions in Manufacturing and Tribological Systems ASME*, New Orleans, Louisiana, December, 1993.
12. Saha, P.K. and Wilson, W.R.D., Influence of plastic strain on friction in sheet metal forming, *Wear*, 172, 167–173, 1994.
13. Saha, P.K. and Wilson, W.R.D., Boundary friction measurements using a new sheet metal forming simulator, *Proceedings of the NAMRC XIX*, Rolla, Missouri, 37–42, May, 1991.
14. Burwell, J.T., Survey of possible wear mechanisms, *Wear*, 1, 119–141, 1957.
15. Schey, J.A., *Tribology in Metalworking*, ASM, Metals Park, Ohio, 1983, pp. 94–101.
16. Kalpakjian, S., *Manufacturing Engineering and Technology*, Addison Wesley, MA, 1992.
17. Saha, P.K., A case study on the abrasive surface finishing of aluminum flat sheet, *Wear*, 258, 13–17, 2005.

8

Cold Forming of Flat Sheet

8.1 Introduction

There is a wide utilization of cold forming of sheet metal for producing traditional fuselage structures of an aircraft. These structures are made mainly from aluminum and titanium alloys due to weight concerns. Titanium alloys generally have poor formability at room temperature so must be hot formed. Aluminum alloy sheet products have more reasonable formability at room temperature so can be cold formed (formed without external heat). Significant numbers of aircraft components are produced mainly from the 2xxx and 7xxx series wrought aluminum alloy sheet products. As discussed in Chapter 6, the fuselage structure has three major components including an exterior skin along with an interior skeleton of frames and stringers. Various cold forming processes have been developed to produce the fuselage components. The wing and empennage use sheet product only for the leading edge and trailing edge details with the structural box consisting of machined plate and extrusion products.

Sheet metal products are available both in flat pieces and also in the form of coil stock produced by the primary sheet metal producers. Aerospace designers select the sheet products according to the type of part designed for the aircraft including part geometry. To satisfy the design requirements for a prospective new metal alloy, the following factors are primary considerations:

- Mechanical properties; strength, facture toughness, fatigue properties
- Environmental compatibility in service
- Metal alloy and temper options
- Product size and thickness availability
- Price

To compete with high demanding lightweight carbon fiber composite structural materials, primary metal producers are always in the mode of developing new alloys with improved density and allowable mechanical and physical properties of metal including strength, damage tolerance, fatigue, and corrosion. To introduce a new metal to the aerospace industry, it is very important to understand the formability of the sheet metal products before making a go-ahead decision of producing sheet metal airplane parts. Basic formability tests are generally performed during the new alloy and sheet product development. Based on the success with basic formability tests, a more detailed producibility study of airframe parts is performed using the major metal forming processes [1].

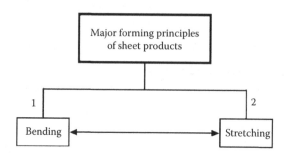

FIGURE 8.1
Sheet metal forming principles.

The fundamentals of sheet metal forming deal with two major forming principles: (1) bending and (2) stretching, or a combination of both as outlined in Figure 8.1.

Simple bend tests to establish the minimum bend radius (MBR) may be considered as a good and quick first indicator of formability of any new alloy. For a given bending operation, the bend radius cannot be made smaller than a certain value; otherwise, the metal will crack on the outer stretching or tensile surface of the sheet metal. Establishment of the MBR for a range of gage thickness for simple bends provides the minimum necessary design guidance for designing a structural part using a new alloy. The MBR varies with thickness, material properties, and the geometry of the bending conditions. The ratio of MBR to gage thickness (MBR/t) is considered to be an important formability index for assessing new alloys.

While many parts are designed with straight bends as assessed in the basic MBR test, the shape of an aerodynamic structure drives to more complex part configurations. In manufacturing production, various forming operations will be used including brake forming, hydro-press forming, stretch forming, deep draw hydroforming, and many more. The major sheet metal part geometry made by various forming technologies have varying combinations of bends of different bend radii and bend angles. The material formability limits can be identified by surface-related quality issues or failures in the process including thinning out, necking and cracks, springback and buckling, and wrinkling of parts. This chapter describes a variety of sheet metal formability tests and introduces an overview of major conventional cold forming processes of flat sheet products as outlined in Figure 8.2 for producing a wide variety of sheet metal parts from small-to-large aircraft components.

8.2 Formability Tests for Sheet Metal

In sheet metal forming operations, the actual forming process required to produce a part may be simple or complex depending on the part geometry and dimensional tolerances. Sheet metal formability is generally defined as the ability of metal to deform into desired part geometry without necking or fracture. Formability of metal plays a very important role to produce a defect-free sheet metal part. Due to the variety of shapes and stress distributions in the forming sheet metal products, formability of metals cannot be predicted only on the basis of simple mechanical properties, such as elongation or tensile strength;

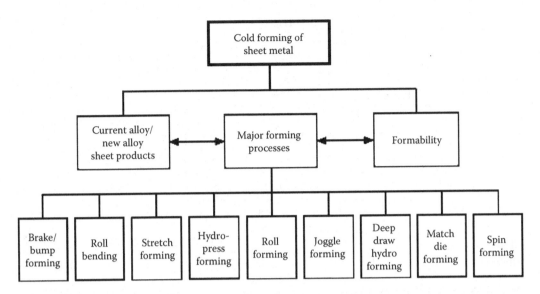

FIGURE 8.2
Major sheet metal forming processes.

or there are no simple evaluation tests that are applicable to all forming operations. As a rule of thumb, the higher the difference between the tensile strength and the yield strength, the strain-hardening capacity of the material becomes proportionally higher. Also, it is generally considered that the higher the strain-hardening capacity, the higher the formability of the material will be. The capacity for strain hardening is described more precisely by the strain-hardening coefficient "n" determined from a true stress–strain plot. The greater the amount of stretching required in a forming operation, the more important is the strain-hardening capacity of the material. The next few sections will highlight some formability test procedures generally used to evaluate flat sheet products of various alloys and tempers before considering as an input raw material in producing large volumes of sheet metal parts for an aircraft.

8.2.1 MBR Test

It was mentioned [2] that the common qualitative test for the evaluation of the formability of sheet metal is the bend test. Test coupons of sheet metal of various gage thickness are bent around smaller radii until cracking appears on the outer surface of the bend radii. The MBR is considered as the smallest radius that can reliably be used without cracking for that particular gage thickness. The value of the MBR depends somewhat on the material property, thickness of the bend specimen, and the method of bending. Bending is the process by which a straight length of flat sheet metal is transformed into a curved length as shown in Figure 8.3. The definitions of terms used in bending are as follows:

1. R_b—inside bend radius, the radius of the arc on the inside surface of the bend area
2. B—bend allowance, the length of the arc through the bend area at the neutral axis
3. W_b—bend width, width at the bend line
4. θ_b—bend angle, the included angle of the arc formed by the bending operation

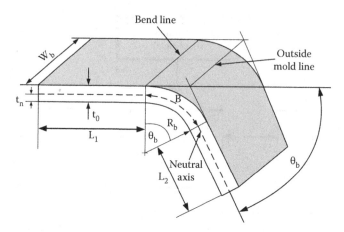

FIGURE 8.3
Forming parameters in bending of flat sheet.

When sheet metal is bent, the outer surface of the bend is stretched and the inside surface of the bend is compressed. Also, its final length tends to increase over its original length in the blank because the metal thickness tends to decrease during bending. The bend radius, R_b, is defined as the inside bend radius of curvature (ROC) on the concave, or inside, surface of the bend. The developed length of the centerline of the bent section is called the bend allowance, B. The included angle of the arc formed by bending is called the bend angle, θ_b. The bend line is the straight line on the inside or outside surfaces of the material where the flange boundary meets the bend area. Mold lines are the straight lines where the flange surfaces intersect in the bend area. This occurs on both inside and outside surfaces of the bend. The neutral axis is the theoretical location at which the material is neither in tension nor in compression. The location of the neutral axis is often referred to as the "K" factor, and it varies depending on

- Material alloy and temper
- Bend radius
- Direction of material grain (L or LT)
- Method of bend
- Ambient temperature

K factor is defined as the ratio between sheet thickness from inside face to the neutral axis, t_n, and the total thickness, t_0, of the sheet material. The bend allowance is useful for determining the length of the blank required for making the bend. The bend allowance B [2] is calculated using a K factor as follows:

$$B = \frac{(R_b + Kt_0)\pi\theta_b}{180} \tag{8.1}$$

The total sheet length needed to have the curved bend

$$L_T = L_1 + B + L_2 \tag{8.2}$$

(a)   (b)

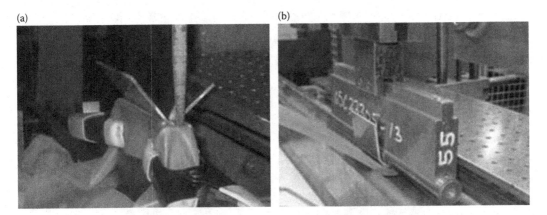

FIGURE 8.4
Straight line bend tests. (a) 90° bend test and (b) 180° bend test. (From The Boeing Company.)

The directionality in mechanical properties produced by rolling, the primary metal working process to produce sheet metal products, can have an important effect on the formability or producibility of the metal. It is generally found that the tensile properties of wrought-metal products are not the same in all directions. The dependence of properties on orientation is called anisotropy. Different mechanical properties in different directions can result in uneven response of the material during forming and fabrication operations. In general, bending is more difficult when the bend line is parallel to the rolling direction than when bend is made perpendicular to the rolling direction. An example of 90° and 180° bend test operations is shown in Figure 8.4. During bend testing, the test coupons were bent parallel (L) and transverse or perpendicular (LT) to the rolling/grain direction (Figure 8.5) to check the effect of grain direction in bending. Figure 8.6 shows examples of post-penetrant bend test coupons bent parallel (L) and also transverse (LT) to the rolling direction. Penetrant inspection is a part of quality control inspection procedures to detect any cracks occurring during forming.

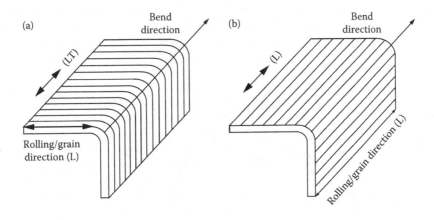

FIGURE 8.5
Bend direction versus material rolling direction, (a) bend direction transverse to the rolling direction (LT) and (b) bend direction parallel to the rolling direction (L).

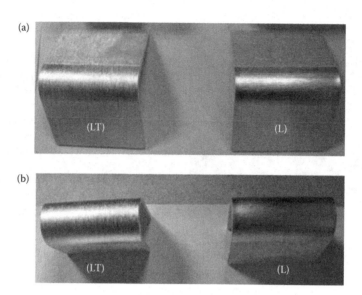

FIGURE 8.6
Straight line bend test coupons. (a) 90° bend test and (b) 180° bend test. (From The Boeing Company.)

As seen in Figure 8.4b, bend testing was performed using a deeper "U" grooved die with rubber pad that allowed the punch to travel deeper into the die to produce an approximately 180° bend angle. 180° bend test allowed to study the deformation pattern at the outer (tension) side of the bend and to measure the bend factor as defined in AMS 4267.

The bend factor as defined below is also employed widely by the primary producers of the sheet products to check the formability of the sheet metal.

$$\text{Bend Factor} = \frac{\text{Bend Diameter, } 2R_b}{\text{Nominal Thickness, } t_0} \tag{8.3}$$

The principal directions in bending are shown in Figure 8.7. The strain in the circumferential or longitudinal direction is the greatest principal strain, and the most widely

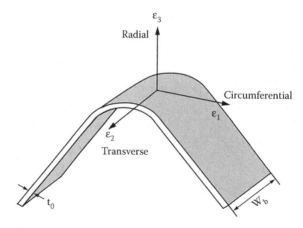

FIGURE 8.7
Principal bending directions.

used experimental results on bending are concerned with the circumferential strain. The circumferential strain distribution depends significantly on the state of stress applied in bending. Three-point bending in a die produces rather nonuniform distribution of strain. The distribution of circumferential strain, ε_1, across the width of the test part is quite uniform except at the edges. It was explained [2] by the fact that an initial rectangular cross section is distorted into a trapezoidal cross section with curled edges during bending. The distribution of transverse strain ε_2 is usually rather nonuniform. Since the stress normal to the free surface is always zero, the edges of a bent specimen are subjected to uniaxial tension. Therefore, because of the Poisson effect a transverse compressive strain is set up. The transverse compressive strain decreases with the distance from the edge in the transverse direction. If the ratio width to thickness is greater than 8, the transverse strain at the center of the width of the sheet is equal to zero. Because of this, transverse strain is important only for narrow pieces.

As explained [2], the ductility of the outer fiber in bending is a function of the state of stress acting on the surface. It is a well-established fact that the occurrence of biaxial state of tension produces a decrease in the ductility of the metal. The biaxiality ratio σ_2/σ_1 of the transverse stress to the circumferential stress increases with increasing ratio of width to thickness, W_b/t_0. Figure 8.8 indicates that, for low value of W_b/t_0, the biaxiality is low because the stress state is practically pure tension, but as the width of the sheet increases relative to its thickness, the ratio σ_2/σ_1 increases until at approximately W_b/t_0 becomes 8 the biaxiality reaches a saturation value of approximately 1/2. Correspondingly, the strain to produce fracture in bending is a reverse function of the width–thickness ratio. In bending sheets with a high width–thickness ratio, the cracks will occur near the center of the sheet when the ductility is exhausted. However, if the edges of the sheet are rough, edge cracking will occur. Frequently, the MBR can be increased by polishing or grinding the edges of the sheet. In bending narrow sheets, the failure will occur at the edges because the biaxiality [3] is quite low at the center of the width.

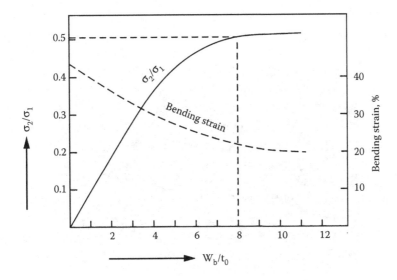

FIGURE 8.8
Effect of biaxiality and bending strain on the ratio of sheet width to thickness in bending.

8.2.2 Limiting Dome Height

The limiting dome height (LDH) test is a sheet formability test introduced by Ghosh [4]. It is based on an earlier cup test suggested by Hecker [5]. Hecker proposed his new cup test because of the inadequacies and lack of reproducibility in two well-established and essentially equivalent tests known as the Erickson and the Olsen cup tests. The LDH test, shown schematically in Figure 8.9, is an important simulative formability test and is commonly used to assess the forming quality of sheet products. The LDH test utilizes coupons of varying widths (Figure 8.10) to provide a range in states of strain from biaxial to uniaxial stretching. The equipment is designed to sense onset of necking and will stop at that point and record the height achieved by the punch. The comparison of dome heights can be used for preliminary assessment. The major process variables affecting LDH test results are mainly:

- Bulk flow stress of workpiece material
- Surface properties of the workpiece

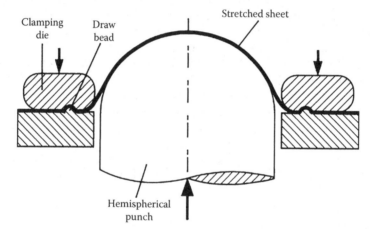

FIGURE 8.9
Schematic of LDH test.

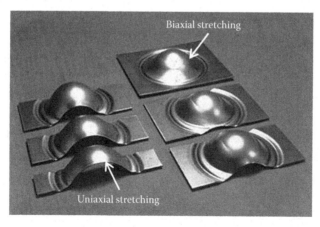

FIGURE 8.10
Different widths of test coupons for LDH tests. (Courtesy of MTS Systems Corporation.)

FIGURE 8.11
LDH test fracture. (From mmri.mcmaster.ca.)

- Punch material and its surface condition
- Punch velocity
- Lubrication at the punch-material interface

These factors indirectly affect the given measure of formability principally through their cooperative influence on the global as well as the local strain path that the workpiece experiences during deformation. Circular grid analysis is normally used in conjunction with the LDH to determine the strain distribution on the workpiece and build a forming limit diagram (FLD) for the material as described later.

Initial bending followed by stretching over the hemispherical punch is commonly used to evaluate formability of sheet metals. The sheet is firmly clamped using the clamping dies with the draw beads to prevent draw-in from the die or the blank holder. As the punch moves upward, the dome shape is developed entirely at the expense of sheet thickness (except narrower specimens which allow draw-in from the side). During the process, necking and finally the fracture of the sheet will eventually occur (Figure 8.11). Thinning of the sheet takes place over the frictional resistance of the sheet on the punch surface. The total depth of material draw before fracture is mainly a function of

- Friction condition at the punch–sheet metal interface
- Material properties
- Strain rate of deformation

For the LDH test to be repeatable study of material properties, care must be taken to standardize on strain rate, tool finish, and lubrication.

8.2.3 Forming Limit Diagram

The LDH increases with the increasing strain-hardening rate "n" value of flow stress, $\sigma_f = K(\varepsilon)^n$. The formability of a material under strain (ε) conditions from biaxial tension to uniaxial tension can be explored with the aid of this test. Each type of sheet metal can be deformed only to a certain limit that is generally imposed by the onset of localized necking, which eventually leads to the ductile fracture. From the onset of necking or fracture, a well-known method of describing this limit, the FLD is obtained. An FLD is a graph of the major strain at the onset of localized necking for range in values of the minor strain. The work of Keeler and Backhofen [6], Keeler [7], and Goodwin [8] has become a widely used method of FLD in the industry. The principal methods of measuring strain or deformation are:

1. Grid method
2. Strain gages
3. Mechanical and optical extensometer
4. Ultrasound thickness and shape measurements

Before constructing the FLD, it is required to obtain the major and minor strains calculated from the gridded circle mechanism. The steps to measure the strain for a sheet metal formed part are as follows:

- Selection of forming tool geometry and test coupon geometries
- Apply grid on the test coupon (Figure 8.12)
- Form the gridded test coupon on the forming tool at various dome heights (Figure 8.13)
- Measure strain by measuring the grid dimensions after deformation (Figure 8.14)
- Plot the strains and construct the FLD

Grid method is one of the commonly used strain measurement methods for strain analysis in sheet metal forming processes [9,10]. There are many types of grid patterns that have been developed including square, circle, and circular dot-shaped grids. Butted- and

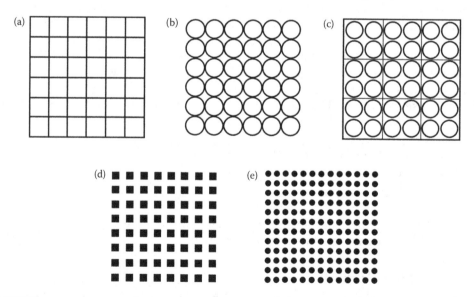

FIGURE 8.12
Examples of commonly used grid patterns, (a) square, (b) butted circular, (c) separated circular, (d) solid square, and (e) solid round.

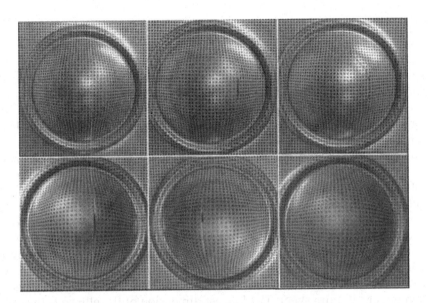

FIGURE 8.13
Example of solid square dot type grids on the test coupon used in LDH test. (From mmri.mcmaster.ca.)

single-circle grids are very common types. There are also various means of applying grids on the test coupon including:

- Electro-chemical etch
- Transferring premade inked circle
- Lithograph

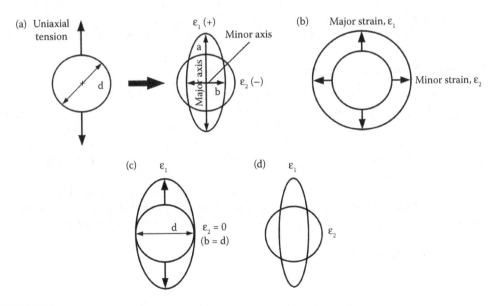

FIGURE 8.14
Schematic of deformation of circular grid in various state of stress, (a) uniaxial tension, (b) pure biaxial stretching, (c) plane strain, and (d) pure shear.

- Photo-sensitive emulsion
- Rubber stamp

When sheet metal is formed, it is subjected to various state of stress. These stresses produce nonuniform strains and might lead to wrinkling or fracturing in the formed part. The sheet is marked with either of the grid patterns as shown in Figure 8.12 before the forming process is carried out. The forming process causes the grid patterns to deform according to the local deformation experienced by the sheet metal in LDH tests as shown as an example in Figure 8.13. A grid of circles will present after deformation as major and minor axes rotated to the orientation of principal strains. A grid of squares will present as components of strain in the established orientation with a shear angle measured between the walls of the resulting box.

Strains are calculated from the grid measurements before and after deformation, which supports the development of the FLD. Figure 8.14a illustrates deformations using the most common circular grid. As shown, a circle becomes an ellipse after deformation unless the deformation is pure biaxial stretching (Figure 8.14b). Figure 8.14c shows the grid deformation in plane strain condition where the minor strain is zero. Figure 8.14d shows the grid deformation pattern in pure shear. The longest dimension of the ellipse is the major axis and the dimension perpendicular to the major axis is the minor axis.

The major strain (ε_1) and minor strain (ε_2) parameters are calculated as follows:

$$\text{Major strain} = \frac{(a-d)}{d} \cdot 100\% \tag{8.4}$$

$$\text{Minor strain} = \frac{(b-d)}{d} \cdot 100\% \tag{8.5}$$

where "a" and "b" are the length of the major and minor axis of the elliptical shape of the grid transformed from the original circular grid whose diameter is d. Figure 8.15 summarizes the grid analysis system of strain measurement and constructing the FLD (Figure 8.16).

8.2.4 Effect of Formability in Sheet Metal Forming

In sheet metal forming, a shape is produced from a flat sheet blank by bending, stretching, and compressing the dimensions of all the metal elements in the three principal directions. The final shape is then the integration of all the local bending, stretching, and compressing of the volume elements. Contoured flanges are widely seen in many sheet metal parts used in the aircraft. Figure 8.17 shows schematic representation of both stretch and compression flange parameters.

Lot-by-lot variations in formability of flat sheet products or variations from different suppliers can impact manufacturing quality for sheet metal formed parts with some kind of forming issues. In sheet metal forming, the most common forming issues are classified as

- Necking and thinning
- Cracking
- Wrinkling
- Elastic recovery/springback
- Surface condition/appearance

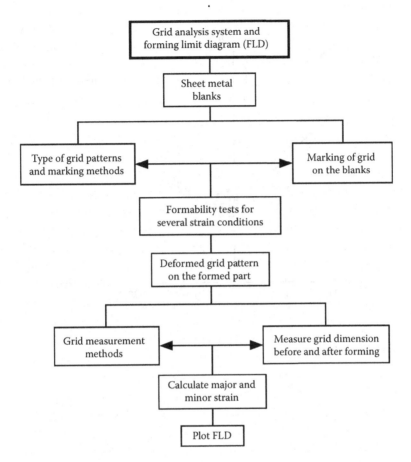

FIGURE 8.15
Grid analysis systems for FLD.

In sheet metal forming, the most common failure is a sequence of thinning, then necking, and finally cracking at the outer surface of the die or punch radius, or closer to the radius where maximum thinning occurs. An example of necking, which occurred in forming a sheet metal part in a hydro-press, is shown in Figure 8.18a. Figure 8.18b is the micrograph of the sheet thickness at section AA of Figure 8.18a to show thinning out during necking. If the radial cracks occur in the flange or the edge of the bend radius, this indicates that the metal does not have sufficient formability to withstand the large amount of circumferential stress in that region for the given die or punch radius. Figure 8.19a shows an example of a crack formed in the flange at the edge of the bend radius during hydro-press forming.

Wrinkling of the flange normally results from buckling of the sheet due to high circumferential compressive stresses. Hydro-press forming is extensively used to make various sheet metal parts for aircraft. Shallow flanged parts with both stretch and compression flanges are readily produced using this process. The compression-flanged parts are prone to wrinkling. A typical example of a wrinkle in a compression-flanged part in regions of compressive stress is shown in Figure 8.19b.

In addition to cracking on the tensile surface during bending, another common forming characteristic is the failure to maintain dimensional tolerances because of springback.

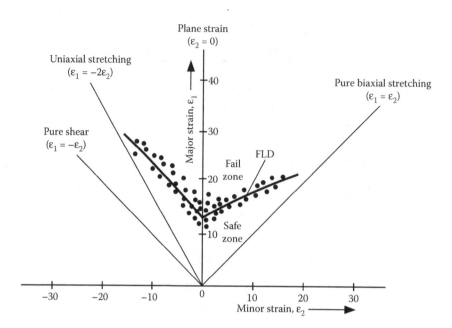

FIGURE 8.16
Schematic representation of FLD.

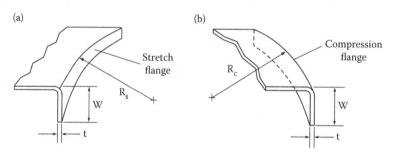

FIGURE 8.17
Contoured flanges, (a) stretch and (b) compression.

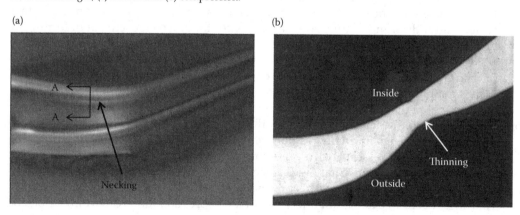

FIGURE 8.18
Example of necking in forming sheet metal part, (a) necking and (b) micrograph showing the thinning at AA. (From The Boeing Company.)

(a)   (b)

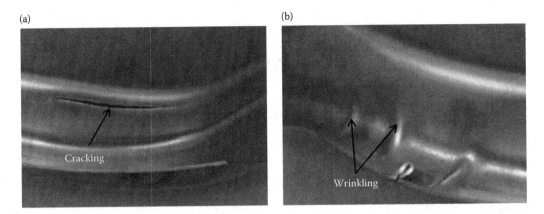

FIGURE 8.19
Example of cracking and wrinkling at the curved bend radius, (a) cracking and (b) wrinkling. (From The Boeing Company.)

Springback is the dimensional change of the formed part after the pressure of the forming tool has been released. It results from the changes in strain produced by elastic recovery as shown in Figure 8.20. When the load is released, the total strain ε_t is reduced to ε_p owing to the elastic recovery ε_r. The elastic recovery, and therefore the springback will be greater, the higher the yield stress and flow curve of the metal, the lower elastic modulus, and the greater plastic strain. For a given material and strain, the springback increases with the ratio between the lateral dimensions of the sheet and its thickness.

Springback is encountered in all metal forming operations, but it is most easily recognized and studied in bending as shown in Figure 8.21. The ROC before release of load, R_0, is smaller than the radius after release of the load, R_f. Since the bend allowance, B, is the same in both conditions, the ROC and bend angle are related as follows [2]:

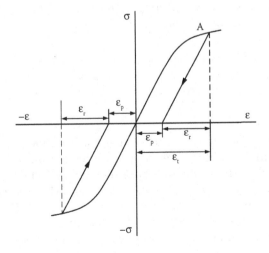

FIGURE 8.20
Schematic of elastic recovery curve.

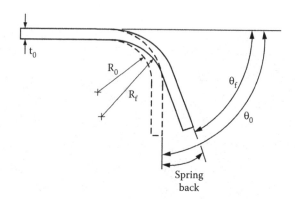

FIGURE 8.21
Springback in bending.

$$B = \left(R_0 + \frac{t_0}{2} \right) \theta_0 = \left(R_f + \frac{t_0}{2} \right) \theta_f \tag{8.6}$$

$$K = \frac{\theta_f}{\theta_0} = \frac{R_0 + (t_0 / 2)}{R_f + (t_0 / 2)} \tag{8.7}$$

The springback ratio, K, defined in this way is independent of sheet thickness and depends only on the ratio of bend radius to sheet thickness.

The most common methods of compensating for springback in sheet metal forming are:

- Bending the part to a smaller ROC than the desired end feature—for brake forming this means selecting a slightly smaller radius punch than specified for the part. For hydro-press forming, which uses a unique tool to provide all contours of the part, this requires manufacturing the tool with adjusted bend angles and radii.
- Iterative contour development—starting with a tool matching the part contour, the contour error due to springback is measured and a safe percentage reflected back to a modified tool surface.
- Predict modified tool surface based on finite element analysis—advanced modeling techniques can provide confidence to go directly to near final tooling solution. This saves much time and effort if reliable model of the specific process is developed.

Springback effects of different alloy and temper combination can be compared in the laboratory by using a universal tensile/compression testing machine with a three-point bending setup. Figure 8.22 shows the schematic representation of the setup. Springback is measured as the difference between peak deflection and permanent set as shown in Figure 8.22. Test coupon with constant width that matches with the width of the three-point bend test tool is a very critical factor for the springback test.

An example of a stretch formed part of complex contoured geometry indicating springback effect after releasing the part from the stretch forming machine jaws is shown in Figure 8.23.

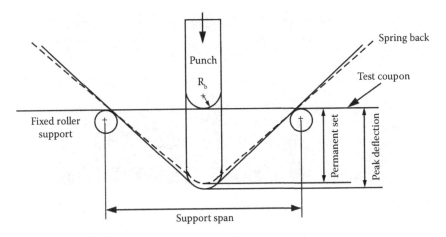

FIGURE 8.22
Schematic representation of springback test.

FIGURE 8.23
Example of springback effect in stretch forming. (From The Boeing Company.)

In addition to necking, cracking, wrinkling, and springback parameters that affect mainly the structural integrity of the sheet metal formed parts of an aircraft, there are few surface-related issues that affect the surface appearance of the part since sheet metal formed parts usually present a large surface area. Pronounced surface roughness in regions of the part that have undergone appreciable deformation is usually called orange peeling. The orange peel effect occurs in sheet metal of relatively large grain size and also when the sheet is deformed under severe stretching or bending (Figure 8.24). It results from the fact that the individual grains tend to deform independently of each other, and therefore, the grains stand out in relief of the surface. This condition is best corrected by using finer grain size sheet metal so that the grains deform more nearly as a whole and the individual grains are difficult to distinguish with the eye.

As explained [2], many metals, particularly low carbon steel and certain Al–Mg, Al–Li alloys, show a localized heterogeneous type of transition from elastic to plastic deformation, which produces a yield point in the stress–strain curve rather than having a flow curve with gradual transition from elastic to plastic behavior. Metals with a yield point have a flow curve or a load elongation curve as shown in Figure 8.25. The load increases

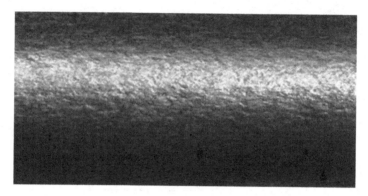

FIGURE 8.24
Example of orange peel surface during sheet bending. (From The Boeing Company.)

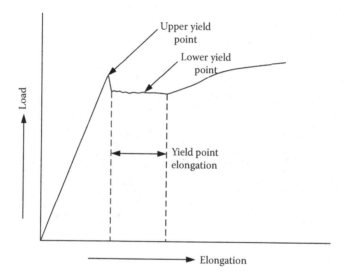

FIGURE 8.25
Typical yield point behavior.

steadily with elastic strain, drops suddenly, fluctuates about some constant value of load, and then rises with further strain. The load at which the sudden drop occurs is called the upper yield point. The constant load is called the lower yield point, and the elongation that occurs at constant load is called the yield point elongation. The deformation occurring throughout the yield point elongation is heterogeneous. At the upper yield point, a discrete band of deformed metal, often readily visible with naked eyes, appears at a stress concentration such as a fillet, and coincident with the formation of the band the load drops to the lower yield point. The band then propagates along the length of the flat sheet workpiece, causing the yield point elongation. In the usual case, several bands will form at several points of stress concentration. These bands are generally at approximately 45° to the tensile axis. They are usually called Luders bands, Hartmann lines, or stretcher strains (Figure 8.26) and appear on the sheet metal surface, which may be a quality issue for the aerospace parts.

FIGURE 8.26
Luders bands appeared on an Al–Mg alloy sheet in stretch forming. (From The Boeing Company.)

8.3 Blank Preparation

The first step in manufacturing a sheet metal part is blank preparation from a large sheet supplied by the flat sheet manufacturers. It is a very important step to cut the blank in a proper geometry to form the desired shape. Shearing and routing are very common methods used in the aerospace industry to produce blanks for making small-to-large sheet metal parts for the aircraft.

8.3.1 Shearing and Blanking

Shearing and blanking mechanisms are shown in Figure 8.27. The workpiece/blank is sheared along the direction of the shearing blade. Since it is a shearing process, shearing

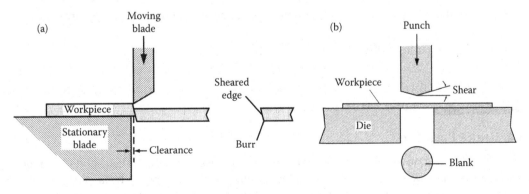

FIGURE 8.27
Shearing and blanking process, (a) shearing and (b) blanking.

(a)

(b)

FIGURE 8.28

Example of CNC route for sheet metal blank preparation. (a) Routing machine in operation and (b) routed blank. (From The Boeing Company.)

along the shear plane, the edge of the cut is left with sharp burrs at the sheared edge. Those sharp edges and burrs are safety hazards for handing the blank in the process of manufacturing parts, and could cause damage to the forming die surface and also burrs left on the part will act as stress risers that are not acceptable for any aircraft parts. It is necessary to break the sharp edges and remove the burrs from the blanks prior to any forming processes.

8.3.2 Routing

A computerized numerical control (CNC) router is widely used in the aerospace industry to cut the sheet metal blank according to a computer-aided design (CAD) program written to match the geometry of the blank needed for the subsequent forming processes including brake press forming, hydro-press forming, hydro forming, spin forming, and more. CNC router performance begins with high-speed spindles, many up to 60,000 rpm, and a vacuum table that makes it extremely easy to set up and machine sheet material blanks in just seconds. Multiple sheets sandwiched together can be routed for mass production in a process called stack routing. Figure 8.28 shows an example of CNC router in operation. The major advantages of using routed blanks include:

- Routed edges are less prone to edge cracking than sheared edges along bend lines
- Optimum material usage since parts can be more effectively nested within the sheet
- Effective forming within the die or punch geometry

8.4 Brake Forming

Brake forming is one of the oldest metal bending processes in which the metal workpiece is placed on an open die and pressed down into the die geometry by a punch that is pressed by the ram of the machine, called the brake press. Both mechanical and hydraulic type brake presses with CNC controls are used on the production shop floor in the

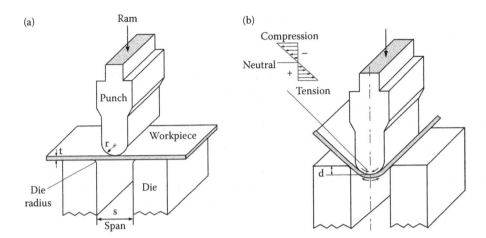

FIGURE 8.29
Brake forming principle, (a) before and (b) after.

aerospace industry. During the process, a piece of sheet metal is formed along the straight axis. This may be accomplished by a "V"-shaped, "U"-shaped, or channel-shaped die and punch set. This process is widely used in aerospace for forming of relatively long, narrow parts. Simple 90° "V"-type bends or more intricate shapes can be formed using brake forming press. In brake forming or in other forming processes when the sheet metal is bent, the metal at the inner bend radius is under compression, whereas the metal on the outer side of the bend is under tension as shown in Figure 8.29. Because of this stress differential, a strain gradient is formed across the thickness of the workpiece. So in brake forming, bending has to overcome both tensile stresses as well as compressive stresses. When the bending is complete, the residual stresses make a part re-bend or springback to its original position, so it is required to over bend the sheet metal keeping in mind the residual stresses. The distance to which punch travel (d) enters the die determines the bend angle. The span width "s" affects the force needed to bend the workpiece. The minimum span width is determined by the sheet metal gage thickness, t, of the workpiece and also by the punch radius "r."

Various types of brake forming can be done using different types of die and punch geometry. Figure 8.30 shows the basic parameters of a brake forming die. Individual types of forming or a combination of various types is needed to make various aircraft parts. Brake formed parts (sections) are also used as initial stock for many other operations like stretch forming, hydro forming, etc. The types of brake form dies include mainly:

- Air bend
- Acute angle
- Bottoming
- Gooseneck
- Flattening
- Offset bending

In air bending (Figure 8.31a), the included angles for both die and punch allow for over bending of sheet metal to compensate for springback. Air bending term is derived from

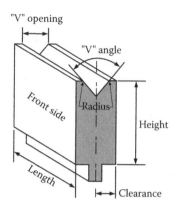

FIGURE 8.30
Brake forming die parameters.

the fact that the workpiece spans the gap between the nose of the punch and the edges of the die allowing enough air gaps between the nose of the punch to the bottom of the die. Angles from very shallow to 90° are formed by adjusting the ram movement of the brake press. If parts are being formed with air bend dies on a brake press, the included angle of the die and punch is adjusted to compensate for all possible springback or to obtain true air bending with 90° bend angle. Acute angle dies and punches (Figure 8.31b) are used for air bends from very shallow angles to 30° angle. The angle formed depends on the depth to which the punch enters the die. Acute angle dies are commonly used to preform hems.

Bottoming dies and punches (Figure 8.31c) are used for making very accurate bends with relatively sharp inside bend radii in comparatively light gage material. Included angle for both the die and punch is kept at 90°. The workpiece is in contact with the complete working surfaces of both punch and die to maintain accurate angular tolerances. As a result, bottoming die requires higher pressure than air bending die. Gooseneck punches (Figure 8.31d) are used for multiple bends of making channels or special shapes with which a straight-sided die would interfere. A deeper throat beyond the punch centerline

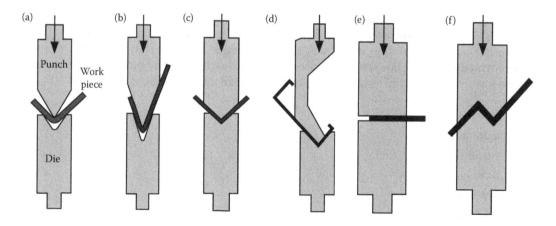

FIGURE 8.31
Common type of brake forming dies and punches, (a) air bend, (b) acute angle, (c) bottoming, (d) gooseneck, (e) flattening, and (f) offset bending.

FIGURE 8.32
Example of a sheet metal part formed in a brake press. (From The Boeing Company.)

increases the width of return flange but reduces the die capacity. Flattening dies (Figure 8.31e) are used for hemming or flattening acute angle bends. Flattening dies and acute angle dies mounted side by side in a press brake can produce a hem with each press brake stroke. In offset bending (Figure 8.31f), both the die and punch have the offset geometry to produce the offset in each press brake stroke. Figure 8.32 shows an example of producing an aircraft sheet metal part.

Bending operation in brake forming is a function of the type of press, the tooling, and the mechanical properties of the workpiece material. Material properties including yield strength, elongation, hardness, and the temper of the material can cause the amount of springback to vary in the material during bending operation.

The overall advantages of brake forming process are as follows:

- No dedicated tooling required; punch and die from selection of standard tools
- Applicable for low and medium production runs
- Suitable for both small and long workpiece

The brake forming process can produce a variety of shaped aircraft parts including small clips, brackets, and shear ties using different types of bend dies and punches to produce various types of bends. To produce long parts, "V," "U," channel-type dies, or other special punch and dies are used. Figure 8.33 shows various production dies and punches used for making a variety of small-to-large sheet metal parts of an aircraft. Brake press forming is also used to preform a shaped part as an initial metal stock for the next forming operation to finish the part to the final geometry. The largest brake form press manufactured by Cincinnati Incorporated is the 50 feet (15.24 m) hydraulic press made for Boeing to form large cylindrical fuel tanks for Delta IV rockets.

8.4.1 Bump Forming

Bump forming is a very well-known process in the aerospace industry, used to produce commercial airplane fuselage skin or other aerospace vehicle skin panels of constant ROC

FIGURE 8.33
Examples of various types of brake forming punch and die. (From The Boeing Company.)

using brake press forming principles. Figure 8.34 shows the fundamentals of the bump forming process as defined by incrementally bending the flat sheet or plate products bump by bump. In bump forming, short length bending takes place over small sequential incremental steps of sheet/plate products along the grain or rolling direction (L), and is used to produce a large panel of constant ROC. A metal blank of flat sheet or plate product is fed between the gap of the punch and the bottom die having width of (w) (Figure 8.34a). The vertical punch travel/depth of penetration (d) determines the bend depth for each bump. The effective bend depth in each step or bump may vary in a complicated way depending on mainly:

- Bend geometry
- Sheet/plate thickness (t)
- Material alloy and temper
- Die width (w)
- Springback (x)

In production practice, the bump forming parameters especially the depth of penetration are optimized by trial-and-error methods. The effective bend depth is roughly calculated from the difference between punch travel (d) and the springback (x). Compensation for springback requires over bending, which means the punch will need to penetrate farther than would appear necessary to produce the desired amount of residual part deformation. Springback may vary depending on these major factors:

- Material properties and behavior
- Sheet/plate thickness (t)
- Die width (w)

Progressive bumping at each incremental step resulted in incremental bending. The sum forming effect of these bumps produces the curved panel with constant ROC (Figure 8.34b).

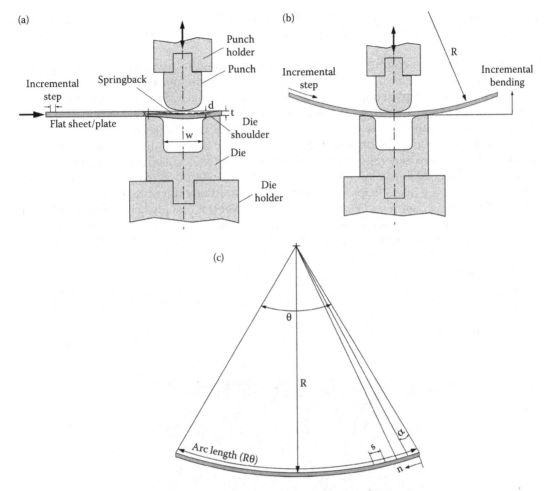

FIGURE 8.34
Fundamentals of bump forming process, (a) adjusting punch travel, (b) incremental bending, and (c) important variables.

Once the panel geometry is defined and the ROC (R) of the panel is known, several important variables will play a very critical role in producing a curved panel:

1. Arc length
2. Distance between step/bump (s)
3. Number of step/bump (n)

The arc length is calculated by using a simple geometrical relation (Rθ), measured on the inside curvature. The distance between steps or bump determines the total number of steps (n) needed to form a panel with an arc length (Rθ) (Figure 8.34c). The distance between the steps (s) is being calculated from the step arc length (Rα). Therefore, the total number of steps is calculated as

$$n = \frac{R\theta}{R\alpha} = \frac{\theta}{\alpha} \tag{8.8}$$

The higher the number of steps utilized (n), the smoother the resultant outside radius would be. But more bumping will make the forming operation more time consuming and less cost effective.

According to common production practice used for manufacturing fuselage skin panels of constant ROC, a large flat sheet or plate blank is initially marked with small graduations of number of steps at opposite edges of the part. This is done so that the operator can feed an equal step distance for each end of the part into the gap between punch and die for every hit or bump by the punch, as shown in an example (Figure 8.35a). In fuselage skin panels, commonly the inner surface of the skin will have number of weight reduction pockets. Pockets are generally numerically controlled (NC) milled or sometimes chemically milled. Special needs are being considered to fill the pocketed areas during bump forming. Use of the fillers helps in maintaining an even amount of material deformation across the varying material thickness caused by the presence of pockets. Pockets are filled using right-size fillers made of different materials, including simple cardboard, plastic, and even aluminum or other metals, depending on the forming conditions including:

- Skin alloy and temper
- Material thickness
- Inner ROC
- Appearance and outer surface condition

(a)

(b)

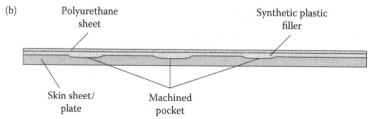

FIGURE 8.35
Production practices of fuselage skin panel of constant ROC, (a) example of bump forming a fuselage skin panel. (From The Boeing Company.) (b) Example of placement of filler on machined pockets (sectional view).

Fillers are generally used on the inner milled pockets prior to forming to avoid some mark off appearance on the outer side of the panel. Visible mark-off impressions may be due to the stiffness variation of the skin material between pocket and the unmachined area that changes the nature of bending in each bump. Placement of fillers on the pocketed areas of the machined skin panel is being done in different ways as shown in the following examples:

1. Individual filler taped in the pocket area (Figure 8.35a)
2. Single piece machined filler taped with polyurethane sheet on the top (Figure 8.35b)

8.5 Roll Bending

Roll bending is another metal forming process utilized to produce sheet metal parts with various contours such as conical, variable, and linear contours. A roll bending process can produce similar parts faster than a bump forming process. Figure 8.36 shows the principles of a roll bending process. The flat sheet skin panel is fed into the gap between the pressure/upper roll and two support/bottom rolls. The amount of bend is controlled by adjusting the height (h) of the pressure roll. The horizontal distance (d), between the support/bottom rolls could be adjusted to vary the bending force. When d increases, the bending force tends to get lower. There are two settings of the roll's direction that determines the forward and backward movement of the sheet relative to the operator's position. An example of roll bending of a flat sheet panel is shown in Figure 8.37 producing a linear contour of constant ROC. This example shows that the panel had machined pockets covered with cardboard fillers on the inner side of the panel as discussed in earlier Section 8.4.1. Fillers are used in the roll bending process for the same reasons as they are used for bump forming, to ensure more uniform contour and thereby less visible markoff on the skin outer surface.

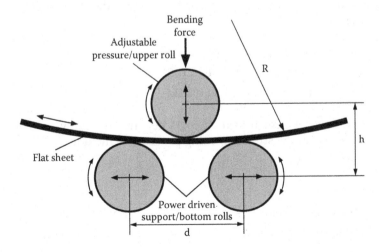

FIGURE 8.36
Roll bending principle.

FIGURE 8.37
Example of roll bending of a sheet metal part of constant ROC. (From The Boeing Company.)

8.6 Stretch Forming

The most basic principle of stretch forming technology is to stretch the metal blank beyond its yield point causing the material to plastically deform within the safe margin of plastic deformation range between yield stress and the ultimate tensile stress of the material being formed. The stretch forming process uses primarily tensile forces to stretch flat sheet or preformed sheet metal profiles over a contoured die or form block to produce parts featuring either simple or complex contoured geometry. Stretch forming of sheet products is widely used by the aircraft industry to manufacture fuselage skin panels, door panels, and doublers as well as some curved wing/stabilizer leading edge profiles. Assemblies made from these parts will include substructures or frames made from aerospace aluminum alloy flat sheet or light gage plate products. The majority of contoured sheet metal "skin-type" parts for aerospace applications are formed by one of these commonly used forming procedures or techniques:

1. Stretch forming (uses flat sheet blank)
2. Stretch-wrap forming (uses preformed profiles)
3. Stretch-draw forming (uses flat sheet blank)

Successful production of stretch formed parts depends on many variables. The major process variables in any of the stretch forming procedures include:

1. Material:
   a. Formability
   b. Material strength properties (determines press power required)
   c. Blank dimension (length and width)
   d. Blank thickness
   e. Profile geometry

2. Press:
  a. Type:
    i. Longitudinal forming—stretches the flat sheet blank along its length
    ii. Transverse forming—stretches the blank along its width
    iii. Profile forming—stretches the preformed profile along its length
  b. Specifications (mainly jaw, die table, and hydraulic system)
3. Stretch form block
  a. Geometry
  b. Material
  c. Lubrication

The stretch forming process mainly involves the following process principles as shown in Figure 8.38:

- Bending
- Stretching/drawing
- Combined bending, stretching, and drawing

Parts made from stretched flat sheet materials are generally produced using large machine/presses consisting basically of a hydraulically driven ram (commonly vertical) that carries/lifts the die or forming block, and two side jaws for gripping the ends of the workpiece. Each gripping jaw is driven by a separate hydraulic cylinder to provide a variety of methods to form a diverse range of shapes. To form a contoured skin panel using a stretch forming machine, the sheet metal blank is first gripped between the jaws and deformed/bent (Figure 8.38a) by moving the jaws inward. The rotational movement of the jaw carriage provides the bending action in the blank and places the jaws in the correct angular position for the desired part geometry. Both the jaw carriages are then moved further outward to exert continuous tensile force at the final jaw position. Then, the vertical ram is slowly moved upward (drawing action) till the deformed sheet blank is stretched around the forming block with a relatively light tensile pull being simultaneously applied by the jaw grip cylinder. The stretching load is increased by extension of the vertical ram while holding the gripping jaws stationary until the blank is strained plastically to its final stretched configuration (Figure 8.38b). Stretch forming of more gentle contours can be accomplished with either extension of the punch or retraction of the side jaws or a combination thereof. Airplane wing leading edge panels are commonly manufactured by using a stretch forming process with a forming block of smaller curvature as shown in Figure 8.38 (with variable contoured surface end to end).

Similarly, large contoured products like fuselage skin panels are produced with large forming blocks having either constant ROC or compound curvature as shown schematically in Figure 8.39. Figure 8.39a shows the flat sheet blank with linear jaw position. Figure 8.39b shows the block position 1 where the blank is pre-stretched (initial tension) and block position 2 where the blank is post-stretched (final tension).

Figure 8.40a shows some examples of stretch forming blocks used for producing various fuselage skin panels of an aircraft. Figure 8.40b shows the large sheet metal blank as it is loaded and an initial deformation or bend is induced by moving the jaw inward. Figure 8.40c shows that the stretch forming in process by moving the forming block upward and stretching the sheet along its length conforming to the block geometry.

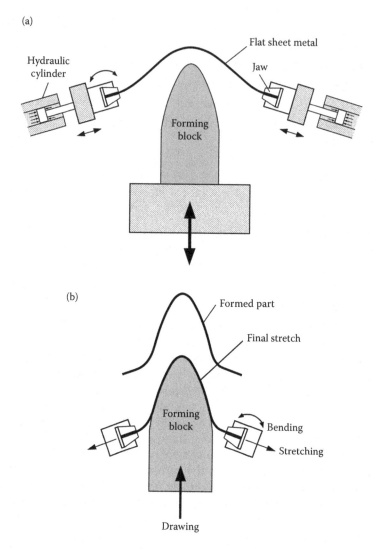

FIGURE 8.38
Fundamentals of stretch forming of flat sheet part with smaller curvature, (a) forming set up and (b) forming complete.

For full conformity, some additional pull is applied by the pulling jaws along with some angular movement of the jaws. Figure 8.40d shows the formed part being unloaded from the stretch forming block.

The fundamentals of a sheet stretch-wrap forming process are shown in Figure 8.41a where generally the preformed sheet profile (roll or brake formed sections) or extrusion is first stretched beyond the yield point and wrapped around the stationary forming block and conformed to the contour of the block. The sheet preformed profile is held in tangency to the front face of the forming block and stretched close to the material yield point. Under this stretched condition, the profile is wrapped around the forming block and a final setting tension is applied to complete forming. Figure 8.41b shows an example of a stretched wrapped sheet metal part from a previously formed 90° angle profile. A wiper block is used at the end of forming cycle to prevent any wrinkling in the part.

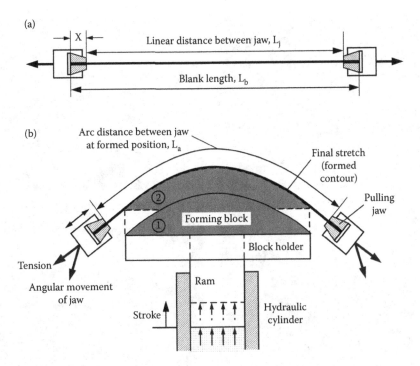

FIGURE 8.39
Schematic representation of stretch forming of flat sheet part with larger curvature, (a) blank with linear jaw position and (b) formed contour at angular jaw position.

In stretch-draw forming, the sheet metal blank is first stretched over the forming block (Figure 8.42a) by stretching the blank with the gripping jaws from each end. Then, the top mating tool is forced down into the stretched blank, thereby drawing the blank into the contour of the forming block to complete conformation of the blank over the forming block contour geometry (Figure 8.42b).

The specific requirements of stretch forming presses are generally:

1. Force or the forming tonnage (Figure 8.43)
2. Distance between jaw (Figure 8.39)
3. Jaw width (Figure 8.43)
4. Stroke (Figure 8.39)
5. Jaw configurations (Figure 8.44)

The amount of force required to stretch form a part determines the power (tonnage) requirement. This is a key parameter used to select a stretch forming press. The required force/pull (T_P) to form a flat sheet blank can be approximately calculated by a simple mathematical relationship:

$$T_P = \sigma_{yp} \times A + T_{FB} \tag{8.9}$$

where σ_{yp} is the yield strength of the pulling blank material, A is the cross-sectional area of the pulling flat sheet blank, and T_{FB} is the additional pull required to overcome frictional resistance (F_{FB}) at the forming block and the flat sheet blank interface (Figure 8.43).

FIGURE 8.40
Example of stretch forming blocks and the forming steps of producing aircraft structural parts. (a) Stretch forming blocks, (b) loading of sheet, (c) stretch forming in process, and (d) formed part. (From The Boeing Company.)

Frictional resistance is dependent on the friction condition of the block material and the lubrication on the block.

Part size is limited by the maximum blank length, which is also controlled by the maximum linear distance between the press jaws as illustrated in Figure 8.39a. Figure 8.39b shows the jaws in angular positions after forming. The arc distance between the jaws (L_a) at final stretched position is estimated with the following relationship:

$$L_a = L_b - 2x + e + E_R \qquad (8.10)$$

where L_b is the blank length, x the length of the blank inserted into each jaw, e the elongation of the blank during stretching, and E_R is the additional allowance for the elastic recovery of the part (springback effect) when the jaws are released.

Maximum distance between jaws will be one of the important machine specifications. Similarly, the width of a part is an important part geometry to be considered while selecting the press jaw width. The forming block is set up on a hydraulic ram between the side jaws providing upward stroke into the flat sheet blank, being stretched and conformed to the forming block geometry. The length of the block determines the maximum width of the part. So, the width of the gripping jaws press will set the limitation on the width of the part.

For sheet stretch forming, two basic jaw configurations are being considered:

1. Curved jaw
2. Straight jaw

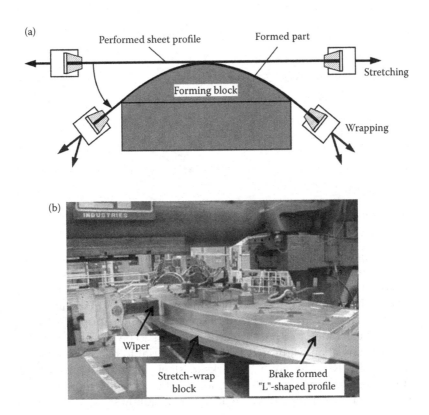

FIGURE 8.41
Stretch-wrap forming, (a) fundamentals of stretch-wrap forming and (b) example of stretch-wrap forming of an airplane part using wiper. (From The Boeing Company.)

Curved jaw configuration means that the jaws can be curved to stretch or pull the sheet material over the ends (end pull) of the stretch form block as shown in Figure 8.44a. Curved jaw configuration is required for long very tight radius fuselage skin panels. In modern presses, the curved jaws are CNC controlled and programmable to eliminate the long, tedious, manual setup.

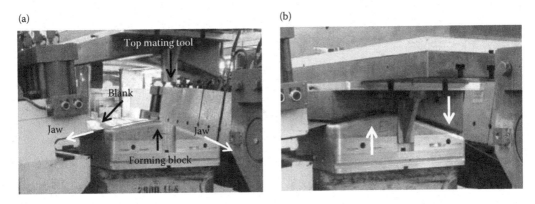

FIGURE 8.42
Stretch-draw forming. (a) Blank is stretched on the forming block and (b) forming complete. (From The Boeing Company.)

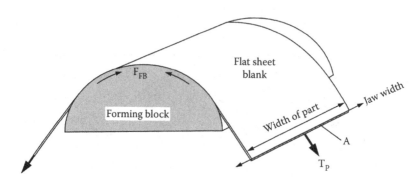

FIGURE 8.43
Stretch forming force/pull.

Straight jaw configuration means that the jaws are oriented only in a straight line to stretch the sheet material over the sides (side pull) of a stretch form block as shown in Figure 8.44b. The straight jaw configuration press is used to form a variety of airplane parts including leading edge skins, cab skins wing tip skins, and many more.

The choice of material from which to manufacture stretch form blocks and mating tools depends on the geometry of part, quantity of parts to be produced, thickness of the sheet blank, and the mechanical properties of the material to be formed. Various materials including phenolic (Richlite, Mycarta), kirksite, steel, aluminum, plastic, and more are used to produce stretch forming blocks depending on its primary requirements. The primary requirements of the forming block material are as follows:

- Maintaining shape under pressure
- Resist abrasion
- Retain smooth finish
- Lighter weight to facilitate handling

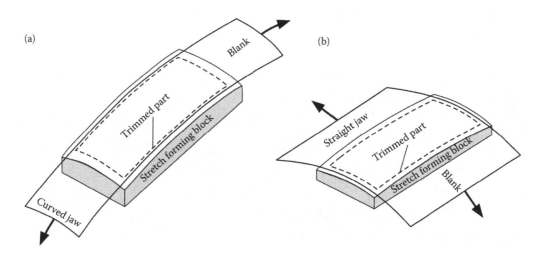

FIGURE 8.44
Stretch forming jaw configurations, (a) end pull and (b) side pull.

It is important to lubricate a stretch forming block to keep the friction low at the forming block and sheet metal interface to minimize the chance of cracking. Various types of lubricants, solid, or oil based, are being explored for use according to complexity of the part and the needs of production.

Developing a stretch form block is a very important consideration for a successful formed part. Major factors to consider when developing stretch form block are:

1. Part geometry
2. Alloy/temper of sheet or extruded product
3. Elastic recovery (variation depends on contour)
4. Contour overbend

Tryout of a stretch form block is required to prove their function and development, and to furnish the production shop with a forming sequence and setup information. Normally, two forming stages are required to make a finished part (1) preform and (2) finish form. Stretch form tryout process steps are generally:

1. Preform—forming in "O." Desirable to form 90% of the contour.
2. Finish form—heat treat preform part to "W." Additional stretch to reach final contour.

Alternative Process Steps:

1. Forming in the "W" condition—forming exclusively in "W" condition. Applicable to shallow or slight contour.

Stretch forming provides a number of advantages over other forming processes. The major advantages are:

- Springback in stretch forming is greatly reduced since the stress gradient is relatively uniform
- Large deformation can be obtained in stretch forming for the material with high ductility because tensile stresses predominate the process
- Wrinkling and buckling are eliminated under tension
- High volume production output and form larger parts with lower cost.

8.7 Hydro-Press Forming

The hydro-press is a special purpose machine, which applies very high pressure on the workpiece by means of the hydraulic fluid bag. Older technology presses operate at 2000–6000 psi (13.8–41.4 MPa). Newer technology presses achieve nearly 30,000 psi (206.8 MPa) forming pressure. The higher applied pressure reduces springback and improves feature definition. Figure 8.45 shows a 20,000 psi (137.9 MPa) hydro-press used to form sheet metal parts with straight, curved compression and stretch flanges, beads, formed cutouts, joggles, or any combination of these features. This machine is used to form small-to-large parts

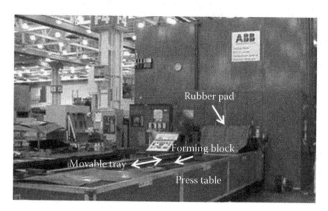

FIGURE 8.45
Hydro form press. (From The Boeing Company.)

including fuselage frames, doorframes, large pans, webs, and more. This press accommodates part with sizes of 150 inch length (3810 mm) × 40 inch (1016 mm) width × 4 inch (101.6 mm) depth.

Figure 8.46 shows a schematic diagram of the process simulation of hydro-press forming. Position 1 is the initial placement of a flat sheet blank on the hydro-block with the protective rubber pad on top of it. In position 2, the flat sheet and the rubber pad are deformed under pressure applied by the hydraulic fluid bag. Position 3 is the final part position after forming.

In hydro-press forming, the lubrication process may be a combination of a boundary and mixed lubrication system. Out of total forming operation in hydro-press forming, some part of the forming is real bending and some part is forming against the contour geometry of the hydro-block (stretching). To do that, the metal is drawn into the forming area by the rubber pad under the influence of fluid pressure generated by the hydro-press. The type of lubricant determines the rate of drawing by controlling the friction between the workpiece and the tooling surface. Figure 8.47 shows an example of the hydro-press forming processes being used to make some frame type part for an airplane.

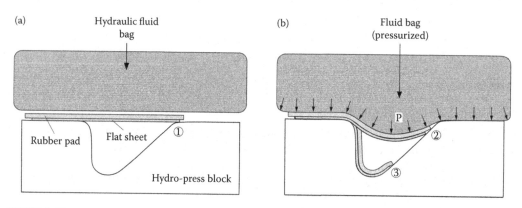

FIGURE 8.46
Schematic of hydro-press forming principle, (a) flat sheet on the hydro-block and (b) forming started.

(a)   (b)

FIGURE 8.47
Example of some hydro-press parts formed. (a) Routed blank on the forming block on the press tray and (b) forming complete. (From The Boeing Company.)

8.8 Progressive Roll Forming

Progressive roll forming is a special type of cold forming process in which a metal strip (or coil) is progressively bent into complex shapes by passing it through a series of driven rolls of defined geometry. Roll forming is ideal for producing parts with long lengths or in large quantities of molded sections (with constant cross sections), such as irregular-shaped channels and trim. Figure 8.48 shows a schematic representation of the progressive roll forming process showing a flat sheet strip being progressively formed by the pairs of rolls at each stage. Strip stock, typically in coil form, is fed through successive pairs of contoured rolls that progressively form the workpiece material to meet the desired specifications. Figure 8.49a shows an example of a progressive roll forming line producing sheet metal stringers for the fuselage structure of an aircraft. Progressive bending of the strip occurred within the geometric gap between the rolls in each stand (Figure 8.49b). The thickness of metal is not appreciably changed during this process.

Design of the rolls is considered with the sequence of profile cross sections for each stand of rolls (Figure 8.49b). The roll cross sections are then derived from the profile contours.

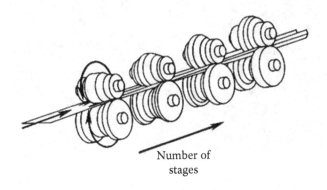

Number of
stages

FIGURE 8.48
Schematic illustration of roll forming process. (From Simply Tool Steel.)

(a)

(b)

FIGURE 8.49
Progressive roll forming line. (From The Boeing Company.)

Because of the high cost of the roll sets, computer simulation is often used to validate the designed rolls and optimize the forming process to minimize the number of stands and material stresses in the final product. To achieve the desired roll formed profile, sets of mated rolls are constructed. Figure 8.50 shows an overview of roll formed stringers in production. The geometry of rolls is dependent on

- Geometry of the stringer profile
- Material alloy and temper to be formed
- Material thickness
- Formability of the material

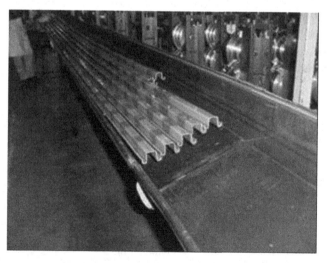

FIGURE 8.50
Roll formed stringers. (From The Boeing Company.)

8.9 Joggle of Sheet Metal Parts

A joggle is a formed feature on a sheet metal or extrusion part. This feature creates an offset bend or transition on the surface of a part. The purpose of this transition is to provide clearance, which allows parts to overlap during assembly of aircraft components. The direction of offset in the joggle is perpendicular to the surface of the part. It creates two parallel planes joined by a sloped or offset plane as shown in Figure 8.51. The depth of joggle and its length are the two design factors measured from the functional surface of the part. The ratio of joggle length and the joggle depth is defined as the joggle ratio.

In a sheet metal part, the joggles are produced by using mechanical or hydraulic presses (Figure 8.52a) with offset and crush joggle tooling for stringer profiles as shown in Figure 8.52b and c. The press ram applies forming pressure to deform the part cross section beyond the yield strength of the part material to create a permanent deformation according to the geometry of the tool. An offset joggle (Figure 8.52d) has the entire section (both top and bottom area) offset while a crush joggle (Figure 8.52e) will offset only one area (top area of the stringer profile). Figure 8.53 shows an example of the difference in deformation pattern between offset and crush joggles formed on a fuselage stringer of an aircraft. Figure 8.53b indicates the material growth in the bottom area (flange) in crush joggle forming as the press compresses the top area (crown) only.

8.10 Deep Draw Hydro Forming

Deep draw forming is a metal working process used for forming sheet metal into deep cup-shaped parts or multiple contour parts by either forming or drawing a defined size blank of sheet material over the shape of a die by applying pressure with a metal punch and resisted by frictional forces around the perimeter of the blank as shown in Figure 8.54.

Deep draw hydroforming process follows the same fundamental means of metal forming as a conventional deep draw process except that the blank is pushed by the punch into

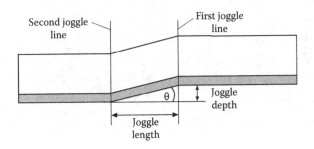

FIGURE 8.51
Fundamentals of joggle geometry.

FIGURE 8.52
Joggle press, tools, and formed geometry. (a) Hydraulic joggle press, (b) offset joggle tool, (c) crush joggle tool, (d) formed offset joggle, and (e) formed crush joggle. (From The Boeing Company.)

(a)   (b)

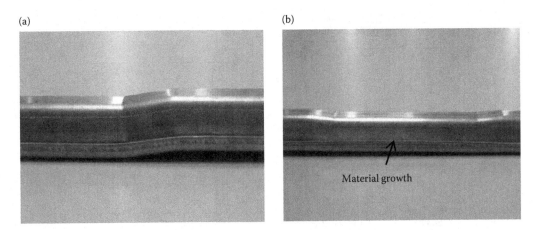

FIGURE 8.53
Difference in deformation pattern between two types of joggle. (a) Offset joggle and (b) crush joggle. (From The Boeing Company.)

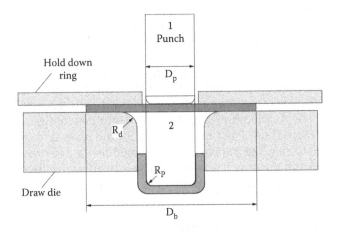

FIGURE 8.54
Fundamentals of deep draw method.

the rubber diaphragm, which is under hydraulic pressure. The major variables in a hydro forming process are as shown in Figure 8.55:

- Forming cavity pressure, p
- Punch force, F_p
- Punch speed, V_p
- Punch travel, d_p

The basic components associated with the hydroforming process are mainly the forming chamber, punch, draw ring, and blank part. The blank is placed on the stationary draw ring and centered over the opening through which the punch will move. The forming chamber is lowered flush with the rubber pad on the top of the blank and the forming pressure is set. The hydraulic piston is activated in an upward direction with the set punch pressure. The punch moves upward through the draw ring opening and forces the

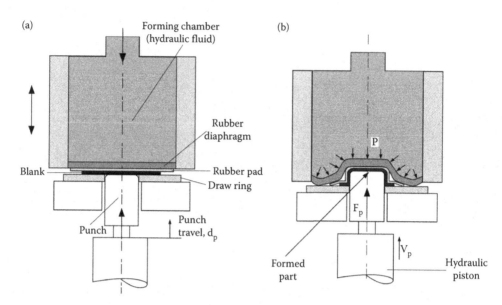

FIGURE 8.55
Principle of hydro forming process, (a) ready to hydro form and (b) after hydro form.

blank and the rubber pad against the outer surface of the diaphragm and upward into the forming chamber. The additional rubber pad is provided in between the blank and the diaphragm mainly to protect against any damage of the diaphragm surface from the edges of the metal blank. The diaphragm forms the blank into the configuration of the punch. In this case, a diaphragm is used as a universal female die instead of using a conventional (method) female die. When the part has been formed at the set pressure, the hydraulic chamber is raised up and the punch is lowered. The formed part is then removed from the draw ring surface. There are a few variations from the standard hydroforming techniques, some of them are highlighted below:

1. Contoured draw ring—contour draw rings are being used when the part with variable depth is required
2. Multiple punch forming—these tools consist of two separate punches mounted on a common base plate and a draw ring with two corresponding central openings
3. Preforming—preforming becomes necessary when the exceptionally deep draw is required in the part which requires to have two separate punches one for preform and other for final form

Variations in the material formability within the same alloy and temper, or materials from different suppliers play a significant role in the forming performance for the same part geometry and forming parameters. Intermediate annealing may be required to increase (restore) the formability of the material to minimize forming defects such as springback thin out, piercing, or tearing.

Lubrication is normally required between the draw ring surface and the part blank due to sliding action, which occurs at that interface. Removal of lubricant using a chemical degreasing process is required before the part is processed through the production line.

The amount of thickness reduction or thin out that may result from hydroforming a part is dependent on

FIGURE 8.56
Hydro forming deep draw production press. (From The Boeing Company.)

- Percentage of reduction—blank size, part size, and depth of draw
- Punch radii
- Material thickness
- Part geometry configuration—circular, square, rectangular, etc.

Figure 8.56 shows a deep draw hydro forming press used in production of mainly fuel dams, seal pans, and baffle types of aircraft parts (Figure 8.57), used generally in the outboard end of the wing box.

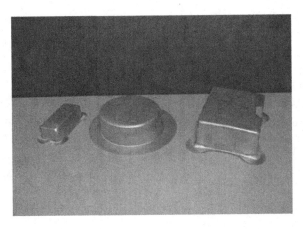

FIGURE 8.57
Smooth corner radii in three different hydro form parts. (From The Boeing Company.)

8.11 Match Die Press Forming

Match die press forming of sheet products basically follows the principle of mating die-punch method (Figure 8.58), which uses two piece die and punch mounted on a single action (Figure 8.58a) or double action (Figure 8.58b) hydraulic press. This method combines the advantages of stretch forming and conventional press forming. When the punch pushes the blank into the die cavity, the blank stretches over the punch radius and sets the contour, and the press forming gives definition to sharply formed contours.

The basic tools used with the matched die metal working press are the punch and die. The punch is the convex "male" tool, which mates with the concave "female" die. Generally, the punch is the moving element. Since accurate alignment between the punch and die is usually required, it is a common practice to mount the tool halves permanently in a sub press or die set, which can be inserted in the press. Figure 8.59a shows an example of a double acting match die forming press being used to form sheet metal parts such as the so-called large "pots and pans." Figure 8.59b shows a closer view of die and punch setup for double action. Figure 8.60 shows an example of a part produced using this technique. Important considerations need to be accounted for in order to form successful parts, which avoid thin out, severe surface damage, cracking, or wrinkling. The primary items to consider include:

1. Die and punch design
2. Blank clamping pressure or hold-down feature
3. Blank thickness, alloy, and temper
4. Proper blank size and geometry
5. Lubrication

Preforming may become necessary in cases where parts include critical angles or complex geometry, or an exceptionally deep draw. The process may also need to gather excess material in a certain area in order to prevent part cracking in the final draw.

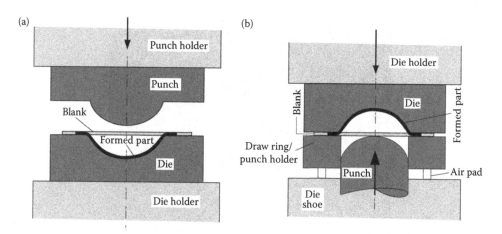

FIGURE 8.58
Fundamentals of match die forming, (a) single action and (b) double action.

(a)

(b)

FIGURE 8.59
Example of double action match die press. (From The Boeing Company.)

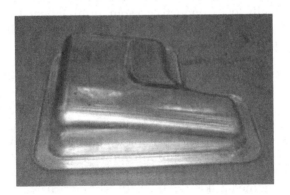

FIGURE 8.60
Formed aluminum part. (From The Boeing Company.)

8.12 Spin Forming

Spin forming is a conventional forming process of an axis-symmetric part over a rotating mandrel. Figure 8.61 shows a schematic representation of the spin forming process. A flat sheet metal blank is clamped against a forming block, which is rotated at high speed. The forming tool with rounded end or a roller applies localized contact pressure to the

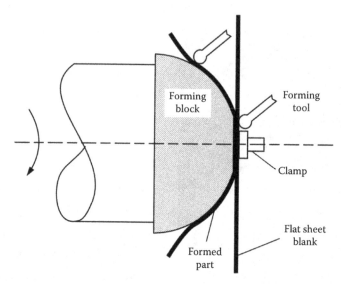

FIGURE 8.61
Schematic representation of spin forming.

workpiece and is gradually moved from the center part to the outer edge of the forming block. This localized contact pressure causes the sheet metal blank to move under plastic deformation and be wrapped over the forming block, thus conforming to the surface of the forming block.

The forming tool in a metal spinning operation is generally made from tool steel. The forming block in some cases may be made of aluminum/wood/Micarta. Setup costs for metal spinning are usually low. Conducting a spin forming operation is relatively expensive on a per unit basis. Deep drawing or match die process can usually be considered as an alternative manufacturing method to metal spin forming, for producing axis-symmetric small parts. In contrast, deep drawing often requires a high initial investment for tooling and set up, while the cost of production per part is low. For these reasons, deep drawing is more useful for mass production of parts. Spinning is recommended for producing smaller quantity of parts, unique, curved, and large parts. Operator skill is an important factor in metal spinning. CNC lathe or CNC spinning machinery is now being used for a high-quality consistent product. Spin forming is generally performed cold, but in some cases, the metal blank may be subjected to warming for hot spin forming. The spin forming process is also being used for manufacturing large parts with diameters as large as the size of an aircraft engine lip skin (Figure 8.62), having a diameter of more than 12 feet (3.66 m). *ASM Handbook* [11] has given an extensive overview of spin forming processes for more information.

8.13 Manufacturing Process Steps for Sheet Metal Forming

Figure 8.63 shows major process steps involved in manufacturing major aluminum alloy sheet metal parts for aerospace industry. Right size blanks are prepared from either clad or bare large aluminum sheets, either in annealed "O" or "T" temper condition according

FIGURE 8.62
Examples of CNC spin forming process to form an aerospace part. (From www.spincraft.net.)

to the part geometry and the forming processes involved to make the part. Depending on the engineering requirements of the parts, some parts are formed directly from the as-received blanks in "T" temper condition and some of the blanks are solution heat treated to "W" temper before forming in "W." Sometimes parts with very complex geometry are formed in "O" for rough forming. Rough formed parts in "O" condition are solution heat treated to "W" prior final forming in "W." Some of the formed parts require additional hand work to deal with springback and other issues associated with the geometry of the part. Excess material is finally trimmed off before the part is sent for cleaning prior to final heat treatment (precipitation hardening or aging) process to achieve the required strength, hardness, and even conductivity of the material.

After aging, the parts are then sent to the quality control inspection area for final measurements to check whether the parts meet the engineering drawing requirements. Sometimes minor hand work is needed to fix minor dimensional or other quality issues including surface defects. Penetrant inspection is often a required inspection method to detect any cracks or other surface defects that are not detected by visual inspection. After

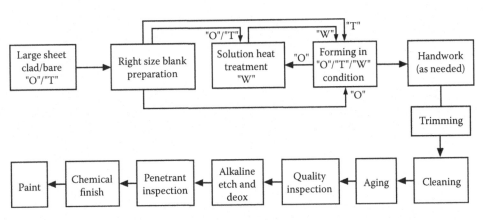

FIGURE 8.63
Manufacturing process flow diagram of aluminum sheet products.

the part is qualified in penetrant inspection, it is sent to the chemical finish line for alodine or anodize treatment according to the engineering requirements to protect the part from corrosion in service. Finally, the part is finished with primer and/or topcoat paint for additional protection from any environmental corrosion.

References

1. Saha, P.K., Cold forming technology vs. aluminum alloy development, *Aluminium Surface Science and Technology, VI International Symposium Presentation,* Sorrento, Italy, May, 2012.
2. Dieter, G.E., *Mechanical Metallurgy,* International Student Edition, McGraw-Hill Kogakusha, Ltd., Tokyo, 1961.
3. Sangdahl, E.L., Aul, E.L., and Sachs, G., Experimental stress analysis, *Proceedings of the Society for Experimental Stress Analysis,* 6(1), 1, 1948.
4. Ghosh, A.K. and Hecker, S.S., Failure in thin sheets stretched over rigid punches, *Metallurgical Transactions A,* 6A, 1065–1074, 1975.
5. Hecker, S.S., A cup test for assessing stretchability, *Metals Engineering Quarterly,* 14, 30–36, 1974.
6. Keeler, S.P. and Backhofen, W.A., Plastic instability and fracture in sheet stretched over rigid punches, *ASM Transactions Quarterly,* 56(11), 25–48, 1963.
7. Keeler, S.P., Determination of forming limits in automotive stampings, SAE Paper No. 650535, 1965.
8. Goodwin, G.M., Application of strain analysis to sheet metal forming in the press shop, SAE Paper No. 680093, 1968.
9. Keeler, S.P., Circular grid system—A valuable aid for evaluating sheet metal formability, SAE Paper No. 680092, 1968.
10. Ozturk, F. et al., Grid marking and measurement methods for sheet metal formability, In *5th International Conference and Exhibition on Design and Production of MACHINES and DIES/MOLDS,* Aydin, Turkey, pp. 1–10, June, 2009.
11. Semiatin, S.L., *ASM Handbook, Volume 14B Metal Working: Sheet Forming,* ASM International, Ohio, 2006.

9

Cold Forming of Plate

9.1 Introduction

The major applications of aluminum plate products, classified as having metal thickness greater than 0.25 inch (6.35 mm), are involved mainly in the manufacturing of aircraft wings and stabilizers. As discussed in Section 6.3, one of the major metal products of the wing structure is the high strength aluminum alloy rolled plate, which is used mainly for wing and stabilizer skins. The wing box normally consists of multiple skin panels integrated together creating a complex aerodynamic shape of the wing. Manufacturing of wing or stabilizer skins from plate products involves various process steps as shown in Figure 9.1.

The major process steps required to form the wing/stabilizer skin to the aerodynamic shape are as follows:

1. Procure rolled and pre-machined wedge-shaped plate, thicker at the inboard end and thinner at the outboard end (Figure 9.2a) from the primary aluminum manufacturer.

2. Perform in-house machining operations to produce the final shape of the wing skin either upper or lower according to the final geometry (Figure 9.2b). OML is the outer mold line, the exterior surface of the wing skin. Similarly, IML is the inner mold line, the interior surface of the skin that incorporates lots of machined steps to produce varying thicknesses and multiple weight reduction pockets.

To create the aerodynamic shape of the wing structure, the OML surface of the wing skins is provided to the required compound curvature. Various metal forming technologies are available to provide the compound curvature on a large skin panel to satisfy engineering requirements of the wing skin. This chapter provides an overview of major cold forming processes as outlined in Figure 9.3, to produce plate product components mainly wing/stabilizer skin of an aircraft.

As discussed in Chapter 8, brake forming (sometimes referred to as bump forming) process is widely followed to form various sheet metal parts from small clips and brackets to large fuselage skin panel of an aircraft. The same process is also used to form aluminum plate products to provide compound curvature as needed in the formed plate using the fundamentals of plate bending technology. Complex curvature cannot be formed with the brake forming process alone. Restrictions are associated with the overall size to fit with press working envelope, material thickness, and material specifications.

Shot peen forming is derived from the fundamentals of a shot peening (SP) process that was originally developed for compression peening, to provide improved fatigue and stress

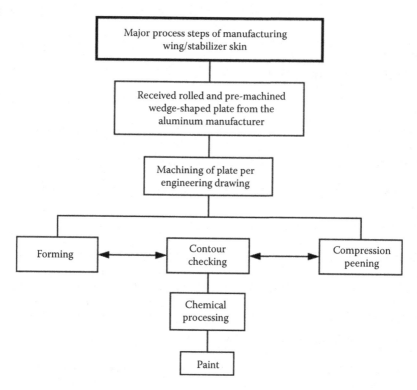

FIGURE 9.1
Major process steps for manufacturing of plate products for aircraft.

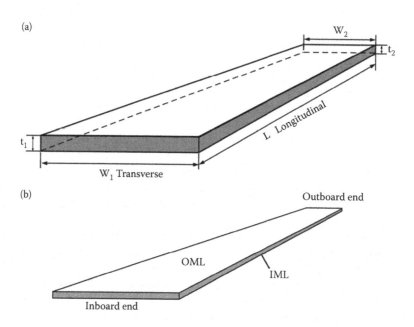

FIGURE 9.2
Schematic of wing skin material geometry. (a) As received and (b) after machining.

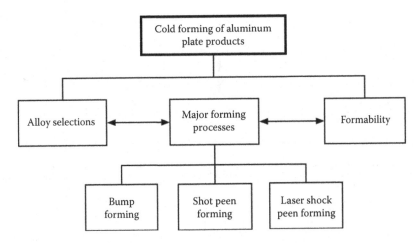

FIGURE 9.3
Major cold forming processes of plate products.

corrosion resistance by imparting a compressive residual stress to a certain depth from the surface of the material. The residual stresses induced in the material resulting from the shot peen forming process causes the original product geometry to transform to a new geometry.

Laser shock peen forming acts in similar fashion to the shot peen forming process but uses pulsed laser energy to produce a shock wave, to impart the compressive residual stress instead of tiny steel shot. The laser shock peen forming process can impart a much deeper compressive residual stress into the metal, thus allowing thicker material to be formed.

9.2 Bump Forming

As discussed in Section 8.4.1, bump forming technology is widely used in the aerospace industry to produce fuselage skin of constant ROC for commercial airplanes. Bump forming technology is equally useful to maintain and correct certain curvature on a wing skin plate product.

Bump forming technology followed the principles of three-point bending. Figure 9.4 shows the principles of the three-point bending process. It is called three-point bending since the metal plate is in contact at three tool point locations. Tool point locations 1 and 2 are the support tool contact locations and location 3 is the applied punch location. Applying the load at location 3 allows the part to form to the desired curvature after bending (Figure 9.4a). The material at the top surface of the plate at location 3 is under compression during deformation while the metal at the bottom surface is under tension. To achieve a specific final contour, the metal will require over bending, accounting for the spring back effect (Figure 9.4b). Moving the span supports/die width, w closer together allows the work piece to be formed to a tighter ROC. Moving the span supports/die width, w further apart allows the work piece to be formed to a larger ROC.

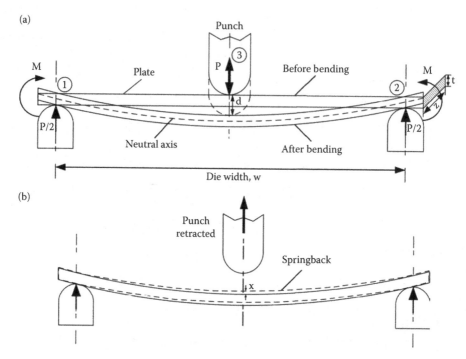

FIGURE 9.4
Three point bending principle. (a) Before and after bending and (b) effect of springback.

Typically in wing skin forming, the curvature that is designed to be formed with this process is primarily in the transverse or chordal direction. During forming, the bend ROC depends on

1. Tool geometry
2. Tool point location
3. Material alloy and temper
4. Part thickness

The bending or flexural stress is calculated by using the standard bending stress equation [1]

$$\sigma = \frac{My}{I} \tag{9.1}$$

where M is the bending moment about the neutral axis, y the perpendicular distance to the neutral axis, and I is the area moment of inertia about the neutral axis.

In this example, the bending moment, M, is given due to the applied load P at tool point location 3 by

$$M = \frac{Pw}{4} \tag{9.2}$$

FIGURE 9.5
Bump forming of wing skin. (From The Boeing Company.)

where w is the support span. The moment of inertia for a rectangular cross section is given by

$$I = \frac{zt^3}{12} \tag{9.3}$$

where z is the width of the rectangular cross section and t is the thickness of the plate or sheet. The perpendicular distance to the neutral axis for the rectangular section would be t/2. Substituting the M, I, and y in Equation 9.1, the bending stress or the flexural stress

$$\sigma = \frac{3Pw}{2zt^2} \tag{9.4}$$

Bump forming is typically used in forming wing skin when shot peen forming is not capable of producing the required ROC on a thicker plate material to a certain location. Figure 9.5 shows an example of bump forming process used to form a specific ROC to a particular span location on a wing skin.

The bump forming operator needs to take special care when forming not to leave any tool marks or dents on the part surface. Marks or dents may occur when using the wrong shape tools or when the distance between the tool point locations is insufficient. It requires the skills of a trained operator, yet is a relatively inexpensive (low tooling cost) approach to form certain sections of the wing skin.

9.3 Fundamentals of SP Process

Starting in the 1920s, the concept of SP technology began being reported in the work as a research paper [2]. General Motors in the late 1920s was involved with blast cleaning of various automotive components to increase fatigue life. A research laboratory of General Motors then continued working on an SP technique that contributed to the understanding and success of the SP process for various applications. Perhaps, one of the greatest demonstrations that highlighted the benefit of SP was noted during World War II. SP started with hand-held equipment using a manually applied technique. Then, a few years later, it was replaced by the mechanized methods. For the aerospace industry, more mechanized approach became the preferred technique. In the 1930s, the chilled iron grit was introduced as a peening media. Then, the industry moved to spherical material using chilled iron shot which was less costly to manufacture as the grit had been manufactured by crushing the shot in the first instance. In the 1950s, cast steel shot was introduced as it was recognized that chilled iron shot shattered too quickly. Cast steel shot became available in many sizes, and it is still the most common SP material available. Starting in the late 1940s, an SP technique was introduced to the forming of aerospace components.

SP is a mechanical cold metal working process in which the spherical media called shot impact the metal surface at high velocity to create small indentations as shown schematically in Figure 9.6a. Upon impact, the material just below the indentation undergoes local plastic deformation. The remaining material tends to push against the plastically deformed zone to restore the shape, resulting in compressive stress as shown in the sectional view AA (Figure 9.6b). The geometry of indentation is dependent on the following major factors:

- Size and geometry of the shot
- Speed of the shot, V
- Angle of impact or attack, θ
- Properties of workpiece material

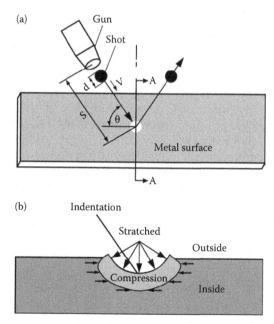

FIGURE 9.6
Principle of SP process. (a) Indentation and (b) section AA.

SP technology was brought into aerospace manufacturing mainly for two different applications:

1. Surface treatment of various structural components
2. Forming large panels; like metal wing skins, to a specific aerodynamic shape

The SP process provides major benefits of surface treatment for aircraft components:

1. Enhances fatigue life (e.g., provides resistance to corrosion and wear enhancement by creating a layer of compressive stress)
2. Relieves induced stress from previous operations such as machining
3. Enhances surface hardness
4. Creates a pleasing isotropic surface texture and also increases paint adhesion

Since SP improves the fatigue strength or working life, this process has gained wide application on aerospace components including aircraft wing skins, stringers, chords and web, landing gear beams, and other components subject to fatigue loads.

The amount of improvement to fatigue strength of the metal part geometry is determined to a large extent by the shot material, size, shape and hardness, as well as the velocity and impact angle. The selection of appropriate size, type of media, and the intensity is determined based on the forming part geometry, material, temper, and thickness. SP media selection is determined by the engineering requirements especially for aerospace applications where the manufacturing drawings or the specifications will indicate the media to be used for specific needs. In order to effectively control the SP process, there should be control on media, intensity, and angle of impingement, coverage, and equipment.

In SP, different peen media are commonly used according to the need of applications:

- Cast steel shot
- Conditioned cut wire shot
- Glass shot
- Ball bearings

Each media has its own advantages and disadvantages. But very common media is the cast steel shot since cast steel shot provides longer life to achieve highest intensity and is easiest to maintain a spherical particle. Peening media should be uniform in size and essentially spherical in shape with no sharp or broken particles. Figure 9.7a shows examples of uniform geometry acceptable media, and Figure 9.7b shows irregular geometry unacceptable media. Broken or sharp-edged media can potentially damage the part surface. Figure 9.8 shows the micrographs of typical surfaces generated by two different types of media. Figure 9.8a shows the micrograph of typical surface generated by uniform geometry media and Figure 9.8b shows the surface created by the poor quality irregular geometry media.

Intensity of shot during SP process is the measure of the energy of the stream of shot. The compressive stress developed on the metal part surface is directly related to the energy of the shot stream. Intensity can be varied mainly with the mass of the shot/media and the velocity of the shot stream. Other variables including angle of impact or attack and the type of media are also considered. Shot intensity is measured by use of an Almen strip.

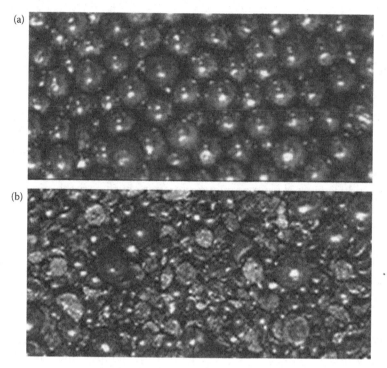

FIGURE 9.7
Shot peen cast steel media. (a) High-quality media (uniform geometry) and (b) poor-quality media (irregular geometry). (From Metal Improvement Company.)

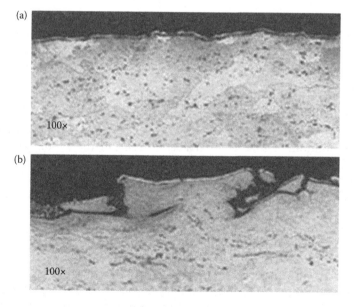

FIGURE 9.8
Surface generated from two different types of media. (a) Typical surface from uniform geometry media and (b) damaged surface from irregular geometry media. (From Metal Improvement Company.)

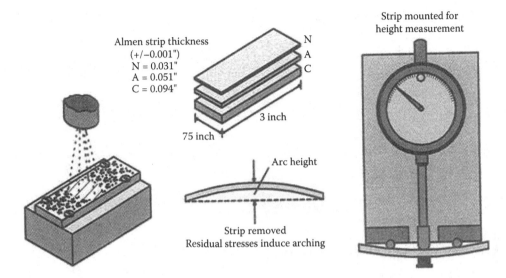

FIGURE 9.9
Almen strip. (From Metal Improvement Company.)

SP was originally transformed from a laboratory technique into a production tool by John Almen during the 1930s when he was working for the Buick Motor Division of General Motors Corporation. He found that the side of the sheet metal that was exposed during shot blasting began to bend and stretch. John Almen also created the Almen Strip, a thin strip of SAE 1070 spring steel as shown in Figure 9.9 used to measure the compressive stresses in the strip created by the SP operation. The residual compressive stress from the peening causes the Almen strip to bend or arc convexly toward the peened side as shown in Figure 9.9.

There are three Almen strip designations that are used depending on the peening application [3,4]:

1. "N" strip: thickness = 0.031 inch (0.79 mm)
2. "A" strip: thickness = 0.051 inch (1.29 mm)
3. "C" strip: thickness = 0.094 inch (2.39 mm)

More aggressive SP utilizes thicker Almen strips.

The intensity of the shot is determined by measuring the arc height of the strip before and after peening and determining the change in the strip's arc height. For the arc height measurement to be valid, the strip must be peened to saturation or beyond and at 100% coverage. A strip that is typically used during compression peening is the "A" strip. If the change in the arc height was 0.006 inch, the intensity would be designated as 0.006 A intensity. If a "C" strip was used, then it would be 0.006 C. The Almen strip is placed in an Almen strip holding block (SAE J442) and the block is located in the area to be shot peened that is representative of the part. In other words, Almen strips must be located in the area or areas where the required intensity can be verified based against the parts configuration and the shot peen equipments configuration. Typically, multiple Almen strips and Almen strip holding blocks are used when determining whether the shot peen parameters will satisfy the engineering requirements. An Almen strip shall not be reused after peening.

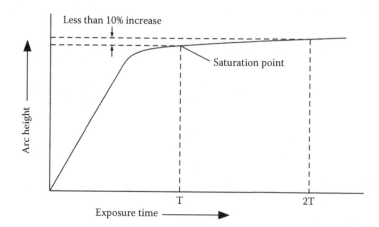

FIGURE 9.10
Saturation curve. (From Metal Improvement Company.)

Initial verification of a process development requires the establishment of a saturation curve shown schematically in Figure 9.10. Saturation is defined as the earliest point on the curve where doubling the exposure time produces no more than a 10% increase in arc height. The process reaches saturation at time equals to T. Saturation curve for any product by SP is developed from a series of Almen strip measurements for a particular machine setup.

Coverage is the measure of original surface area that has been regenerated by SP dimples. Complete coverage of a shot peened surface is the qualifying factor of an SP process. Coverage should never be less than 100%. Figure 9.11 shows some photographs of both complete and incomplete peening coverage.

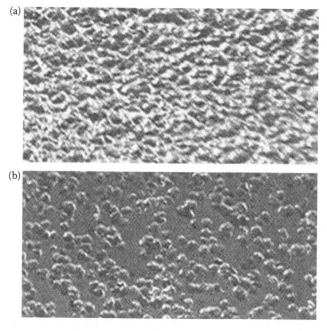

FIGURE 9.11
SP coverage. (a) Complete and (b) incomplete. (From Metal Improvement Company.)

Controlled automated peening machinery assures process repeatability, which is very important to control the percentage coverage and other process variables including shot velocity, angle, distance, and feed rate of the part.

9.4 Shot Peen Forming

SP technology is widely used to form single piece or integrally stiffened wing skins [5,6]. No forming dies are required for this process, except small round steel shot impacting the surface of the metal work piece. Shot peen forming is performed by peening one or both sides of the part with varying intensities so that the part gets the right contoured shape. The use of SP technology to form large metal parts has been proven to be capable of providing compound, complex curvatures within the part.

The shot may be propelled against the metal work piece by means of

- Air nozzle
- Centrifugal wheel

SP using an air nozzle involves compressed air being used to propel shot media through a nozzle in the direction of the metal part to be treated. Air nozzle peening is generally selected over wheel type peening (which uses a rotating, centrifugal wheel to propel shot) when the direction of the media must be more accurately focused. Frequently, multi-nozzle setups are utilized to shot peen complex parts that have more surface area than one nozzle can cover. The velocity of the shot determines the energy of the shot stream and is adjusted by modifying air pressure and nozzle settings.

During peening, the surface of the workpiece is subjected to compressive stresses due to the indentations caused by the shot on the surface layer of the material. Since there is no loss of material during the process, the surface layer tends to expand/stretch under plastic deformation in all three directions X, Y, and Z (Figure 9.12a) from the volume constancy relation. As shown along section AA, the width, W, of the plate or sheet metal expanded/stretched to width, W′, of the workpiece (Figure 9.12b). Since the layer of material below the peened surface layer remains undeformed, the surface stretching causes a stress differential between peened surfaces to the unpeened surface of the plate/sheet to develop a curvature, R, as shown in Figure 9.12c.

The amount of deformation caused by the shot peen forming process, as explained in Figure 9.6, to generate an ROC R on a plate product is mainly dependent on

1. Energy of shots ($1/2 \, mV^2$), where $m = 2/3\pi d^3\rho$, ρ is the density of shot material
2. Distance of shot gun from the metal surface, S
3. Angle of attack, θ
4. Material properties of the work piece (σ, ε, $\dot{\varepsilon}$)

Figure 9.13 shows a schematic representation of peen forming of a compound convex curvature wing skin panel before (Figure 9.13a) and after forming (Figure 9.13b). Figure 9.13b shows the formed curvature along the chordal and span directions of the wing.

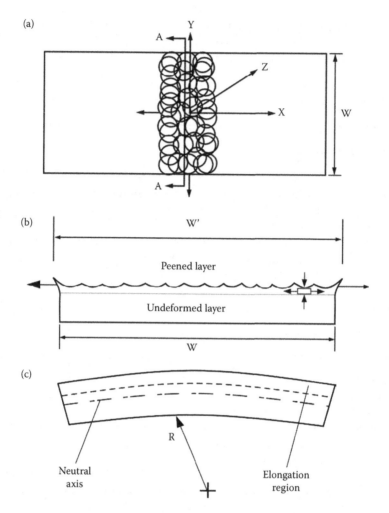

FIGURE 9.12
Shot peen forming principle. (a) Shot peened area, (b) section AA, and (c) formed ROC, R.

Wing skin panels are normally fed vertically hanging on its longitudinal axis (Figure 9.14a) into the shot blasting zone of the machine using a special type of moving system, having the flexibility to pass a wide variety of wing components including spar chords, stringers, spar webs, etc. The machines may have large banks of centrifugal wheels or multiple nozzles that spray shot to produce the peening coverage and intensity for the desired part geometry. In manufacturing operation, the wing skin is first processed in the span peen forming machine. After completion of the span peening, the skin is then passed through the chordal peen forming machine (9.14b).

As shown in the manufacturing process steps in Figure 9.1, after completion of both span and chordal forming, the wing skin is placed on the check fixture as shown in Figure 9.15 to check the spanwise and chordal curvature of the skin panel. Certain procedures and standards are followed to inspect the skin on the check fixture. After the skin panel is passed in the check fixture inspection, the part is transported to the sanding machine to sand those areas where the surface is roughened more than the standard roughness value due to SP process. The sanding process brings the overall roughness requirement of the

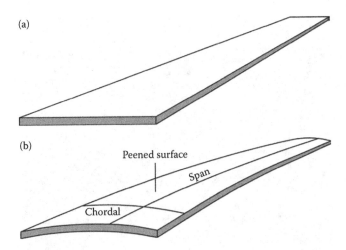

FIGURE 9.13
Schematic of compound convex curvatures on the peened panel surface. (a) Flat machined skin before forming and (b) curved skin after peen forming.

skin surface into compliance before compression peening of the entire skin surfaces to further improve fatigue and corrosion cracking resistance. In the compression peening process, the wheels or nozzles on both side of the machine are utilized to complete the process. Final contour check is performed after completion of compression peening before the skin is being delivered to the tank line for chemical processing, followed by corrosion inhibitor primer application and painting. Regular process control checks are made of the short intensity in both span and chordal machines. Almen strips are used as a control check. Certain arc height at a specified wheel or nozzle controls and shot flow rates in a certain time must be met for a specific part number.

To minimize or eliminate trial-and-error testing, computer modeling work has been developed to model SP as well as shot peen forming process. The computer model is used to determine the peening intensities across the wing skin and the initial size of the flat skin. To support the computer modeling, the relationships among the thickness of the plate material, the stress induced into the material, and the intensity of the SP were developed [7]. Numerical model of wing skin peen forming process was also developed [8]. The model is used to predict the initial size of the flat skin given an arbitrary aerodynamic contour requirement.

9.5 Fundamentals of Laser Shock Peening

Laser shock peening (LSP) is another surface treatment technique that uses a laser pulse to improve fatigue performance of metallic components. The application of a high-energy laser pulse generating shock waves for plastic deformation of metallic materials was first recognized and explored in the 1960s [9]. Conventional SP processes have existed in industry for over six decades. The LSP process is envisaged as a substitute for the conventional SP process. LSP process was initially developed for investigating fastener hole applications during 1968–1981 at Battelle Columbus Laboratory. Since 1986,

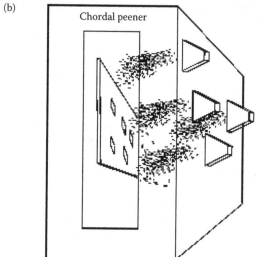

FIGURE 9.14
Type of shot peen forming. (a) Span peening and (b) chordal peening. (From The Boeing Company.)

more systematic research on LSP applications has been carried out outside the United States including France, China, and Japan [10,11]. Since the development of LSP, a number of patents [12,13] have been issued addressing its strong commercialization interest. After LSP was invented in the early 1960s, the ensuing studies mainly focused on the basic process development, understanding mechanisms, and the use of high laser power density to achieve high pulse pressures [14] and development of physical models to characterize LSP processes [10]. A lot of research work on LSP has been done on treating aluminum alloys [15–18].

The demand for industrial laser systems drove several of the laser shock peening researchers from Battelle to start LSP Technologies, Inc. in 1995 as the first commercial supplier of laser peening equipment [19]. In the late 1990s, Metal Improvement Company (now part of Curtis Wright Surface Technologies) partnered with Lawrence Livermore

FIGURE 9.15
Shot peen formed wing skin on a check fixture. (From The Boeing Company.)

National Laboratory to begin developing laser peening capabilities. The growth of industrial suppliers and commercial proof of laser peening technology via published literature convinced many companies to adopt laser peening technology for various applications. Some of the companies that have adopted laser peening included GE, Rolls-Royce, Boeing, Pratt & Whitney, and more.

In the LSP process, a compressive residual stress is induced beneath the treated metal surface by a high magnitude mechanical shock wave that is created by a high-energy laser pulse. The principle of the laser shock peening process is shown by the schematic diagram in Figure 9.16. The technique involves producing mechanical shock waves through the expansion of generated confined plasma. The metal surface area required to be shock peened is coated by an overlaying sacrificial material, commonly a black paint or opaque tape. Overlaying this sacrificial layer may be a tamping layer of de-ionized water, which is transparent to laser radiation but acts as a confinement medium for the generated plasma. The high-power laser vaporizes the sacrificial ablative material creating plasma.

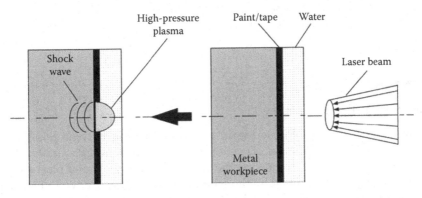

FIGURE 9.16
Laser shock peening principle.

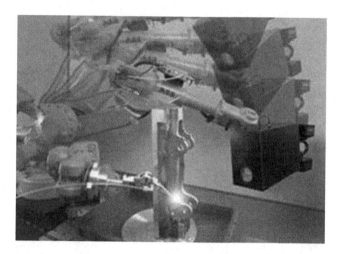

FIGURE 9.17
Laser fired at the metal surface. (From Metal Improvement Company.)

The plasma rapidly expands but is trapped between the workpiece and the tamping layer resulting in an extremely high pressure. This leads to a mechanical shock wave that is driven into the component. The shock wave pressure can be large enough to overcome the yield strength of the work piece material causing plastic deformation to the material to a depth where the shock wave pressure is lower than the yield strength. The undeformed material attempts to restore the original shape of the surface, which in turn creates in-plane compressive stress fields in the near surface region of the metal work piece.

When the laser is fired at the metal surface as shown in Figure 9.17, it generates the pressure pulses of significant magnitude that send shock waves through the part. Multiple firings of the laser in a predefined surface pattern will impart a layer of residual compressive stress on the surface.

The primary benefit of laser peening is a deep compressive layer with minimal cold working, which increases the component's resistance to failure mechanisms such as fatigue, fretting fatigue, and stress corrosion. A secondary benefit is that thermal relaxation of the residual stresses of a laser peened surface is less than a shot peened surface due to the reduced cold work that is involved with the laser peening process [3]. Figure 9.18 shows the improvement of laser shock peening over conventional SP process. Fatigue life comparison was done on 6061 –T6 aluminum SAE key hole specimens [3,4].

9.6 Laser Shock Peen Forming

Laser peen forming is another plate forming process that has become popular in recent years. It uses a high-power laser in place of tiny shot to create a high amplitude stress or shock wave on the metal surface causing the surface layer to yield and plastically deform. The forming principle is explained by the schematic diagram as shown in Figure 9.19.

Deeper compressive stresses achieved by laser peening allow the formation of thick plate sections and tight curvatures. Laser peening was able to form a 9-inch radius (230 mm) in

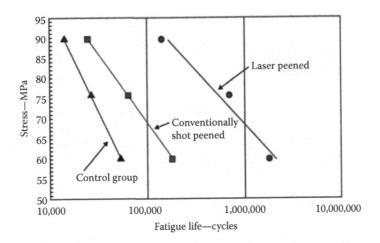

FIGURE 9.18
Fatigue comparisons between SP and laser shock peening. (From Metal Improvement Company.)

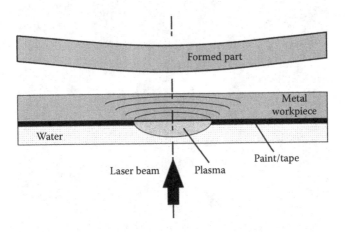

FIGURE 9.19
Laser shock peen forming principle.

a 0.4-inch (10 mm) thick piece of Aluminum 7050 as shown in Figure 9.20. Compared to other methods, laser peening can provide a cost-effective alternative for the formation of complex aerodynamic shapes.

Laser peen forming of large wing skin components requires preparing the parts for laser peening. This includes applying the ablative layer to the surface to be laser peened along with proper indexing of the part to the laser peen system to ensure the part is oriented to the laser. Ensuring the laser will impact the surface where it is programmed is critical to the process. Also, monitoring and recording the energy that is imparted into the part is critical to the ensuring quality of the process. If the ablative layer were to fail and the laser light were to strike the part, there may be cause for part rejection. Typically, the ablative layer needs to be visually inspected prior to removal from the work piece. This allows the quality inspector to verify spot locations and ensure there was no failure of the ablative layer. Figure 9.21 shows a large wing skin panel part entering into a laser shock peen forming machine.

FIGURE 9.20
Laser shock peen formed part. (From Metal Improvement Company.)

FIGURE 9.21
Wing skin entering into laser shock peen forming machine. (From The Boeing Company.)

References

1. Timoshenko, S. and Young, D.H., *Elements of Strength of Materials*, 5th Edition, Van Nostrand Reinhold Co., New York, 1968.
2. Marsh, K.J., *Shot Peening: Techniques and Applications*, Engineering Materials Advisory Services Ltd., England, UK, 1993.
3. http://www.metalimprovement.com
4. *Shot Peening Applications*, 9th Edition, Metal Improvement Company, NJ, November, 2005.
5. Yamada, T., Takahashi, T., Ikeda, M., Sugimoto, S., and Ohta, T., Development of shot peening for wing integral skin for continental business jets, *MHI Technical Review*, 39(2), 57–61, 2002.
6. Ramati, S., Levasseur, G., and Kennerknecht, S., Single piece wing skin utilization via advanced peen forming technologies, *International Shot Peening Conference*, Warsaw, Poland (ICSP-7), pp. 1–17, September, 1999.

7. VanLuchene, R.D., Johnson, J., and Crapenter, R.G., Induced stress relationships for wing skin forming by shot peening, *Journal of Materials Engineering and Performance*, 4(3), 283–290, 1995.

8. VanLuchene, R.D and Cramer, E.J., Numerical modeling of a wing skin peen forming process, *Journal of Materials Engineering and Performance*, 5(6), 753–760, 1996.

9. Ding, K., Ye, L., *Laser Shock Peening*, Woodhead Publishing Limited, Cambridge, England, 2006.

10. Peyre, P. et al., Laser shock processing of materials, physical processes involved and examples of applications, *Journal of laser Applications*, 8, 135–141, 1996.

11. Chaojun, Y., Yongkang, Z., Jianzhong, Z., Fang, Z., and Aixin, F., Laser shock wave and its applications, *3rd International Symposium on Advanced Optical Manufacturing Technologies: Advanced Optical Manufacturing Technologies*, 6722(68), Chengdu, China, 2007.

12. Mallozi, P.J. and Fairand, B.P. *Altering material properties*, US Patent 3,850,698, November 26, 1974.

13. Clauer, A.H. Fairand, B.P., Ford, S.C., and Walters, C.T. Laser shock processing, in shock waves and high-strain-rate phenomena in metals, US Patent 4401477, August 30, 1983.

14. Fairand, B.P., Wilcox, B.A, Gallagher, W.J., and Williams, D.N. Laser shock-induced microstructural and mechanical property chances in 7075 aluminum, *Journal of Applied Physics*, 43(9), 3893–3895, 1972.

15. Tucker, T.R. and Clauer, A.H., Laser Processing of Materials MCIC Report, MCIC-83–48, Metals and Ceramics Information Center, November, 1983.

16. Clauer, A.H., Walter, C.T., and Ford, S.C., The effects of shock processing on the fatigue properties of 2024-T3 aluminum, *Lasers in Materials Processing*, ASM International, Ohio, pp. 1–26, 1983.

17. Ford, S.C. et al., Investigation of Laser Shock Processing, Report AFWAL-TR-80-3001, August 2, 1980.

18. Clauer, A.H., Laser shock processing increases fatigue life of metal parts, *Materials and Processing Report*, 6(6), 1–4, 1991.

19. www.lsptechnologies.com

10

Cold Forming of Extrusion

10.1 Introduction

Light, medium, and heavy extrusion shapes are widely used as structural components in various locations within an aircraft. Extrusions are used in as-produced straight configurations as well as in subsequently formed shapes with certain contour geometry depending on the application. The major application of formed extrusion includes wing stiffeners, channel vents, spar chords, fuselage frames, stringers, body chords, and side of body chords. Significant numbers of aircraft structural components are produced mainly from the 2xxx and 7xxx series wrought aluminum alloy extrusion products. To introduce new aluminum alloy extrusions to the aerospace industry, it is very important to understand the formability of the alloys in extrusion products before making a go-ahead decision to produce airplane parts from them. Basic formability tests are generally performed during the new alloy and extrusion product development in coordination with production manufacturing experts.

This chapter introduces an overview of major conventional cold (ambient temperature) forming processes of extrusion products as outlined in Figure 10.1 for producing a wide variety of extrusion parts from small-to-large component of an aircraft.

10.2 Stretch Forming

As discussed in Chapter 8 about stretch forming of sheet metal parts of an aircraft, the same fundamentals of stretch forming technology are also applied for extrusion shapes using the extrusion jaws to form a contour over a stretch form block. The extrusion is stretched just beyond the yield point to produce plastic deformation into the metal to conform it to the geometry of the forming block extending outward. The position of the jaws is controlled to maintain the required minimum tensile stress during stretching. Figure 10.2 shows the schematic representation of the fundamental components of stretch forming of extrusion shapes. Depending on the size and shape of the extrusion, the right size stretch-forming machine [1–4] is chosen to form the part. The fundamentals of stretch forming of an extrusion involve:

1. Grip extrusion at each end by a jaw and apply initial stretch by pulling the jaw from each end and engage the extrusion as properly aligned with the block (Figure 10.2a).
2. The ram is gradually moved upward and the grip jaws are moved radially outward and proceed to stretch the extrusion beyond the yield point.

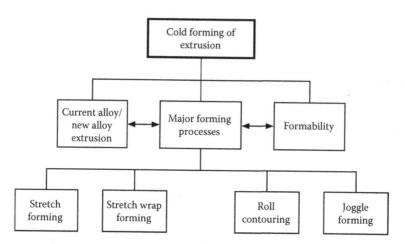

FIGURE 10.1
Major cold forming processes of extrusion.

3. Increase tension while the part is stretched around the block contour geometry (Figure 10.2b) for final setting.

To maintain very tight dimensional tolerances on the part geometry, the forming parameters including jaw tension, ram pressure, and the forming block geometries are very

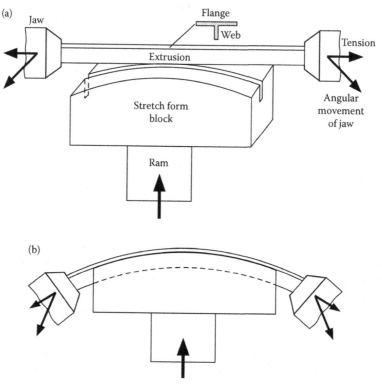

FIGURE 10.2
Schematic of stretch forming of extrusion, (a) before forming and (b) after forming.

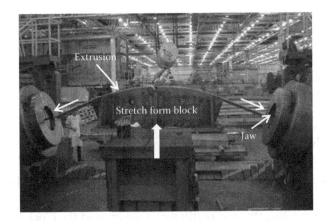

FIGURE 10.3
Example of stretch forming of large size extrusion. (From The Boeing Company.)

critical. Figure 10.3 shows an example of stretch forming of a short web and long flange "T"-shaped large size extrusion shape used for manufacturing the fuselage body frames of an aircraft. Another example of stretch forming of a very large-shaped extrusion to manufacture a side of body chord, which integrates wing to fuselage structure, is shown in Figure 10.4.

FIGURE 10.4
Example of stretch forming of very large extrusion parts for an aircraft. (a) Stretch forming and (b) extrusion after forming. (From The Boeing Company.)

10.3 Stretch Wrap Forming

The fundamentals of stretch wrap forming of extrusion, which is similar to the stretch wrap forming of sheet-shaped products, are shown in the schematic diagrams (Figure 10.5). Basic operations of stretch wrap forming of extrusion involve:

1. Grip extrusion at each end by the jaws and apply initial stretch by pulling the jaw from each end. Engage the extrusion with the block to check the alignment of extrusion with the block (Figure 10.5a).
2. Wrap extrusion around the stretch wrap block by applying angular movement of the jaws from the swing arm of the stretch wrap machine.
3. Increase tension while the part is still wrapped around the block (Figure 10.5b) for final setting of the part geometry.

Figure 10.6 shows an example of stretch wrapping of a "T"-shaped extrusion. For any given part geometry, the right size forming block, jaw inserts, and other supporting tooling are set on the stretch wrap machine before the first part is tried out. The initial stretch and final setting tension are initially set, based on experience with similar parts and adjusted during development prior to first production. The initial stretching load must be sufficient to avoid any distortion of the workpiece during forming and also to minimize springback after wrapping is complete. The pulling tonnage will be dependent on

- Cross section of the part geometry
- Contour to be formed
- Extrusion material and temper

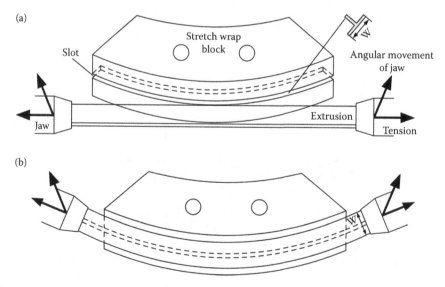

FIGURE 10.5
Schematic of stretch wrap forming of extrusion. (a) Before stretch wrap and (b) after stretch wrap.

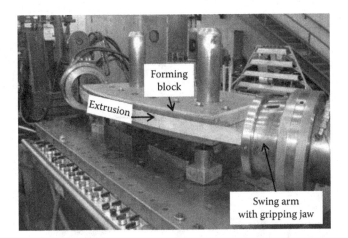

FIGURE 10.6
Example of extrusion stretch wrap forming. (From The Boeing Company.)

10.4 Roll Contouring

Roll contouring is an alternative to stretch forming and is generally used for low-quantity production of contoured parts, parts that are not subsequently machined extensively, parts with very gentle contour, or parts in which cross-sectional tolerances are not that critical. Parts required in large quantities or with complex contour geometry are generally stretched formed or stretch wrap formed as already discussed. Roll contouring equipment is sometimes used for altering angularity of the extruded or flat sheet section profiles (roll or brake formed sections) and correcting sometimes heat treat distortion remaining in the parts. In a roll contouring or bending process, a set of rolls is used to exert opposing compressive forces to create a bend or contour in a part. The roll contour equipment is generally designed to apply constant and compound curvature to form aircraft body stringers and frames made of extrusion and flat sheet section profiles.

Roll contouring machines generally employ the basic principle of bending the metal as the part enters into the driving rolls and bending occurs beyond its elastic limits upon contact with the forming rolls. There are two basic types of roll contouring equipment configurations:

1. Pyramid rolls (Figure 10.7)
2. Pinch roll (Figure 10.8)

In a pyramid roll configuration, there are three rolls. The fundamentals of pyramid roll contouring are shown in Figure 10.7a. The two lower rolls are aligned on their fixed centers at the same elevation. The third or upper roll called the bending roll is placed above and between the lower rolls and is generally adjustable up and down in vertical relation to the lower rolls. All three rolls are generally power-driven. A part is placed across the two lower rolls and as it passes between them it is bent down by the upper roll, midway between the two points of support. Gradual bending of the part resulting from the upper roll position determines the contour produced. The upper roll is never set very close to the lower rolls to pinch the part. The gap between either of the lower rolls and the upper roll is

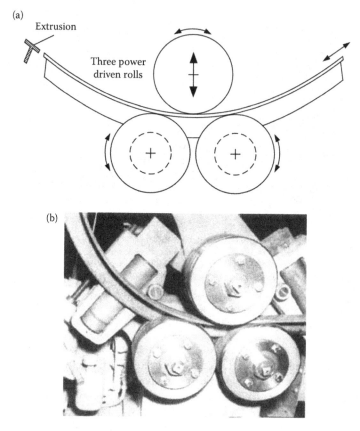

FIGURE 10.7

Pyramid roll contouring of extrusion. (a) Fundamentals of pyramid rolls and (b) extrusion contouring machine. (From Buffalo Machines, Inc.)

always more than the thickness of the forming part profile. Figure 10.7b shows an example of roll contouring of an extrusion.

Whereas, in a pinch roll configuration, the machine utilizes two opposed driving rolls, mounted vertically, plus one or two yoke or forming rolls are located on one or both sides of the pinch rolls. One drive roll (usually the upper one) is adjustable to obtain a pinching action and both yoke rolls are adjustable to vary the radius of curvature (ROC) of bend. Typical roll contouring production equipment uses "pinch" rolls to drive the part and restrain it from buckling during the forming process. The roll adjustment produces a variable degree of pressure or "pinch" on the part. Both pinch rolls are power-driven. Yoke side rolls are usually arranged within adjustment guides at varying angles of inclination toward the vertical centerline of the pinch rolls. Yoke rolls on either side of the pinch rolls are progressively raised to give contour to the part as it is rolled back and forth between the "pinch" rolls. Figure 10.8 shows schematics of the fundamentals of a roll contouring process using a "T" extrusion profile as an example. The "T" extrusion is firmly gripped between the main rolls and pinching pressure is applied. Bending or contouring of the part is produced when the yoke roll position is changed in relation to the opening between the pinch rolls. Contouring proceeds by rolling the part back and forth as the yoke rolls are progressively adjusted. Figure 10.8a shows an example of contouring of "T" extrusion keeping the leg down, whereas Figure 10.8b shows the contouring of same "T"

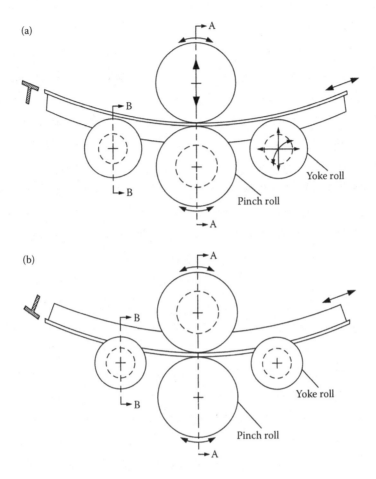

FIGURE 10.8
Schematic diagram of pinch roll contouring of extrusion. (a) Leg out "T" and (b) leg in "T."

extrusion with the leg up. Successful roll contouring of various extrusion shapes is very much dependent on the geometry of pinch and yoke rolls and their arrangements. The roll arrangements for contouring "T" extrusion are shown in Figure 10.9.

Since roll contouring is essentially a bending process, the estimates of machine capacity for various section shapes, sizes, and materials are based on analysis of the bending stresses. It is assumed that the neutral axis is near the center of the part and the metal inside of the neutral axis (toward the center of theoretical contouring radius) is subjected to compression while the metal outside of the neutral axis is stretched. When the section to be roll formed is not symmetrical, the neutral axis is assumed to move toward the heavier part of the section as shown in Figure 10.10.

Since the forming loads in roll contouring are highly concentrated by the pinch rolls, heavy sections may be contoured on smaller equipment than that used for stretch forming. Figure 10.11 illustrates the pinch roll contour forming process in operation. A high degree of operator skill is required to achieve maximum process capability of roll contouring equipment to form extrusions and sheet metal formed sections of varying geometry.

Roll contouring uses low-cost standard tooling and is therefore more economical than stretch forming, where there is a need of an individual stretch form block per part for

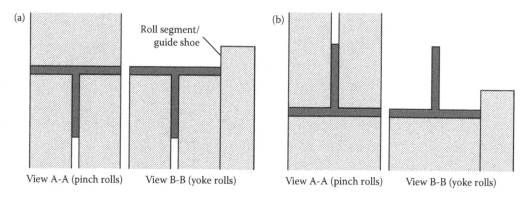

FIGURE 10.9
Roll arrangements for contouring "T" extrusion. (a) Leg out "T" and (b) leg in "T."

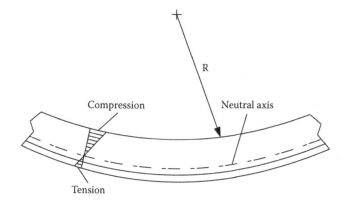

FIGURE 10.10
Distribution of bending stresses.

FIGURE 10.11
Example of an extrusion contouring process. (From The Boeing Company.)

FIGURE 10.12
Example of NC roll contouring of extrusion. (From The Boeing Company.)

producing small quantities of parts. In addition, stretch forming press would normally require very large floor space compared to the roll contouring machine.

Figure 10.12 shows a multiaxis NC roll contouring machine generally used to contour extrusion and roll formed body stringers. The equipment can operate in three planes of space. The upper pinch roll has an adjustable movement up and down. Each yoke roll has generally three movements (1) up and down, (2) in and out, and also (3) twist in and out.

10.5 Joggle Forming of Extrusion

The fundamentals of joggle forming related to sheet metal parts have been discussed in Section 8.9. The same fundamentals are also applied to the joggle forming of extrusion. A joggle is a very common feature to provide offset in structural shapes to accommodate doublers, straps, or adjoining members. The offset is placed in a direction normal to the plane of the original surface so that two parallel surfaces could be joined by a short transition. Offset areas are generally produced by joggling.

Based on the geometry of the extrusion and the type of joggle geometry, the basic principle of joggle tooling (a combination of blocks and punches) is shown schematically in Figure 10.13. Figure 10.14 shows an example of joggle forming of "T"-shaped extrusion where the punch comes down and pushes the flange against the moving block to form an offset joggle.

Joggle ratio is the key design factor in the joggle forming processes for making airplane parts. Different combinations of joggle length and joggle depth could be used to produce the same joggle ratio. Depth of joggle determines the amount of metal deformation in joggle forming, the higher the joggle depth, the higher the amount of deformation that may cause necking and cracking along the first joggle line. An example has been shown in Figure 10.15 with two joggle ratios 6:1 (Figure 10.15a) and 5:1 (Figure 10.15b). The extrusion with a joggle ratio of 5:1 has developed a crack in forming (Figure 10.15b). The material is deformed in both the flange and web of the extrusion.

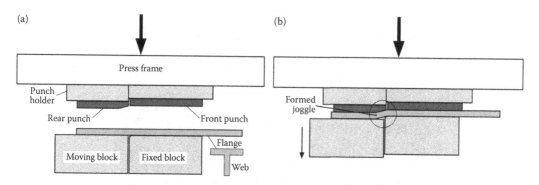

FIGURE 10.13
Basic principle of joggle forming of extrusion. (a) Press frame moving down and (b) joggle complete.

FIGURE 10.14
Example of joggle forming of "T" extrusion. (From The Boeing Company.)

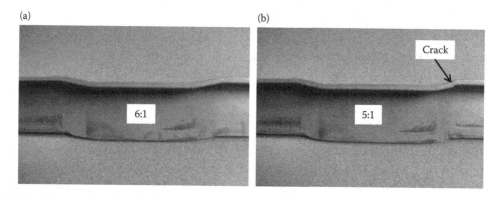

FIGURE 10.15
Depth of deformation for two different joggle ratio. (a) Joggle at 6:1 ratio without crack and (b) crack along the first joggle line at 5:1 joggle ratio.

Typical vertical type joggle presses for aircraft part applications are typically designed in the range of 100–250 t, capable of forming joggles with various extrusion geometries of various alloys and tempers with various cross sections. A joggle could be performed in cold and also at elevated working temperature, without changing the temper of the material, using heated tooling depending on the joggle ratio to be produced in the part.

References

1. www.cyrilbath.com
2. www.eriepress.com
3. www.stretchformingcorp.com
4. www.bmt-usa.com

11

Hot Forming of Flat Sheet, Plate, and Extrusion

11.1 Introduction

In the last three chapters, some of the major conventional cold forming technologies for manufacturing various components of airframe structures, mainly with aluminum alloys flat sheet, plate, and extrusion products were discussed. Manufacturing complex light-weight aerospace structural high strength metal components is quite challenging. Sheet metal parts of complex geometry and difficult-to-form alloys are sometimes made by hot forming processes using powered presses at elevated temperature. There are several hot forming processes available to produce heat-resistant very high strength-to-weight ratio aerospace components, mainly from titanium sheet, plate, and extrusion products. This chapter introduces an overview of major hot forming processes of flat sheet, plate, and extrusion products as illustrated in Figure 11.1. These processes are used for producing a wide variety of components utilized at various locations including engine aft fairing heat shields, tail cone muffler, auxiliary power unit exit, and side of body joint chords of an aircraft.

Hot forming processes including superplastic forming (SPF), either alone or in combination with diffusion bonding (DB), and hot die forming processes are commonly used to fabricate titanium sheet metal parts. The main objective is to reduce the cost and weight of part components, which enhances the performance of an aircraft. Performance of some parts made by SPF or SPF/DB process is sometimes a more significant factor, even though in some cases, there may be a slight increase in overall manufacturing cost.

11.2 Superplastic Forming of Flat Sheet

Superplasticity of metals is the capability at the proper temperature and strain-rate to deform crystalline solids in tension to unusually large plastic strains, often well exceeding 1000%. An SPF process [1–8] is a hot working process where the workpiece metal temperature is brought to a certain stage where the flow stress of the metal is sufficiently low to enable large plastic strain. Large plastic strain can create localized deformation that allows successful forming of a complex contoured part in a single press cycle. During the SPF process, the sheet metal blank is heated to the required SPF temperature within a sealed die. Generally argon gas pressure is then applied at a controlled rate (strain-rate), forcing the sheet material to slowly take the shape of the die geometry. During deformation, the flow stress of the material increases rapidly with increasing strain-rate.

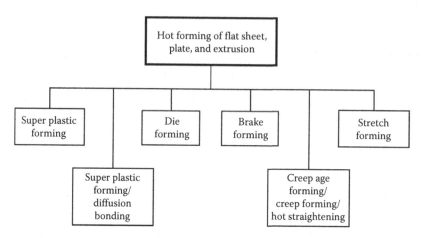

FIGURE 11.1
Major hot forming processes of sheet, plate, and extrusion products.

Sheet metal parts of complex shapes that are difficult to form at room temperature are mainly formed using an SPF process from the materials with superplastic properties. SPF process steps have been illustrated by the schematic diagram as shown in Figure 11.2. Typically, the die is placed in a press and heated to the required temperature. The flat sheet metal blank is placed on the lower half of the SPF die and the upper die is placed directly on the top of the bottom die (Figure 11.2a). The sheet metal and die are heated to a specified temperature (T) appropriate for the metal. There is an inlet (shown in the upper die), to purge atmospheric air, and input argon gas to apply pressure (p) on the sheet metal for a prescribed amount of time (t). There is an air vent in the lower die to remove the air pressure from the bottom die as the forming cycle starts. Once the set temperature for the forming

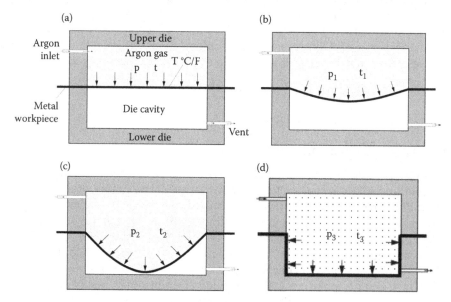

FIGURE 11.2
Schematic of superplastic forming process steps: (a), (b), (c), and (d).

metal is reached and controlled, the argon gas pressure (p_1) slowly pushes the workpiece into the die cavity for a given time (t_1) as shown in Figure 11.2b. The gas keeps pushing the workpiece for another set of pressure and time (p_2, t_2) to form further into the die cavity (Figure 11.2c). For the final set of gas pressure and applied time (p_3, t_3), the final part is formed by conforming the workpiece fully into the die cavity as shown in Figure 11.2d.

The major variables in SPF process include:

- Workpiece temperature, T
- Argon gas pressure, p
- Time cycle, t or strain-rate ($\dot{\varepsilon}$)

SPF is a process designed to form sheet metals that allow elongation of several hundred percentage. The materials developed for superplastic forming include the following:

1. Aluminum (2004, 2419, 5083, 7475)
2. Aluminum–lithium (2090, 2091, 8090)
3. Zinc–aluminum
4. Titanium (CP, 6Al-4V, 6Al-2Sn-4Zr-2Mo)
5. Inconel 718
6. Stainless steel (2205 series)

The major benefits of the SPF process over the conventional metal forming processes are as follows:

- Produces complex part in one operation
- Produces larger parts to eliminate assemblies or reduce weight
- Provides fine surface finish
- Does not induce spring back or residual stresses
- Enhances design freedom
- Minimizes the amount of scrap
- Reduces the need for machining
- Lowers overall material costs

SPF presses (Figure 11.3) are high-temperature presses with precision gas management systems for forming exotic metals with superplastic properties into complex geometries. The general features of the presses include:

- Down-acting and up-acting press frame (Figure 11.3a) constructions with optional lower platen shuttle (Figure 11.3b)
- Operating temperatures up to 1000°C (1832°F)
- Precise control of closing speeds, position set-points, gas pressure forming set-points, ramp rates, and real-time trending of all process parameters
- Metallic and ceramic platen options are available (Figure 11.4)
- Hot platen working area is fully enclosed by retractable, insulated heat shields

(a)   (b)

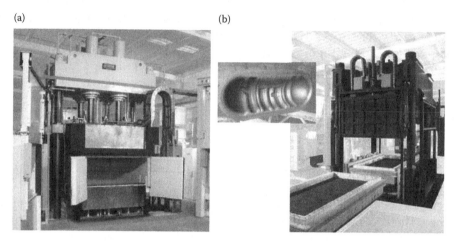

FIGURE 11.3
Examples of SPF presses. (a) Up- and down-acting press frame and (b) lower platen shuttle. (From Accudyne.)

Metallic platens as shown in Figure 11.4a are drilled for radiant, multi-zone heating elements and ground to precision flatness and parallelism specifications, and are offered in a number of different chromium–nickel high-temperature alloys. Ceramic platens as shown in Figure 11.4b are cast with multiple rows of heating elements near the working surface and are ground to precision flatness and parallelism specifications, and are offered in a number of reinforced ceramic compositions.

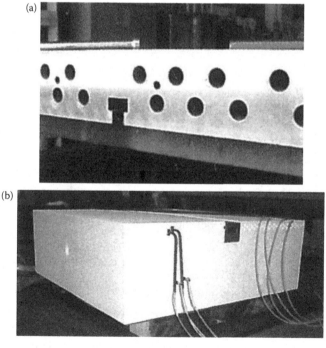

FIGURE 11.4
Platen segment. (a) Metallic (Supertherm®) and (b) ceramic. (From Accudyne.)

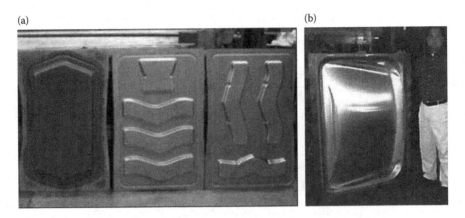

FIGURE 11.5
Examples of SPF parts: (a) Fine grain 6Al-4V titanium sheet metal part (Adapted from Hefti, L.D., *SAE International Journal of Aerospace*, 3(1), 2010-01-1834, 173–179, 2010) and (b) aluminum part. (From Accudyne.)

The main disadvantage of an SPF process is having a slower forming rate. Forming cycle time varies from a few minutes to hours depending on the geometry of the part being produced. The SPF process is mainly used for lower volume products of special need. The technology is very useful for fabricating complex parts from special metal like titanium especially for various aerospace applications. Figure 11.5 shows examples of a few SPF parts made from two different metals.

11.3 Diffusion Bonding

Diffusion bonding (DB) is a solid-state joining method of bonding a wide range of metals under controlled temperature and pressure in a controlled atmosphere. The major factors involved in the process are as follows:

- Time
- Pressure
- Temperature
- Heating method
- Protective atmosphere or vacuum

The process is carried out mainly in a vacuum or a protective atmosphere with heating being applied by radiant, induction, direct or indirect resistance heating. Once the desired elevated temperature is reached at the interface between two pre-machined metal surfaces, pressure can be applied uniaxially or isostatically. The DB process requires a good surface finish of the mating surfaces to have proper plastic yielding contributing to bonding. Typically surface finishes of better than 0.4 μm Ra are recommended, and in addition the surfaces should be as clean as practicable. Temperature is usually maintained between 0.5 and 0.08 times the melting temperature in °C or °F, and time can range from 1 to more than 60 minutes depending upon the materials being bonded, joint properties, and the remaining bonding parameters [9].

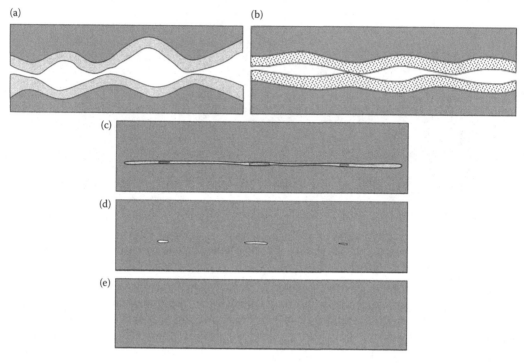

FIGURE 11.6
DB mechanism: (a), (b), (c), (d), and (e).

An example of sequence of DB mechanism illustrated in Figure 11.6 emphasizes the importance of the original surface finish [9] as follows:

1. Initial point contact—showing residual oxide contaminant layer (Figure 11.6a)
2. Yielding and creep—leading to reduced voids and thinner contaminant layer (Figure 11.6b)
3. Final yielding and creep—some voids remain with very thin contaminant layer (Figure 11.6c)
4. Continued vacancy diffusion—eliminates oxide layer, leaving few small voids (Figure 11.6d)
5. Bonding is complete (Figure 11.6e)

For DB, it is necessary for two clean and flat surfaces to come into atomic contact, with micro-asperities and surface layer contaminants being removed from the bonding faces during preparation for bonding. Various models have been developed to provide an understanding of the mechanism involved in forming a bond. First, it is considered that the applied load causes plastic deformation of surface asperities reducing interfacial voids. Bonding development continues by diffusion-controlled mechanisms including grain boundary diffusion and power-law creep.

This technology has been widely developed in the aerospace industry for meeting special needs of an aircraft part [5,6]. The process is used commercially for titanium and titanium alloys, especially 6Al-4V (usually called 6–4) for its superplastic properties at

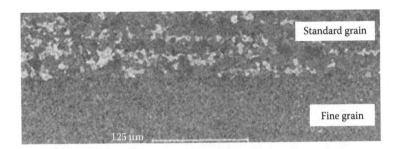

FIGURE 11.7
Micrograph of standard and fine grain materials joined in DB. (Adapted from Hefti, L.D., *SAE International Journal of Aerospace*, 3(1), 2010-01-1834, 173–179, 2010.)

elevated temperatures within defined strain-rate condition. The unique combination of temperature and pressure coincides with the conditions required for bonding and therefore two processes have been combined into one manufacturing operation either in sequence or together. Development continued in DB of similar or dissimilar materials, or of the same alloy with two different grain sizes. This is done to establish cost savings in manufacturing various aircraft components, which are exposed to hot air from the engine or auxiliary equipment and also engine components including engine blades.

The world's largest titanium manufacturer located in Russia yielded a fine grain version of 6–4 titanium with a grain size of about 1 μm in order to lower the SPF temperature to around 775°C (1425°F). The chemistry of the alloy was kept within the standard grain 6–4 alloy within the limits of the AMS-4911 specification. SPF at the lower temperature reduces the amount of alpha case (a brittle oxygen-rich layer) on the surface of the fine grain 6–4 titanium sheet metal workpiece. A chemical milling process is generally used to remove the alpha case from the aircraft parts produced by SPF and SPF/DB to avoid any risk of fatigue failure of the part in service. The lower the amount of alpha case produced, the lower the cost of chemical milling. Figure 11.7 shows an example of a micrograph of a diffusion bond between fine and standard grain 6Al-4V titanium flat sheets. A test was conducted bonding a standard grain alpha–beta 6–4 titanium alloy with fine grain 6–4 titanium at 775°C using the same time and pressure profiles used for 6–4 fine-grained titanium. Standard grain alpha–beta alloys generally require 900–925°C (1650–1700°F) for complete DB [5,6]. In 2009, a US patent was granted [10] covering a topic on SPF and SPF/DB of fine grain 6–4 titanium.

Typical SPF or SPF/DB applications are made out of titanium alloys used for heat-resistant parts around engines or handling hot air including pylon panels, nacelle panels, wing access panels, and also jet engine components used for commercial aircraft. Titanium heat shields are also used to protect mainly the wing structure above the engine from the high-temperature exhaust gases through the exhaust sleeve of the engine. In earlier single aisle commercial aircraft models, the heat shield components were fabricated typically from titanium castings. In recent aircraft models, the heat shields are fabricated from titanium sheet metal details produced by hot die forming and or SPF or SPF/DB to save both the cost of manufacturing and overall weight of a heat shield assembly. An example of SPF/DB part used in commercial aircraft applications is shown in Figure 11.8. Figure 11.9 shows an example of a heat shield assembly for installation in an aircraft. Figure 11.10 shows the heat shield assembly installed on the aircraft.

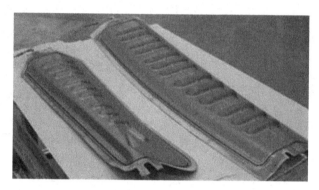

FIGURE 11.8
SPF/DB fabricated panel for engine heat shield. (Adapted from Hefti, L.D., *SAE International Journal of Aerospace*, 3(1), 2010-01-1834, 173–179, 2010.)

FIGURE 11.9
Computer-aided design (CAD) drawing of engine heat shield assembly with SPF and SPF/DB components. (Adapted from Hefti, L.D., *SAE International Journal of Aerospace*, 3(1), 2010-01-1834, 173–179, 2010.)

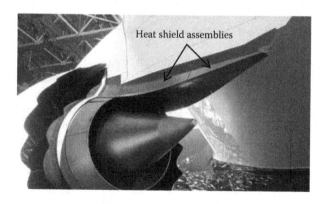

FIGURE 11.10
Example of SPF/DB heat shield assemblies installed on an airplane. (Adapted from Hefti, L.D., *SAE International Journal of Aerospace*, 3(1), 2010-01-1834, 173–179, 2010.)

11.4 Hot Die Forming

The hot die forming process is similar to traditional cold die forming with the exception of heating tool and the workpiece and forming at elevated temperature. In general, hot forming temperatures fall within the range of 600–950°C (1112–1742°F), depending on the part configuration and the material used. Hot forming processes are carried out within the enclosed hot chamber of the press as shown in Figure 11.11. Down-acting four post-frame presses (Figure 11.11a) for hot forming stainless steel, titanium, and inconel type of metals have certain manufacturing features as follows:

- The press frame working areas fully enclosed by retractable, insulated heat shields providing full access for both front and rear
- Operating temperatures up to 1010°C (1850°F)
- Heated platen material options include metallic or ceramic with multi-zone temperature profiling for optimum uniformity and control flexibility, same as specified in SPF presses
- Press control of closing speeds, position set-points, forming rates, and forming tonnage
- Custom engineered tooling solutions including draw forming pins, part eject circuits, and platen shuttles (Figure 11.11b) for ease of tooling change

While SPF of titanium is performed at 750–925°C (1382–1697°F), hot forming is typically performed around 730°C (1346°F). In hot forming of a sheet metal part, the heated metal blank is brought into contact with the hot die that incorporates the part geometry in the inner surface while the hot punch that represents the outer surface of the part descends into the die and conforms to the part geometry. The part is then held under forming pressure for a certain period to eliminate any spring-back effect. Figure 11.12 shows an example of a hot forming steel die used for making a titanium sheet metal part. Figure 11.13 shows a representative hot formed aerospace component, and as attached to the heat shield installed on an airplane (Figure 11.14). Typically formed radii in the part (Figure 11.13) are

(a)   (b)

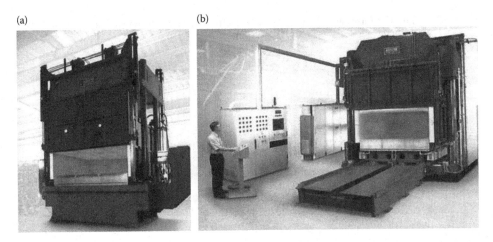

FIGURE 11.11
Examples of hot forming presses: (a) Four post frame and (b) platen shuttle. (From Accudyne.)

FIGURE 11.12
Hot forming steel die set. (Adapted from Hefti, L.D., *SAE International Journal of Aerospace*, 3(1), 2010-01-1834, 173–179, 2010.)

FIGURE 11.13
Part produced from the hot forming die. (Adapted from Hefti, L.D., *SAE International Journal of Aerospace*, 3(1), 2010-01-1834, 173–179, 2010.)

FIGURE 11.14
Hot die forming titanium part in heat shields installed in airplane. (Adapted from Hefti, L.D., *SAE International Journal of Aerospace*, 3(1), 2010-01-1834, 173–179, 2010.)

about two to four times the starting thickness of the sheet product. Preformed starting part blanks are used to get the initial shape as close to the final configuration as possible. This way, during final forming, there is not much material stretching that occurs and the thickness of the material remains fairly constant. Typically, hot forming uses sheet material (<4.75 mm thick).

Metal plate product thicker than 0.187 in (4.75 mm) can be formed by the same hot die forming process and is often used to pre-form plates that will be subsequently machined. For certain shapes of parts, titanium plate is used and then machined to the required configuration. Some of the more simple shapes lend themselves to being formed to a given contour and then machined to the final configuration. The forming process allows a thinner plate of starting material to be used, which lowers the buy-to-fly ratio (how much material starts the process to how much material is contained in the final part), as well as reduces the amount of time for machining, which leads to lower cost products. Figure 11.15 shows a ceramic hot forming die with a 1.6 in (40 mm) thick piece of plate material. Figure 11.16 shows the material after it has been formed. Machining of this formed plate occurs next to achieve the final part configuration. If a formed plate is not used for this component, a starting thickness of approximately 11 in (380 mm) would be required in order to machine the final part. This represents a significant cost savings in raw material plus the time required to machine away the extra material.

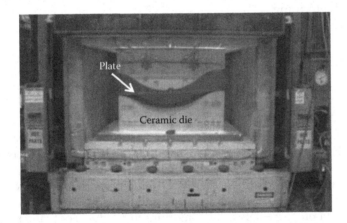

FIGURE 11.15
Hot forming ceramic die. (Adapted from Hefti, L.D., *SAE International Journal of Aerospace*, 3(1), 2010-01-1834, 173–179, 2010.)

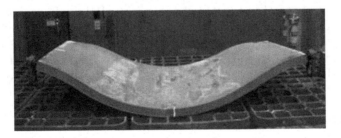

FIGURE 11.16
Hot formed plate from the ceramic die. (Adapted from Hefti, L.D., *SAE International Journal of Aerospace*, 3(1), 2010-01-1834, 173–179, 2010.)

11.5 Brake Forming

Fundamentals of the brake forming processes have been discussed in Chapter 8, cold forming of flat sheet. Hot brake press forming is similar to the cold brake forming of aluminum flat sheet products at room temperature with the exception that preheated blanks are removed from the oven before pressing in the brake press. The hot brake forming process is generally used to produce titanium parts with relatively simple shapes, such as brackets and clips for aerospace applications. Hot brake forming is a cost-effective method for forming titanium that does need standard brake press die and punch type tooling, but can produce high-quality, close-tolerance parts. Brake press forming is also used to pre-form parts for subsequent processes like hot die forming operations.

Hot forming processes including die forming and brake forming provided significant advantages over cold forming and SPF as follows:

- Higher surface profile tolerances
- Consistent thickness control
- Resistance to crack
- Forming accuracy
- Complex part forming
- Lower cost

11.6 Creep Age Forming

Creep age forming (CAF) process [11,12] was developed to form large heavy metal panels with varying thickness utilizing the heat treating technology. The CAF process is generally applicable to aluminum sheet, plate, and extrusion products. Aluminum products of solution heat treated "W" condition are placed onto a contour tool and loaded into an autoclave under tightly controlled temperature and pressure settings. A prescribed combination of heat and pressure deforms the aluminum products onto the tooling geometry, during the final heat treat (aging) cycle.

In aerospace manufacturing, there is always demand for producing metal aircraft panels with improved performance including enhanced strength, lower weight, reduced fabrication costs, and increased resistance to fatigue and corrosion. The CAF technique, which is a combination of mechanical and heat treatment process [11], has been in use for some time to shape large structural airframe components like wing skins, chords, etc. In the CAF process, the metal panel is deflected toward the formed tool under the pressure differential created with the vacuum bag and is heated at a specified heat treatment temperature for the specific metal alloy as shown in Figure 11.17. CAF could be applied to any component whose final design properties require material in an artificially aged temper with higher tensile strength. Forming of wing skin components of commercial aircraft is a complex operation due to the following factors:

- Physical size of component
- Distribution of thickness variation

- Assembly tolerance
- Aerodynamic shape requirements

CAF process steps are shown in Figure 11.18. In the first step, a machined panel is loaded and located onto a form tool with given contour. The next step is to use the vacuum bagging technique to seal the panel against the contoured surface of the formed tool. Air is removed under the vacuum bag, creating an atmospheric pressure differential forcing the panel to form onto the contoured tool surface. Once completed, the panel with the tool assembly is loaded into a thermal chamber for a thermal aging cycle recommended for the material to achieve the design mechanical properties of the panel. This loading is maintained during the heat treatment cycle, which may be completed by using multiple temperature and dwell periods activating creep (stress relaxation) and aging mechanisms. Once complete, following removal of the force constraints from the vacuum bag after heat treatment, the panel may have some permanent deformation, which remains along with some elastic deformation recovery that is termed as elastic spring back. The tool contour may need some design corrections to compensate for the elastic spring back to attain the final shaped components.

The CAF process is relatively simplified in set-up and operation as compared to other forming techniques such as shot peen forming, bump forming, and roll forming. A particular feature of the CAF process is that the initial deformation experienced by the panel during forming onto the tool surface is elastic in nature. The elastic recovery, spring back constraint after CAF is generally high compared to other metal forming processes. This drawback can often be compensated for with the necessary design consideration of the tool surface. Finite element modeling (FEM) may help to reduce the design correction time to compensate the

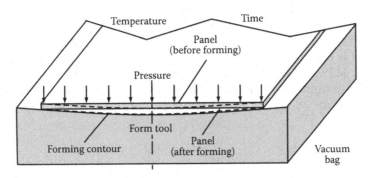

FIGURE 11.17
Schematic of CAF.

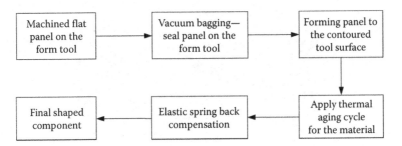

FIGURE 11.18
CAF process steps.

spring back effect of the part after heat treatment. An advantage of a CAF process is that the final structure has lower residual stresses and strains as compared to structure formed by conventional means such as roll forming, shot peen forming, etc. [12]. The long-term performance of skin panels made by CAF is improved due to its enhanced resistance to both fatigue and stress corrosion cracking. Another advantage of CAF is that the forming and aging portion of the heat treatment are done simultaneously as the forming.

The choice of thermal chamber for CAF (oven or autoclave) is primarily dictated by the ability of holding the atmospheric pressure differential created by vacuum bagging to hold the panel in intimate contact with the forming tool surface during heat treatment. Additional force may be required in extreme circumstances and that can be applied using a pressurized oven or autoclave. The main advantage of using an autoclave is the improvement of heat transfer coefficient into the large panel because of the increased density of the pressurized atmosphere.

11.7 Creep Forming/Hot Straightening

Creep forming/hot straightening is done by applying force in the form of a weight block or vacuum pressure on the part over a period of time while the part is heated at an elevated temperature preferably in a vacuum furnace in the case of a titanium part. This creep effect is to force the part into the correct geometric configuration while also stress relieving. This technique is typically used to straighten parts that did not meet contour requirements in prior forming followed by a machining operation.

A case study is shown of a machined titanium extruded part that has certain inaccuracies in shape and dimensions. The part is placed on the tool surface having the correct geometry and it is found that there exists a gap between the part and the tool surface as shown in Figure 11.19. Weight blocks are then placed on the part (Figure 11.20), in the areas where additional forming needs to occur to bring the part as close to the tooling surface as possible. The entire set-up is then placed into a heating chamber, which is typically a

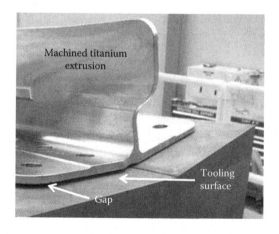

FIGURE 11.19
Part on creep forming tool showing gap. (Adapted from Hefti, L.D., *SAE International Journal of Aerospace*, 3(1), 2010-01-1834, 173–179, 2010.)

FIGURE 11.20
Part on creep forming tool with weights applied. (Adapted from Hefti, L.D., *SAE International Journal of Aerospace*, 3(1), 2010-01-1834, 173–179, 2010.)

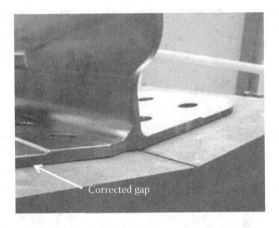

FIGURE 11.21
Part on creep forming tool after thermal cycle was run and the weights removed. (Adapted from Hefti, L.D., *SAE International Journal of Aerospace*, 3(1), 2010-01-1834, 173–179, 2010.)

vacuum furnace. At an elevated temperature similar to hot forming, 730°C (1346°F), the weights force the material down to the tool, if it has not already reached there prior to the thermal cycle, and the temperature stress relieves the part so that it conforms to the shape of the tool surface geometry after it cools down and the weights are removed, as shown in Figure 11.21. Since the only forming force comes from the applied weights in this case study, only simple shapes of similar kinds can be achieved by this process.

11.8 Hot Stretch Forming

The fundamentals of stretch forming of aluminum extrusion at room temperature (cold) were discussed in Chapter 10. The stretch forming at elevated temperature (hot) will follow the same principle except at an elevated temperature as required to bring the flow

(a) 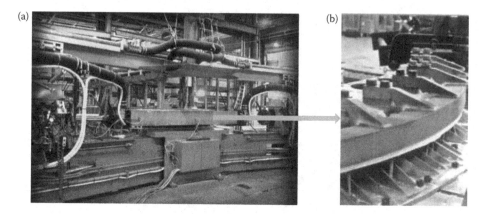 (b)

FIGURE 11.22
Hot stretch forming of titanium extrusion. (a) Stretch forming and (b) after forming. (From the-pilot-group.com, RTI International Metals, Inc.)

FIGURE 11.23
Example of stretched formed titanium extrusion. (From RTI International Metals, Inc.)

stress of the material into the comfortable range to match the power of the stretch forming machine. Figure 11.22a shows a stretch forming press in operation for forming titanium extrusion needed for aerospace applications; and Figure 11.22b shows the contoured extrusion after stretch forming. Figure 11.23 shows an example of hot stretched formed titanium extrusions for aerospace applications.

References

1. Ghosh, A.K. and Hamilton, C.H., Mechanical behavior and hardening characteristics of a superplastic Ti-6Al-4V alloy, *Metallurgical Transactions*, 10A, p. 699, 1979.

2. Xun, Y.W. and Tan, M.J., Applications of superplastic forming and diffusion bonding to hollow engine blades, *Journal of Materials Processing Technology*, 99(1–3), 80–85, 2000.
3. Wenbo, H., Kaifeng, Z., and Guofeng, W., Superplastic forming and diffusion bonding for honeycomb structure of Ti-6Al-4V alloy, *Journal of Materials Processing Technology*, 183, 450–454, 2007.
4. Beal, J.D., Boyer, R., and Sanders, D., Forming of titanium and titanium alloys, in Semiatin, S.L. (ed.), *ASM Hand Book Metal Working: Sheet Forming*, 14B, pp. 656–669, 2006.
5. Hefti, L.D., Fabrication of titanium aerospace hardware using elevated temperature forming processes, *SAE International Journal of Aerospace*, 3(1), 2010-01-1834, 173–179, 2010.
6. Hefti, L.D., Elevated temperature fabrication of titanium aerospace components, *Key Engineering Materials*, Vol. 433, Trans Tech Publications Ltd., Switzerland, 49–55, 2010.
7. www.technet.pnnl.gov/dme/materials/superplastic.stm
8. Hariram, S., Philipp, P., and Dummeyer, D., Fire Protection: Engines and Auxiliary Power Units, www.boeing.com/commercial/aeromagazine, Aero Quarterly, QTR_04| 10
9. www.twi.co.uk
10. Comley, P.N. and Hefti, L.D., Superplastic forming and diffusion bonding of fine grain titanium, US Patent 7,533,794, 2009.
11. Holman, M.C., Autoclave age forming large aluminum aircraft panels, *Journal of Mechanical Work Technology*, 20, 477–488, 1989.
12. Zhu, A.W. and Starke Jr., E.A., Material aspects of age forming of Al-xCu alloys, *Journal of Material Processing Technology*, 117, 354–358, 2001.

12

High Energy Forming and Joining

12.1 Introduction

Conventional cold and hot forming processes of flat sheet, plate, and extrusion products were discussed in the earlier chapters. In conventional forming processes, parts were mainly formed on the forming tools, which are of a male or female type depending on the part geometry and forming processes used to produce parts using mechanical or hydraulic presses and other types of forming machines. Conventional forming processes are normally low-strain-rate deformation processes, whereas high energy forming and joining processes are very high-strain-rate deformation processes. In high energy forming processes, very high pressures are applied on the workpiece for a short time interval. The workpiece is formed at a rapid strain rate on the order of microseconds, as induced by high velocity energy sources. This chapter will discuss three major high energy forming and joining processes, as illustrated in Figure 12.1.

12.2 Explosive Forming

Explosive forming technology is a high-strain-rate deformation process used to form a variety of metal products including flat sheet, plate, and tube including materials from aluminum to high strength alloys. As compared to the conventional metal forming process, the press punch is replaced by an explosive charge in explosive forming. The name of the forming process has been derived from the fact that the forming energy is generated due to the detonation of an explosive. Chemical energy from the detonating compound is used to generate shock waves through an energy-carrying medium, generally water, that are directed to deform the workpiece material at very high velocity (strain rate). The use of water also muffles the sound of the explosive blast. This technology has been explored by the metal forming industry and the research institution and these have published useful documents [1–5] for continuation of this technology especially in the aerospace industry.

In the 1880s, explosive forming technology was used in the engraving of iron plates. During that period of time, research was performed in developing military applications. During World War I and II, many programs were investigated aimed at developing torpedoes and other weapons. The development of explosive chemicals has been intimately tied to military weapon development. In the early 1950s, aerospace companies in the United States, such as Rocketdyne, Aerojet General Corporation, and Ryan Aeronautical, were using explosive forming for the manufacture of complex curved aerospace components.

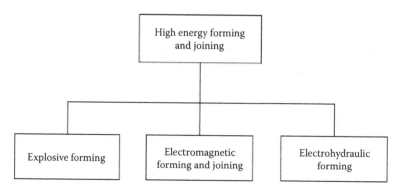

FIGURE 12.1
Classification of high energy forming and joining.

Explosive forming was especially important in the development of short-production-run missile components—particularly for the curved domes of missiles and rocket nose cones. Other aerospace components were produced through explosive forming: complex corrugated panels for aircraft, and fuel filters and asymmetrical exhaust tubes for jet engines. During this time, the Soviet Union also began using explosive forming in their rocket industries for large curved panels. The process was actively used in the 1960s–1970s. Especially in the 1960s, there were at least 80 programs financed by the U.S. Government running simultaneously. In the 1970s, TNO Prins Maurits Laboratorium (TNO), Holland, experimented with explosive forming. However, it was not until 1992 that TNO defined a research project with six Dutch companies. From the project, undertaken between 1995 and 1997, it was concluded that explosive forming was a useful technology for small batches, complex shapes, and difficult-to-form materials.

Explosive forming technology is generally classified into two methods depending on the position of the explosive charge with respect to the workpiece materials as follows:

1. Standoff method
2. Contact method

In the standoff method, the explosive charge is placed at certain predetermined distance from the metal workpiece within the energy-transmitting medium. The standoff distance and the amount of charge determine the amount of pressure transmitted to the metal. Other factors including explosive type, shape of explosive, and type of medium also affect the forming pressure. Water that is readily available and inexpensive has provided excellent results by producing high peak pressures out of other medium including air, oil, gelatin, and liquid salt. Under normal operating conditions it is recommended to detonate explosive charges as far below the surface of the water as possible. This reduces the amount of water that will be thrown by the explosion and the amount of energy lost through venting the gas bubble to the atmosphere. Higher efficiency would be obtained if the distance from the charge to the surface of water is at least twice the standoff distance.

In the contact method, the explosive charge is placed directly in contact with the metal workpiece. This method requires specific studies of the interactions between the metal and the explosives. If not predicted and monitored properly, there may be an increased risk of fracture and failure involved with the contact method due to localized high stress

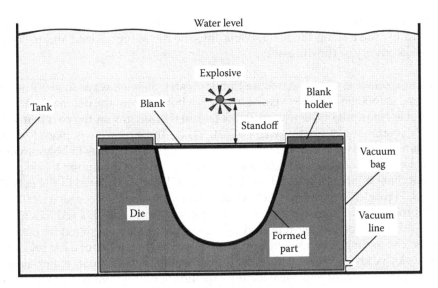

FIGURE 12.2
Fundamentals of explosive forming process.

for a very short duration. Because of the increased risk of failure involved with the contact method, the standoff method is more commonly used in explosive forming operations. In standoff method, the shockwave energy transferred to the metal workpiece is typically lower than in the contact method. Larger parts can be produced with a smaller amount of explosives. Depending on the part geometry and the material properties of the part, the explosive charge (detonation chords) can be placed in one spot or distributed such that the energy release is spread over the metal blank evenly. The fundamentals of the standoff method are illustrated in Figure 12.2.

The major variables in explosive forming are as follows:

1. Tooling geometry and design
2. Blank geometry
3. Material and thickness
4. Type and quantity of explosive (weight per unit length)
5. Distance between workpiece and explosive—standoff
6. Vacuum in the die

The major process steps of explosive forming used for the standoff method are as follows:

- Metal sheet blank is placed on the die with the blank holder
- Die assembly with the metal blank is placed inside the vacuum bag
- Explosive charge is placed above the blank at a predetermined distance (standoff)
- The assembly is lowered to the bottom of the water tank
- Explosive is detonated—a pressure pulse of high intensity is produced. A gas bubble is also produced and expands spherically and vents out at the surface of the water

- Explosion transfers energy through water to a metal blank that is accelerated toward the die forcing the metal blank into the die at high speed and to conform to the geometry of the die cavity

When an explosive charge is detonated under water, a shock wave moves out from the explosive and is followed by the expansion of a gas bubble from the detonated explosive. If the gas bubble vents out of the water surface before it impinges on the part being formed, a considerable amount of energy will be lost. When the explosive is placed sufficiently deep in water, several cycles of expansion and contraction of the gas bubble can be noted. Each time a bubble expands, an additional pressure pulse is transmitted to the part being formed. But these bubbles may not be severe enough to be considered in the actual forming process. Figure 12.3 shows an example of explosive formed sheet metal part.

Dies and tooling for explosive forming can be made simple, with a reduction of cost of up to 80% over that for tooling required to perform the same operation by conventional techniques. The tooling is simplified since only the female member of a die set is required with the explosive shock waves acting as a punch (male die). Different materials are used for manufacturing dies for explosive forming including high-strength tool steel, cast steel, cast iron, plastics, fiberglass, and concrete. Relatively low-strength dies are used for short run products. The shock waves provide a compressive load to a greater extent than tension in the die. Since most die materials can withstand much greater load in compression than in tension, the lighter dies can be used.

Explosives are chemicals that undergo rapid chemical reaction, which generates heat and large quantities of gaseous products. Explosives could be solid (TNT—trinitro toluene), liquid (nitroglycerine), or gaseous (oxygen and acetylene mixture). Explosives are generally classified into two classes as follows:

1. Low explosives: Used for propellants in guns, rockets, and missiles
2. High explosives: High rate of reaction with a large pressure build up, used for explosive forming

Table 12.1 shows properties including detonation velocity, energy, and maximum pressure of some selected high explosives commonly used for metal forming. The cost of

FIGURE 12.3
Example of an explosive formed ring type part of 42 inch (1067 mm) diameter from stainless-steel sheet product of 0.16 inch (4 mm) thick. (From 3D Metal Forming.)

TABLE 12.1

Properties of High Explosives

Explosive	Relative Power, % TNT	Form of Charge	Detonation Velocity (ft/s)	Energy (ft-lb/lb × 10^5)	Storage Life	Maximum Pressure (Million psi)
TNT	100	Cast	23,000	2.62	Moderate	2.4
RDX	170	Pressed granules	27,500	4.25	Very good	3.4
PETN	170	Pressed granules	27,200	4.35	Excellent	3.2
Pentolite (50/50)	140	Cast	25,000	3.17	Good	2.8
Tetryl	129	Pressed granules	25,700	–	Excellent	–

Source: Adapted from Noland, M.C. et al., High velocity metalworking, *A survey by N.A.S.A.*, SP-5062, 1967.

explosives is normally very small in comparison with die costs. Holding the explosive charge in position is a very important consideration. Normally, wire or masking tape is used to hold the explosive charge in proper position with respect to the workpiece while it is placed in the water. The only requirement is to maintain the explosive in a predetermined location without displacing it by water flow when the die is put into the water tank.

The explosive forming processing methods present a series of important advantages, being recommended for wide-scale applications in the aerospace industry [5]. By using the high explosives, parts with different shapes and almost unlimited overall size are formed using heavy duty materials provided that the material is sufficiently ductile at high strain rates. The major benefits of explosive forming are as follows:

1. Low tooling cost
2. Short prototyping and delivery time
3. Special shapes in special metals in lower quantities
4. Unlimited power and size
5. Capable of forming complicated double curved parts with metal in the age (hardened) condition
6. Ability to compensate thickness variation
7. High-strain-rate deformation process, highly reproducible
8. Reduced spring back and in some cases increasing the formability

The peak pressure generated in water can be calculated by using the following equation [3,7]:

$$p = K\left(\frac{\sqrt[3]{W}}{R}\right)^a \tag{12.1}$$

where p is the peak pressure in MPa, K the constant depending on the type of explosive, W the weight of explosive in kg, R the distance of the explosive from the workpiece (standoff) in m, and a the constant, generally taken as 1.15.

An important factor in determining the peak pressure is the compressibility of the energy-transmitting medium (such as water) and its acoustic impedance (defined as the

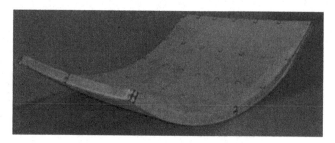

FIGURE 12.4
Example of an aerospace developmental part from plate products. (From 3D Metal Forming.)

product of mass density and sound velocity in the medium). Thus, the lower the compressibility of the medium and the higher its density, the higher is the peak pressure.

Explosive forming technology has gradually returned to the manufacturing industry, especially for aerospace applications due to the addition of powerful simulation analysis methods to the forming analysis prior to actual forming. Figure 12.4 shows an example of a compound curvature explosive formed developmental aircraft part made from an aluminum plate product for a monolithic structural part.

Figure 12.5 shows an example of a machined monolithic developmental part for the nose landing gear bay of an aircraft. Thick plate product was formed by using explosive forming technology, and it is then machined to the engineering drawing to meet the structural requirements of the part. The main advantage of this process would be to

- Reduce weight
- Reduce cost

Numerical simulations provide a better understanding of the effect of many complex process variables in actual forming. The results of the forming trials have become more predictable due to the application of numerical simulations [8,9]. The explosive forming process is normally simulated using the hydroform module from Autoform. The hydroforming process is much like the explosive forming process. The major difference is the forming

FIGURE 12.5
Example of a monolithic developmental part for nose landing gear bay. (From 3D Metal Forming.)

speed (strain rates) of explosive forming ranges from about 10–100 s^{-1}. It is also known that certain materials show increased ductility at the higher strain rates, which drives one to determine forming limit diagrams (FLD) for explosive forming applications. The determined FLDs are used as an input in the hydroforming software of Autoform and enable the software to be applied for explosive forming.

12.3 Fundamentals of Electromagnetic Forming and Joining

Electromagnetic forming is also called magnetic pulse forming and was developed in the 1960s for shaping, forming, and joining circular metallic parts [10–12]. In this process, the energy required to form the part is derived from the electrical energy by means of a magnetic field that exerts force on a current-carrying conductive workpiece. The EMF process uses a direct application of pressure created in an intense, transient magnetic field. Without mechanical contact, a metal workpiece is formed by passing a pulse of electrical current through a forming coil. The conductivity of the workpiece and the eddy current that interacts with the magnetic field of the coil result in a net pressure on the surface of the workpiece. As the workpiece surface moves inward under the influence of this pressure, it absorbs energy from the magnetic field. The electromagnetic pulses can be precisely controlled in both magnitude and time of pulse. Controlling the magnitude of the pulses affords excellent shape reproducibility and maintains close dimensional tolerances. In most forming applications, pulses have duration in between 10 and 100 μs. A high-pressure magnetic pressure between the coil and the workpiece is produced due to the interaction between the magnetic field produced by the induced current and the magnetic field produced by the coil current as illustrated in Figure 12.6. The force from the magnetic pressure is sufficient to stress the workpiece metal beyond its yield strength, resulting in a permanent deformation.

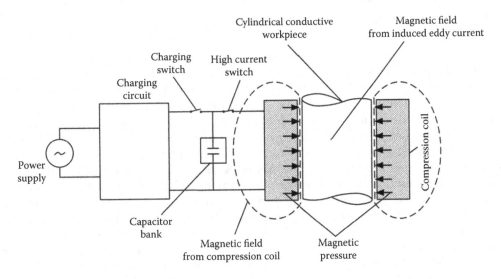

FIGURE 12.6
Fundamentals of electrical circuit of EMF process.

When the energy storage capacitor is discharged through the forming coil, the discharge is a momentary oscillating current. The frequency of this oscillation is often called the ringing frequency. Electrically, the forming coil is an inductor (L), along with a small amount of series resistance (R). The oscillating current has a damped sine wave form [10]. The damping of the current is due to the loss of energy for the mechanical work in forming, as well as the resistive losses in both the coil discharge circuit and the workpiece itself. The frequency (ω) is inversely proportional to the square root of the product of the capacitor (C) of the storage bank and the inductance (L) of the coil. The pressure produced by the coil is proportional to the square of the current; thus, the pressure pulse is essentially positive and approximates a damped sine square wave form [10]. The pulse is characterized by its peak pressure, which is that of the first wave. Peak pressure is

- Directly proportional to the energy of the electrical pulse from the capacitor bank
- Inversely proportional to the resistivity of the workpiece and coil
- Inversely proportional to the total volume of the workpiece and the field shaper penetrated by the electromagnetic field (skin effect) and the volume between the coil and the workpiece surfaces

Forming coils are generally designed to be mounted directly on the machine or positioned in a separate forming location away from the machine. Depending on the part geometry, material to be formed, and coils are constructed either with a single winding of heavy cross section or with multi-turn windings. The coils are generally constructed to provide a certain clearance with the workpiece without causing an excessive loss in intensity of magnetic field. The radial clearance is generally kept in the range of 25–30 thousand of an inch (0.63–0.76 mm). The coils are generally of three types as follows:

1. Cylindrical coil for expansion forming
2. Flat spiral (pancake) coil for flat sheet forming
3. Field shaper for compression forming of tube

Figure 12.7 illustrates the application of expansion, flat, and compression coils used to form tubular and flat sheet products based on the end use of the part. Figure 12.7a shows an expansion coil inserted into the tubular part to expand the tube radially to form inner grooves as provided in the forming die. Expansion coils are primarily used to provide bulge, shape, and flange in tubular products. Figure 12.7b shows a flat coil or pancake coil placed above the flat sheet blank to form a bell mouth shape. Flat coils are generally used to form coin, blank, and dimple using a forming die. Figure 12.7c shows the compression coil with field shaper used to swage the tubular workpiece outwards to join with a fitting. A field shaper is used in conjunction with the primary coil to achieve a variety of forming applications. The field shaper is a single-turn coil split into two halves, usually made from beryllium copper that has lower electrical resistivity and higher tensile properties than high-strength aluminum alloy. Field shaper is inserted between the work coil and the workpiece is placed into the field shaper. Compression coils with or without field shapers are generally used to produce structural joints, hermetic joints and seals, metal to non-metal seals, swaged torsion shaft, hydraulic connections, and uniformly reduce the cross sections of tubular shapes.

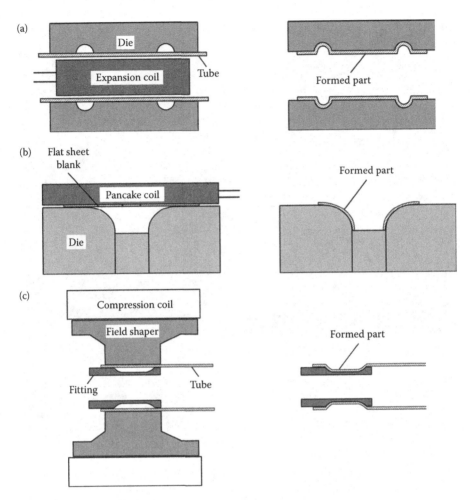

FIGURE 12.7
Type of coils for forming tubular and flat sheet products: (a) Expansion coil, (b) pancake-type flat coil, and (c) field shaper.

A few examples of expansion, flat, and compression type formed products are shown in Figure 12.8.

The following workpiece conditions must be met in order to be successful using the electromagnetic forming process:

1. Workpiece must have good electrical conductivity
2. Workpiece must provide continuous path for the induced current
3. Less conductive materials having higher resistivity such as stainless steel and titanium can be successfully formed by using an intermediate driver of conductive materials such as aluminum and copper tube and sheet place between the magnetic work coil and the workpiece. The driver deforms and it transfers the pressure to the high resistivity workpiece material

There are good conductors such as copper, aluminum, brass, or low carbon steel. Table 12.2 lists the values of electrical resistivity, ρ, for some materials.

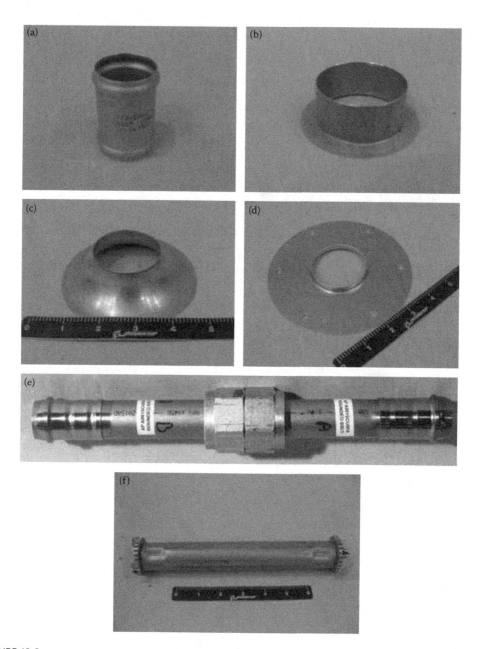

FIGURE 12.8
Examples of few EMF aircraft test parts: (a) Beaded duct, (b) 90° flange, (c) bell mouth, (d) 90° collar, (e) hydraulic joint, and (f) torque tube. (Adapted from Saha, P.K., *SAE Technical Paper 2005-01-3307*, 2005.)

12.3.1 Electromagnetic Forming of Tube

The knowledge of compression forming is used to form tube products. A few case studies of compressive-type joining processes applied on both aluminum and titanium or stainless tubes for aerospace applications were presented [13]. The major applications for aerospace industry include:

TABLE 12.2

Electrical Resistivity of Some Materials (ρ), Micro-ohm-cm

Materials	Resistivity, ρ
Aluminum, 2024-T3	5.8
2024-T6	4.5
6061-T6	4.3
7075-T6	5.7
Beryllium–Cu	7.9
Brass	4.7–6.6
Copper	1.72–1.77
Steels	12–30
Stainless steels, austenitic	69–79
Ferritic	60–67
Martensite	40–72
Titanium, CP	48–57
6Al-4V	172
8Al-1Mo-1V	199

Source: Adapted from Wick, C., Benedict, J.T., and Veilleux, R.F., *Society of Manufacturing Engineers Hand Book*, Vol. 2, Chapter 19, SME, Dearborn, MI, 1984.

- Torque shaft assembly
- Control rods
- Linkages
- Engine push rods

In the first case study, tests were conducted to join aluminum tubes with steel end fittings to form torque tubes. Torque tubes are hollow seamless drawn tubes (generally 2024-T3 alloy) connected with end fittings that are splined for connection to transmissions or other tubes. Torque tubes are used for driving the leading edge and trailing edge of the flight control system of an aircraft. Torque tubes/torque rods are used on all aircraft models in varying lengths, diameters, and type of joints. Figure 12.9 is an example of an EMF joint used in a trailing edge flap drive for an airplane wing.

The overall look of an EMF machine used for forming tube products is shown in Figure 12.10. The major components of an EMF machine are as follows:

- Power supply (1)
- Energy storage capacitors (1)
- Control circuitry (2)
- Switching devices for discharging the capacitors (2)
- Transmission lines and busses (3)
- Permanent coil (4)
- Field shaper (5, 5A)
- Work bench (6)

Figure 12.11 shows the process flow diagram of compressive forming of aluminum tube products onto an end fitting. Each half of the field shaper (Figure 12.11) is insulated with a

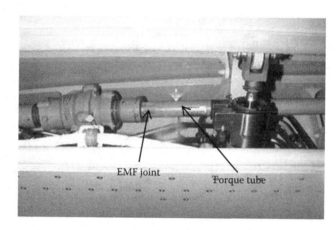

FIGURE 12.9
Training edge flap drive showing the torque tube with an example of EMF joint. (Adapted from Saha, P.K., *SAE Technical Paper 2005-01-3307, 2005.*)

layer of fiberglass or other insulated tape with a maximum separation of 0.762 mm between the two field shaper halves. The workpiece (tube + end fitting) is installed between the field shaper halves (lower and upper) lined on the work bench (6) and the field shaper assembly (5A) inserted into the main coil (4). Then the machine is ready for discharging the power kilo-joules (kJ) into the main coil to form the part. The formed joint is shown in the lower half of the field shaper. Figure 12.12 illustrates the sectional view of the forming joint area within the main coil.

EMF machines are typically rated according to their energy storage capacity. The energy storage capacity is usually given in kJ. The stored energy (E) is shown in the relationship with the supplied voltage (V) and capacitance (C).

$$E = \frac{CV^2}{2} \tag{12.2}$$

Commercial machines are available in sizes ranging from 2 to over 150 kJ. The maximum operating voltage is normally held to a maximum of around 10 kV. The maximum

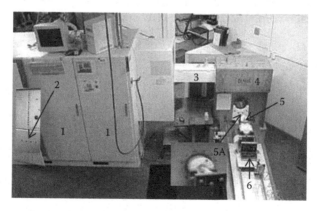

FIGURE 12.10
EMF machine components. (Adapted from Saha, P.K., *SAE Technical Paper 2005-01-3307, 2005.*)

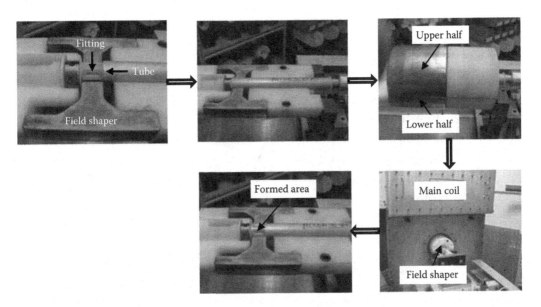

FIGURE 12.11
Process flow of compressive forming of aluminum torque tube. (Adapted from Saha, P.K., *SAE Technical Paper 2005-01-3307*, 2005.)

safe current is usually established by the limitations of the switching devices. The discharge current in electromagnetic forming is usually between 100 and 400 kA. The rated energy capacity of the machine used for production of torque tube is generally 60 kJ. The machine power setting is chosen in 12 kJ increments, and then fine-tuned by adjusting the (%) settings. Figure 12.13 shows an example of a damped current waveform used in forming a 2024-T3 aluminum torque tube joint. The peak pulse current (I_{max}) and the pulse width (T/2) are 391 kA and 86.2 μs, respectively.

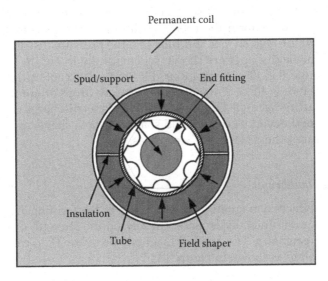

FIGURE 12.12
Cross-sectional view of forming area.

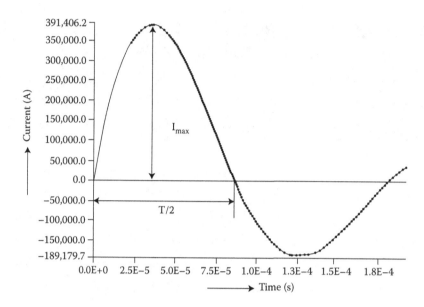

FIGURE 12.13
Example of a damped current waveform at a discharged energy 36 (97.5%) kJ. (Adapted from Saha, P.K., *SAE Technical Paper 2005-01-3307*, 2005.)

12.3.2 Kinematics of Electromagnetic Forming

12.3.2.1 Pressure Distribution

As previously shown in Figure 12.10, the field shaper assembly (5A) is used to adapt a large diameter compression coil to a smaller diameter workpiece and to concentrate the pressure at a specific location on the workpiece. When the capacitor is discharged through the permanent coil, the induced current in the field shaper produces a magnetic pressure on the conductive tube. The amount of discharged power has to produce sufficient magnetic compressive pressure to conform the tube wall into the fitting geometry. The pressure distribution between the field shaper and tube and the pressure distribution between the primary coil and the field shaper are shown schematically in Figure 12.14. The pressure distributions are quite flat, or a rectangular distribution, except at the ends within the distance S. The distance is about 2 mm and the pressure drops around one hundredth of the uniform pressure in that area. Distance "S" depends on discharge frequency and electrical conductivity of the workpiece and field shaper, or field shaper and primary coil. As frequency (ω) increases, the distance "S" decreases.

12.3.2.2 Forming Equilibrium

The field shaper produces a magnetic pressure pulse in the forming area toward the end fitting. The resulting deformation depends entirely on the strength of the tube material and the end-fitting geometry. The applied magnetic pressure, P_{mag}, can be calculated by the equilibrium relation as shown in Figure 12.15:

$$P_{mag} = P_{inertia} + P_{def} \tag{12.3}$$

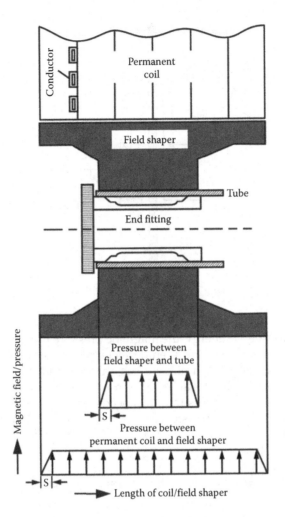

FIGURE 12.14
Schematic diagrams of the pressure distribution in EMF of torque tubes.

where the inertia pressure $P_{inertia}$ is a multiplication of material wall thickness, material density, and acceleration and the resultant deformation pressure is P_{def}.

P_{def} is a function of yield stress (σ_{yp}), strain (ε), and strain rate ($\dot{\varepsilon}$). The deformation pressure may be derived from hoop–stress relationship as shown

$$P_{def} = \frac{\sigma_{yp} \cdot t}{r} \qquad (12.4)$$

where r is the outside radius of the tube and t is the tube wall thickness.

The required magnetic pressure P_{mag} may be defined as a factor (n) of P_{def}

$$P_{mag} = n \cdot P_{def} \qquad (12.5)$$

The initial gap between the tube and the end-fitting surface is called the fly distance as shown in Figure 12.15. The fly distance has to be large enough to allow the material to

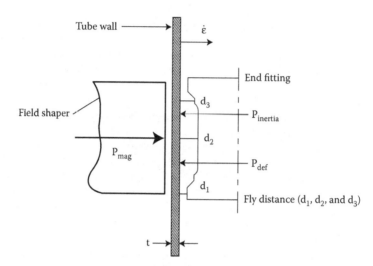

FIGURE 12.15
Force equilibrium in EMF of tube.

deform plastically. To balance energy, the magnetic pressure pulse will accelerate the tube material to a maximum velocity and then decelerates to a zero velocity by the time the tube covers the full fly distance. In many cases, the required deformation varies within the deformation zone as the fly distance varies according to the geometry of the end fitting. Accordingly, the working pressure (P_{mag}) needs to be adjusted to the correct amount of deformation with a varying fly distance within the end fitting. If the applied pressure pulse is too small, the tube wall does not have sufficient energy to reach the end-fitting surface. Conversely, if the applied pressure pulse is too high, it will impact and rebound from the end-fitting surface.

The major variables associated with the forming of tubes onto the end fitting are as follows:

- Fitting geometry
- Power setting kJ (%)
- Material formability

End-fitting geometry [14–16] plays a critical role in forming drawn aluminum tube, generally 2024-T3. Instead of forming in solution heat treated condition in "W," forming of tube in "T3" temper minimizes fabrication cost by reducing the process cycle time involved in heat treatment, freezer storage, and any straightening required due to heat treat distortion.

The common joining methods between fittings and hydraulic tubes for aircraft applications are as follows:

1. Roller swaged
2. Elastomeric
3. Welding

Figure 12.16 shows an example of each process. Each process involves many process steps with different issues like capacity, a low tool life, degreasing, heat-affected zone,

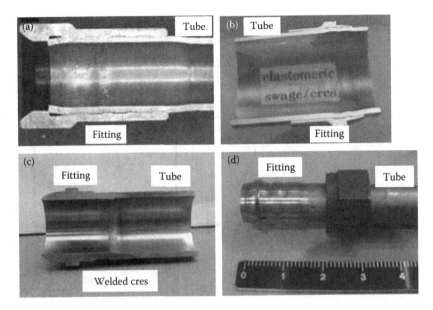

FIGURE 12.16
Different joining processes for hydraulic tubes: (a) Roller swaged, (b) elastomeric, (c) welded, and (d) EMF. (Adapted from Saha, P.K., *SAE Technical Paper 2005-01-3307*, 2005.)

and others that involve additional production costs, including quality issues. The EMF process was developed [13] to produce a stronger joint to withstand very high hydraulic pressure. This involved designing of groove geometry on the tube outside diameter and fitting geometry [17].

Every manufacturing process has its own safety concerns. In EMF, production and maintenance personnel need to know the proper operational and safety procedures, since the process deals with high voltages and currents. The major safety measures include:

1. Insulation between coil and workpiece must be adequate
2. Main coil insulation should be visually examined daily
3. Field shaper insulation should also be examined before using
4. Main coil, field shaper, and the forming workpiece must be free from any metallic dust to avoid arcing

12.4 Electrohydraulic Forming

Electrohydraulic forming (EHF) [3,18] also known as the electro spark forming process, which is similar to an explosive forming process, involves using electrical energy instead of igniting of explosive energy. In EHF, the capacitor bank discharges the high density current across a gap between two electrodes connected with a thin wire (initiating filament) that provides a path along which an arc channel develops. As a result, it instantly vaporizes the wire and generates a high level explosion within water which is the most

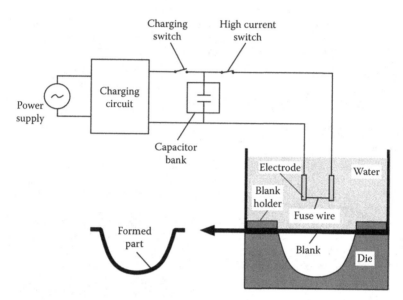

FIGURE 12.17
Schematic illustration of EHF.

suitable medium to generate shock waves that travel in radially outward directions from the wire at high velocity and push the metallic blank into the die cavity to form a part. The workpiece is set up similarly to the standoff system described earlier in explosive forming. The fundamentals of EHF are illustrated in Figure 12.17. The wire needs to be replaced after every operation. Since the high velocity shock wave propagates radially outward, bulging of tubular parts into conical, spherical, or cylindrical shapes is the most common application in terms of energy utilization from this process. Another application is to form shallow or deep drawn flat sheet metal parts (Figure 12.17). The performance of EHF is very much dependent on the selection of fuse wire and its placement with respect to the required forming geometry. The pressure pulse geometry could be cylindrical or planar based on the type of wire and the electrode placement configurations. The energy level in EHF may be relatively lower than that of the explosive forming. That limits the use of the process from small to intermediate sized parts.

References

1. Strohecker, D.E. et al., Explosive forming of metals, Battelle Memorial Institute, DMIC Report 203, May, 1964.
2. White, T., Materials and manufacturing for tomorrow's engineer, *3rd Annual KU Aerospace Materials and Processes "Virtual" Conference*, The University of Kansas, 2003.
3. Mynors, D.J. and Zhang, B., Applications and capabilities of explosive forming, *Journal of Materials Processing Technology*, Vol. 125–126, 1–25, 2002.
4. Carton, E.P., Stuivinga, M., and Verbeek, H.J., Explosive forming of aerospace components, Public release report from TNO Defence, Security and Safety, PO Box 45, 2280 Rijswijk, the Netherlands, 2005.

5. Ghizdavu, V. et al., Explosive forming—Economical technology for aerospace structures, *Incas Bulletin*, 2(4), 107–117, 2010.
6. Noland, M.C. et al., High velocity metalworking, *A survey by N.A.S.A.*, SP-5062, pp. 179–183, 1967.
7. Kalpakjian, S. and Schmid, S.R., *Manufacturing Processes for Engineering Materials*, 5th Edition, Pearson, Upper Saddle, NJ, 2008.
8. Groeneveld, H.D. et al., Deformation control in explosive forming, *Numisheet*, Zurich, pp. 1–5, 2008.
9. Groeneveld, H.D., Photogrammetry applications for explosive forming, *The Paton Welding Journal*, 11, 52–56, 2009.
10. Wick, C., Benedict, J.T., and Veilleux, R.F., *Society of Manufacturing Engineers Hand Book*, Vol. 2, Chapter 19, SME, Dearborn, MI, 1984.
11. Guenter, Z., Getting the most out of electromagnetic metal forming, Reprinted from September 1976 issue of *Assembly Engineering*, Hitchcock Publishing Company, 1976.
12. Plum, M.M., Electromagnetic forming, *Metals Handbook*, Vol. 14, ASM, 1988.
13. Saha, P.K., Electromagnetic forming of various aircraft components, *SAE Technical Paper 2005-01-3307*, pp. 999–1009, 2005.
14. Dolan, L.E. et al., Tube forming on an end fitting, US Patent, 5,983,478, November 16, 1999.
15. Saha, P.K., Berden, M.J., Low Chamfer angled torque tube end fitting metal, US Patent, 6,932,118, August, 2005.
16. Saha, P.K. and Berden, M.J., Low Chamfer angled torque tube end fitting with elongated overflow groove, US Patent, 7,363,945, April, 2008.
17. Saha, P.K. et al., Electromagnetic mechanical pulse forming of fluid joints for high pressure applications, US Patent, 7,954,221, June, 2011.
18. Aluminum Company of America, *Forming Alcoa Aluminum*, Pittsburgh: Alcoa, PA, 1974.

13

Tube and Duct Forming

13.1 Introduction

The applications of metal tubes and ducts in an aircraft are significant and many are associated with the manufacture and installation of the environmental control systems in the aircraft, including heating, ventilation, air conditioning, oxygen, and pressurization and depressurization systems. In addition to the environmental control system, there are applications in the auxiliary power unit, air handling unit ducts, and wire management conduits, as well as fluid containment tanks. The applications of hydraulic tubing in the aircraft are mainly in the landing gear operating system, the wing trailing edge flap, and the leading edge slat actuation systems, as well as engine operating systems. There are also applications of metal tubes as control rods, struts, actuation and push rods, links, and more. The metal tubes for aerospace applications are mainly seamless hollow products made with various metal-forming technologies discussed in Sections 4.6 through 4.8. The ducts are generally fabricated using sheet metal forming and welding technologies. This chapter introduces an overview of major tube- and duct-forming processes as illustrated in Figure 13.1.

13.2 Bending

Tube bending is a metal-forming process used to bend straight generally round tube stock in multiple directions and angles for various industrial applications especially in the aerospace industry. The most common simple tube bending operation consists of forming elbows with angles varying within 90°. Another bend is the "U" type 180° bend. Tube bending is an important technology designed to produce shapes, which allow routing of tube and ducting through tight envelopes throughout the aircraft. Bends are generally classified as (1) two-dimensional (2D) and (2) three-dimensional (3D). In 2D tube bending, the tube openings are on the same plane (Figure 13.2a), whereas in 3D bending, the openings are on different planes (Figure 13.2b). In tube bending operations, fundamentally the inner layer of the tube is in compression, whereas the outer layer of the tube is in tension as illustrated in Figure 13.2a. As a result, strain varies in both inner and outer walls, causing the wall along the inner radius of the tube to become thicker and the outer radius wall to become thinner. Owing to the nature of this strain, there may be some process issues including buckling, wrinkling, and cracking of tubes depending on bending effects of bend angle, tube material, tube wall thickness, and the bending process. To avoid those

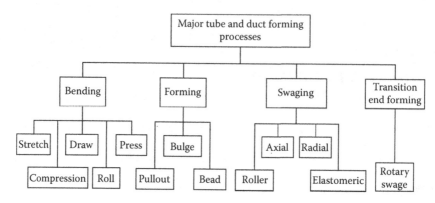

FIGURE 13.1
Major tube- and duct-forming processes.

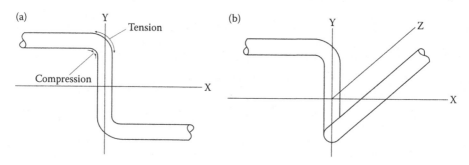

FIGURE 13.2
Fundamentals of tube bending: (a) 2D bending and (b) 3D bending.

issues, the tubes may be supported internally or externally to preserve the cross section of the tube during bending.

13.2.1 Bending Methods

The oldest method of bending tubes or pipes is to fill or plug inside with loose particles mainly sand and then perform the bending operation. This method prevents the tubes or pipes from buckling by pressurizing the inside to distribute compressive strain uniformly over the forming layer of the tube or pipe. Tubes or pipes can also be plugged inside with various flexible type mandrels for certain types of bending operations. There are several processes of bending tubes or pipes for various industrial applications including aerospace. The most common type processes are (Figure 13.3) as follows:

- Ram/press
- Roll
- Rotary-draw
- Compression
- Stretch

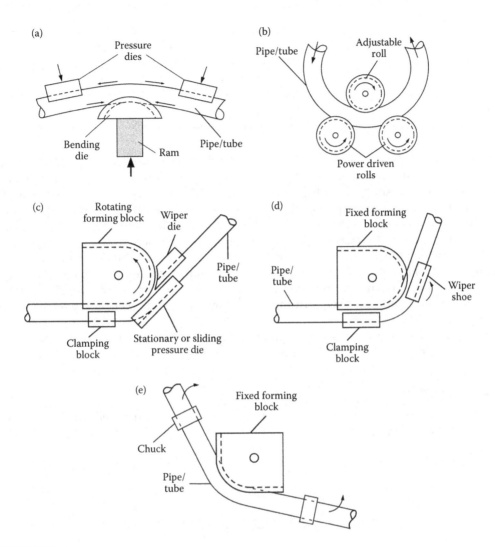

FIGURE 13.3
Most common type pipe/tube bending processes: (a) Ram/press, (b) roll, (c) rotary-draw, (d) compression, and (e) stretch.

The ram/press type tube or pipe bending method as shown in Figure 13.3a pushes a bending die against the tube or pipe by a mechanical force provided by the press ram, forcing the pipe or tube to conform to the contours of the bending die.

In a roll-type bending process (Figure 13.3b), pipes/tubes are fed through a series of spool-like rollers with inside grooves to support the outside diameter of the pipe or tube. The fundamentals of a roll bending process as discussed in Section 8.5 were applied to roll bending of tube and pipe. The pressure is exerted by the powered rolls to the pipe/tube to start bending. The relative location of the rollers determines the radius of a desired bend. The roller type pipe bending process is ideal for forming helical form of pipe/tube coils for heat transfer applications.

The rotary draw-type tube bending process (Figure 13.3c) is widely used to produce high-quality and tight radius bends for various applications in some industries, especially for aerospace. A steel mandrel (not shown in Figure 13.3c) that supports the pipe/tube internally

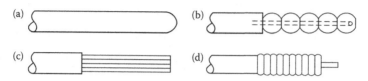

FIGURE 13.4
Internal mandrels for tube bending: (a) Plug, (b) balls, (c) laminated, and (d) cable.

to maintain mainly the tube cross section at the bend radius moves along the rotating bend die to bend the tube to the radius of the die. The bend die has two functions: (1) define the bend radius and (2) support the tube cross section at the inner bend. The bend die along with the mandrel and the wiper die determines the quality of the inner bend. The sliding pressure die provides the forces on the pipe/tube to cause it to conform to the die radius.

Figure 13.3d illustrates compression bending process [1], where the tube is clamped on the fixed forming block while the wiper shoe wraps the tube in compression against the forming block. Figure 13.3e illustrates the stretch bending process [1], which follows the fundamentals of stretch-forming process (as discussed in Section 8.6), which stretches the tube and wrap around fixed forming block to get the bend according to the radius of curvature of the block.

Figure 13.4 illustrates various flexible internal mandrels [1] needed to plug the tube prior to bending using any of the bending processes shown in Figure 13.3.

13.2.2 Bending Tools

The basic tooling system in tube bending operation is illustrated in Figure 13.5. The major components are:

1. Bend die
2. Clamp die

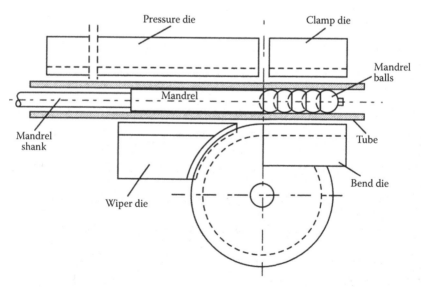

FIGURE 13.5
Schematic of tube bending process tooling system. (From Sandvik Special Metals Corporation, USA.)

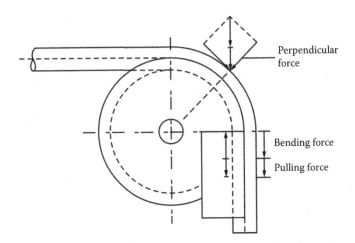

FIGURE 13.6
Forces on tube in bending. (From Sandvik Special Metals Corporation, USA.)

3. Mandrel

4. Pressure die

5. Wiper die

During bending, the tube is longitudinally pulled while the tube is firmly fastened in the clamp die. As the bend is in progress, the pulling direction becomes diagonal causing the deformation of the tube cross section as shown in Figure 13.6. As shown, the ball mandrel provides the support to the tube cross section during the bending process. There will always be an unsupported length of tubing due to limited number of balls provided in the plug. The unsupported length may be flattened due to diagonal forces.

Since higher bending forces are required for titanium tubing as compared to CRES, titanium tube has a greater tendency for changing the round cross section to oval shape. A method of using bending dies with extended radius grooves (Figure 13.7), which keeps minimizing the flattening effect. However, the extended groove distance should not be more than 0.020–0.060 in (0.5–1.5 mm) [2]. The pressure die must have a smaller groove when offset dies are used. It is advantageous if the movements of the pressure die relative to the tube are CNC controlled.

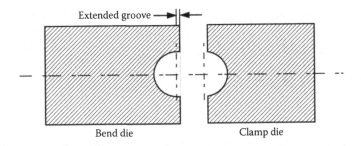

FIGURE 13.7
Bending dies with offset grooves. (From Sandvik Special Metals Corporation, USA.)

13.2.3 Springback

Titanium and stainless-steel tubes are widely used in various bent shapes for both commercial and military aircraft. The tubes are used for both high- and low-pressure hydraulic and pneumatic lines to operate various mechanical systems for the operating systems of an aircraft. A comparison of mechanical properties between 3Al-2.5V Ti and 21-6-9 CRES is shown by a schematic representation of stress–strain curve (Figure 13.8). The curve shows the difference in bending range between two metals, titanium and CRES. Since titanium has a lower bending range than CRES, more care needs to be taken during bending operations, mainly by

- Slowing the bending speed
- Controlling lubrication

The smallest radius of bend required for the hydraulic system for normal aircraft is three times the nominal tube diameter. The elastic region of the stress–strain curve results in retention forces in bending, which may cause a springback effect in tube bending. The springback effect in tube bending is illustrated in Figure 13.9.

13.2.4 Lubrication

The selection of proper lubricant in titanium tube bending operation is very important. Since very high pressure is developed between tube and tooling, extreme pressure lubricant is necessary. Every interruption of lubricant film results in chattering that stops smooth flow of metal during bending to cause surface defects. The lubricant must be free from chlorine or chlorine compounds since they can leave residues on the tube surface, which may embrittle the tube if it is later exposed to higher temperatures. Tube bending lubricants could be categorized generally into two types as follows:

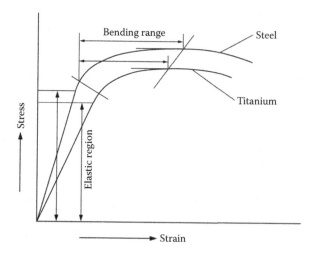

FIGURE 13.8
Schematic representation of stress–strain curve for two different tube metals. (From Sandvik Special Metals Corporation, USA.)

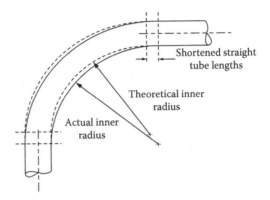

FIGURE 13.9
Springback in tube bending. (From Sandvik Special Metals Corporation, USA.)

1. Water-soluble
2. Non-water-soluble

On the basis of the performance testing of each type of lubricant on specific metals for intended applications, a series of lubricants from different lubricant manufacturers have been qualified for use in tube bending applications of aluminum, titanium, and steel products by various manufacturers.

13.2.5 Speed

The bending speed of titanium must be lowered compared to CRESS due to low ductility of titanium. With proper tooling combinations, bend speeds of 20–25° per second can be used to obtain good quality bends.

13.3 Forming

Besides bending of tubes/ducts, several other metal forming processes are utilized; including bulge forming, draw formed pullout, and bead-forming technologies. These are commonly used to produce various tube/duct products as needed in the environmental control system of an aircraft.

13.3.1 Bulge Forming

Isostatic bulge forming utilizes a liquid-filled chamber or pressure vessel contained within a static structural support. A tooling die holding a blank tube is placed inside the chamber, and the chamber is pressurized causing the blank tube to form to the die geometry as illustrated in Figure 13.10. Figure 13.11 shows an example of bulge formed aluminum duct.

The axial bulge-forming process consists of forming a tubular sheet metal blank into a die cavity by applying a combination of internal hydraulic pressure and axial compressive force. Figure 13.12 shows some examples of axial bulge formed parts.

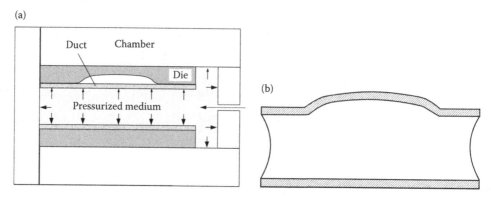

FIGURE 13.10
Fundamentals of isostatic bulge forming: (a) Forming process and (b) bulge formed.

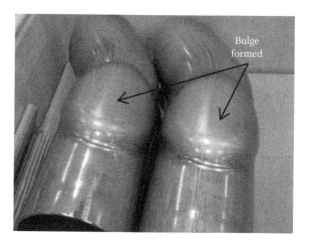

FIGURE 13.11
Example of isostatic bulge formed part. (From The Boeing Company.)

FIGURE 13.12
Examples of axial bulge formed part. (From The Boeing Company.)

13.3.2 Draw Formed Pullout

Draw formed pullout of duct in the form of an angular "T" provides an attach point to braze/weld a branch tube to the parent duct assembly as illustrated in Figure 13.13. Pullout process performance is very much dependent on the following parameters:

- Type of material
- Wall thickness
- Diameter of duct

The major process steps involved in a pullout process are illustrated in Figure 13.14 as follows:

1. Cutout hole in the parent duct by using an electrical discharge machine (EDM), abrasive water jet, or laser (Figure 13.14a)
2. Polishing of cutout hole periphery to remove the heat-affected zone
3. A ball punch is placed inside the tube perpendicular to the tube axis and attached to a piston (Figure 13.14a)
4. Piston pulls the ball punch through the tube cutout and forms a T branch (Figure 13.14b)

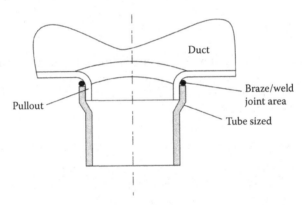

FIGURE 13.13
Schematic of tube size over pullout.

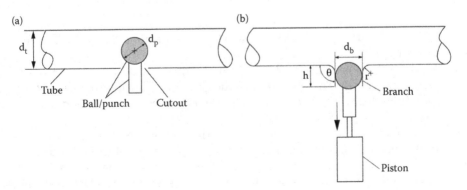

FIGURE 13.14
Fundamentals of tube/duct pullout: (a) Before and (b) after forming.

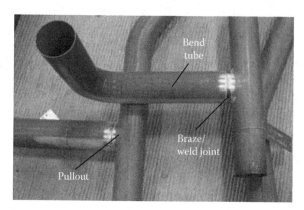

FIGURE 13.15
Example of pullout and joint. (From The Boeing Company.)

The basic engineering dimensions for the pullout process are:

1. Branch diameter, d_b
2. Branch height, h
3. Bend radius, r
4. Angle, θ

Pullout branch angle generally stays between 45° and 90°. The recommended angles are 45°, 60°, 75°, and 90° with standard increments of 5°. The pullout angle is very much dependent on tube/duct material. Figure 13.15 shows an example of a tube pullout branch brazed or welded to a straight or bends tube.

13.3.3 Bead Forming

The main objective of a bead-forming process is to create a smooth ridge of material around the circumference of a tube/duct end for joining to other ducting runs. Bead forming is generally performed by two major processes as follows:

1. Roller
2. Elastomeric

Electromagnetic forming processes can also be used forming beads in the tube/duct as explained in Section 12.3.

Roller bead forming is performed by placing a duct inside two die halves with an offset arbor in the center. The arbor forces the duct material into the die grooves as it rotates around the duct ID and forms close to the bead geometry in the die as shown in Figure 13.16.

Elastomeric bead forming follows the same principle of elastomeric internal swaging illustrated in Section 13.4.2. This is accomplished by placing a duct into a die set over a draw bolt. The draw bolt is fitted with an elastomeric sleeve. The ram on the machine pulls the draw bolt down against a stop in the die, compressing the elastomeric expander against the inside duct wall, forming the bead into the die bead groove. The elastomeric beading process provides a higher quality bead over roller beading with minimal work

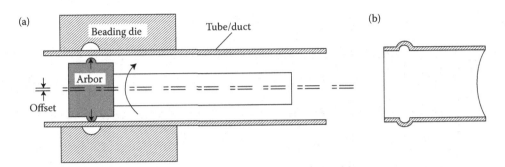

FIGURE 13.16
Fundamentals of roller bead forming: (a) Roller bead forming and (b) bead formed tube/duct.

hardening and little to no sizing requirements after forming. This is due to the uniform forming action of the elastic expander.

13.4 Swaging

Joining hydraulic end fittings to the end of a tube is done by using swaging processes. Depending on the type of hydraulic joint, attaching of an end fitting is accomplished by a few different type of swaging processes, which include:

1. Roller
2. Elastomeric
3. Axial
4. Radial

Roller and elastomeric processes are accomplished by forming the tube wall outward into the grooves within the end fitting. In the case of axial and radial swaging, fittings are mechanically attached to the outer diameter of the tube by deforming the fitting into the tube. Fittings are generally made of high strength alloys including 15-5 CRES, 7075 aluminum, and 6Al-4V titanium.

13.4.1 Roller Swage

Internal roller swaging is used for installing ferrules, couplings, unions, and other miscellaneous fittings to the tube ends. The fittings are permanently attached to the appropriate tube by placing a rotary expander tool into the end of the tube assembly and applying a predetermined torque to the expander. The tube wall material is then forced into the serrated grooves of the fitting as illustrated in Figure 13.17.

13.4.2 Elastomeric Swaging

Elastomeric swaging is a cold forming process, which utilizes extremely high radial internal swaging forces for joining of end fittings at the end of the tube. In this process, the

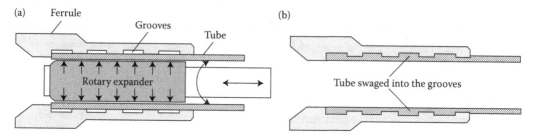

FIGURE 13.17
Schematic of internal roller swaging: (a) Before forming and (b) after forming.

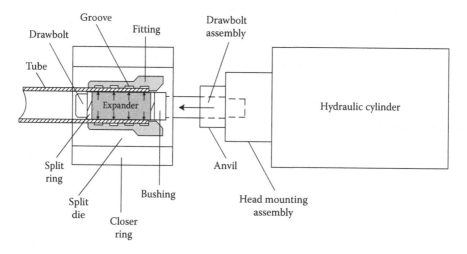

FIGURE 13.18
Schematic of internal elastomeric swaging.

tube outside wall is deformed into the grooves on the inside of the fitting geometry by an elastomeric expander.

The internal swaging forces are produced by the controlled linear travel of a high-strength draw bolt, which compresses an elastomeric expander against a static retainer ring surrounding the draw bolt as illustrated in Figure 13.18. During the swaging cycle, controlled compression of the elastomeric material forces the tube wall radially into internal circumferential grooves in the end fitting skirt, producing a strong, leak-proof mechanical attachment. During forming, the tube end fitting is contained within a precision matching split die cavity to provide a reactionary force to the forming operation and prevent distortion of the fitting.

This process is widely used in the commercial and military aircraft industry to fabricate tubing assemblies for hydraulic, fuel, oxygen, and similar fluid distribution systems. Currently, elastomeric swaging is used to fabricate tubing assemblies for both low- and high-pressure fluid systems made from aluminum, steel, and stainless-steel alloys. Applications include high-temperature nickel-based alloys and titanium.

13.4.3 Axial Swaging

In axial swaging process, the fitting is mechanically attached to the tube by deforming the fitting into the tube by moving a ring (a component of Permalite/AS fitting) axially along the fitting length using a Permaswage Axial swage Tool (DAT). Axial swaging produces an interference fit between fitting and tube resulting in a leak-proof metal-to-metal seal and structurally sound joint [3].

13.4.4 Radial Swaging

In radial swaging, the permaswage system, a tube connecting system, mechanically attaches a Permaswage fitting to a tube. In this process, the metal fitting is plastically deformed onto a tube radially (360° around the circumference) using the Permaswage Lightweight Tool (DLT). Permaswage system is used to make permanent and separable connections between metal tubes [3].

13.5 Transition End Forming

Transition ends of tubes/ducts are required to change the end geometry for attaching mechanical fittings or connectors or for joining with another tube or duct. The right type of forming technology is essential to meet the specific needs of the product. Rotary swaging technology is considered to have a wide variety of transition end-forming capabilities.

13.5.1 Rotary Swaging

The rotary swaging process is used to reduce the cross-sectional area of rods and tubes at their ends. Swaging is often called as cold forging process because the forming of metal takes place under hammering impact of die sections at room temperature. The fundamentals of rotary swaging are shown in Figure 13.19. The machine consists mainly of a hollow spindle, which carries the die sections and rollers. The die is inserted in the slot

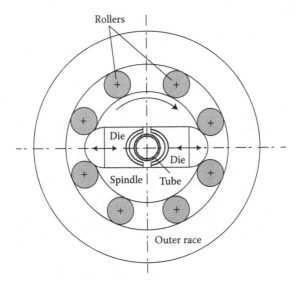

FIGURE 13.19
Fundamentals of rotary swaging process.

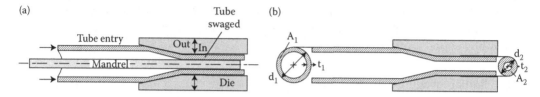

FIGURE 13.20
Schematic representation of taper swaging of tube: (a) Swaging with mandrel and (b) dimensional change.

in a spindle, rotated, and forced together repeatedly by the rollers around the periphery, as much as several times a minute, forming the taper on the workpiece. In this process, the diameter of the tube or the rod is reduced by forcing the workpiece into confined die geometry. A set of reciprocating dies provide radial force to cause the metal to flow inward and acquire the form of the die groove geometry. The die movements may be in and out type (Figure 13.20a) or rotary type. The rotary type is obtained with the help of set rollers in the cage, which is similar to a roller bearing. The workpiece is held stationary and the dies rotate, the dies forge the workpiece at a rate as high as 10–20 strokes per second.

In tube swaging, the tube thickness and or internal diameter of the tube can be controlled with the use of internal mandrels (Figure 13.20a). For small diameter tubes, a smaller diameter rod can be used as a mandrel. In tube swaging, the tube thickness grows as the tube is taper swaged as shown in Figure 13.20b. Since there is no loss of material in the process, the volume constancy relationship is being maintained through the swaging process, starting from the tapered entry to the final reduction in diameter.

$$A_1 t_1 = A_2 t_2 \tag{13.1}$$

As a tube is being pushed through the tapered entry of the die, there is a reduction of tube cross-section, $A_2 < A_1$ with some increase of tube thickness as well as the overall length of the tube after swaging. Figure 13.21 shows an example of the overall growth of length after swaging, $L_2 > L_1$.

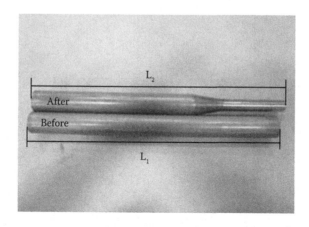

FIGURE 13.21
Growth of aluminum tube length after taper swaging. (From Primus International, University Swaging Division, WA, USA.)

FIGURE 13.22
Rotary taper swaging machine and forming die. (From Primus International, University Swaging Division, WA, USA.)

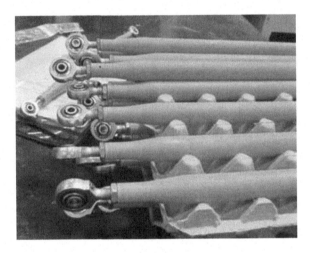

FIGURE 13.23
Example of rotary swaged aluminum parts for an aircraft. (From Primus International, University Swaging Division, WA, USA.)

Rotary swagers are available in two types: (1) two-die, (2) four-die configurations to meet the need of swaging harder materials and generally achieve greater reduction per pass. Swaging dies are the most critical elements of the swaging machines. Production of high precision parts for aerospace requires high-quality dies. Production machines can swage from a needle to larger diameter tubing. The length of a swaging die is dependent on the taper length and taper angle of the product. Aerospace products generally have long, shallow tapers. Figure 13.22 shows an example of a rotary swaging machine in a two-die configurations with long shallow taper angle.

There are wide applications of rotary taper swaged parts in an airplane, including flight control rods, engine/nacelle struts, door actuation rods, landing gear push rods, links, and more. An example of a flight control actuation rod (taper swaged, machined inner thread, chemical finish, painted, and mechanical threaded attachment to the taper swaged ends) of an aircraft is shown in Figure 13.23.

References

1. Kalpakjian, S. and Schmid, S.R., *Manufacturing Processes for Engineering Materials*, 5th Edition, Pearson, Upper Saddle River, NJ, 2008.
2. Forney, C.E. and Meredith, S.E., *Ti-3Al-2.5V Seamless Tubing Engineering Guide*, Sandvik Special Metals Corporation, WA, USA, 1990.
3. www.permaswage.com

14

Welding Technology in Aerospace

14.1 Introduction

Welding is most commonly a metal-joining technology. Welding technology mainly involves four fundamental factors, which are (1) the metals to be joined, (2) an energy source, (3) filler material (if used), and (4) a kind of gas shield (if used). The metal is heated to its welding temperature while at the same time an inert gas from the air shields in order to avoid oxidation, and a filler metal is also often added to the joining area when producing a welded joint.

In 1750 AD, a process known as forge welding was first described. In this process, the joining area of two metal pieces was heated to a temperature below the material melting point and then mechanically forged. This joining process continued until 1876. In 1877, Elihu Thomson invented resistance welding technology and encouraged inventors to explore many different joining technologies to start the modern welding age. Welding technology made rapid advancement during the early twentieth century as World War I and World War II increased the demand for reliable and inexpensive metal joining methods. During this period, several welding techniques were developed, including shielded arc welding, gas metal arc welding, submerged arc welding, flux-cored arc welding, etc. Developments continued with the invention of laser beam welding (LBW) and electron beam (EB) welding in the latter half of the twentieth century. Today, researchers continue to develop new welding methods and automation to meet the industrial need for greater control of weld quality and properties.

Welding technology is a major need for most metal manufacturing industries. Industries from underwater construction to space exploration depend on welding technology. Welding has been used in the automotive industry for quite some time for joining metal frames and sheet metal parts for the car chassis. Application of welding technology in the aerospace industry is still a big opportunity for various structural applications. Buy-to-fly ratio (weight ratio between the input raw material used and the final component installed in the aircraft, which describes the cost associated with the amount of input raw materials to the finished product) is the driving factor when considering welding technology in various aerospace applications for both military and commercial aircraft. Welding technology can reduce usage of fastener to join metal parts without making any holes for the fasteners. Welding technology may reduce the buy-to-fly ratio by joining metal parts of near net shape geometry followed by machining to the final geometry of the part.

The American Welding Society defined a welding process as "a materials joining process which produces coalescence of materials by heating them to suitable temperatures with or without the application of pressure or by the application of pressure alone and with or without the use of filler material." Many different energy sources are used to

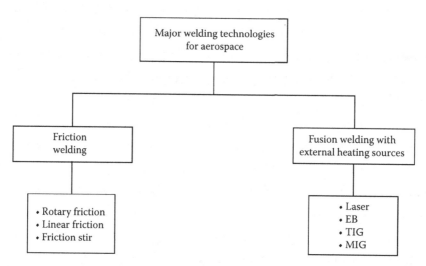

FIGURE 14.1
Major welding technologies for aerospace applications.

heat the metals for major welding processes including a gas flame, laser beam, EB, electric arc including tungsten inert gas (TIG) and metal inert gas (MIG), etc. Among all the major welding processes, few processes are being considered to be successful in aerospace applications due to certain key factors such as fatigue performance, defects, and quality assurance (QA) and process qualification. This chapter focuses on major welding and joining technologies being considered in the aerospace industry as illustrated in Figure 14.1. The basic principles of each process and the associated variables in the process will be discussed.

The properties of a weld joint must satisfy mainly the following requirements for aerospace applications:

1. Mechanical strength
2. Fracture toughness
3. Damage tolerance
4. Fatigue strength

In some cases

1. Formability
2. Corrosion performance
3. Electrical conductivity

14.2 Type of Weld Joints

Weld joints can be prepared geometrically in many different ways. There are five basic types of weld joints as shown in Figure 14.2 for joining metal components:

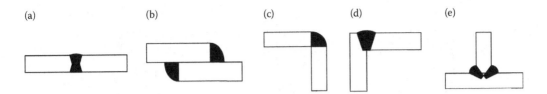

FIGURE 14.2
Basic weld joints: (a) Butt, (b) lap, (c) corner, (d) edge, and (e) tee.

1. Butt joint
2. Lap joint
3. Corner joint
4. Edge joint
5. Tee joint

14.3 Friction Welding

Tribology and thermodynamics and their relationship play an important role in friction welding technologies. Tribology is the science of surface topography and its effect on friction between two rubbing surfaces in relative motion. Thermodynamics in friction welding is the effect of heat generation due to friction between the workpiece surfaces or workpiece and tool surfaces in relative motion under contact pressure. The frictional heat brings both the joining metal objects into the plastic deformation phase before the welding pressure is applied to join in solid state. No additional material or filler material is normally needed and there are typically no harmful emissions from the process. As illustrated in Figure 14.1, friction-assisted welding processes are classified into three major types including rotary friction, linear friction, and friction-stir welding (FSW).

14.3.1 Rotary Friction Welding

Rotary friction welding (RFW) follows the principle of frictional heat generation at the interface of two objects in rotary motion [1]. In RFW, one object is rotated against the other. RFW is most commonly used for joining round objects of similar or dissimilar materials as solid or hollow as shown in Figure 14.3a. Object 2 is stationary, while object 1 is rotating at high speed at a revolution N per unit time and is brought into contact with object 2 under pressure P_C. The pressure at the interface and the resulting friction force F_f generate heat to raise sufficient temperature T_f to bring both the metal objects into a plastic deformation stage before the rotating object is brought to a quick stop for the welding to take place within a very short time period t_w before applying the weld pressure P_w. It is basically a solid state joining process. Due to the combination of frictional heat and joining pressure P_w, which is slightly greater than the contact pressure P_c at the interface, a forging flash is developed by plastic deformation within the heated zone as shown in Figure 14.3b. If the flash is not acceptable with respect to product quality, then it can easily be removed by machining. Dissimilar metals with two different flow stresses σ_1 and σ_2 would be joined together using the same principle.

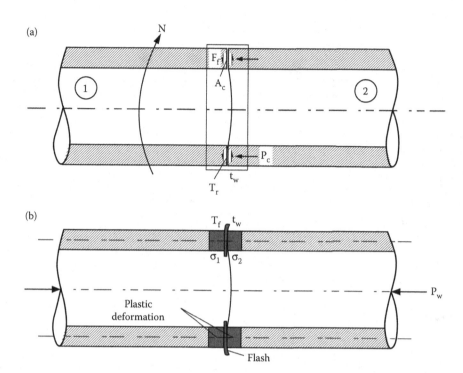

FIGURE 14.3
Schematic diagram of RFW: (a) Rotating object 1 is brought in contact with stationary object 2 and (b) develops flash in forging.

The principle variables applied to RFW are:

1. Joining metal contact geometry and the area of contact, A_c
2. Rotation of object, N
3. Contact pressure, P_c
4. Frictional force at the interface, F_f
5. Joining pressure, P_w

Interface friction can be represented in a functional form as

$$F_f = f(A_c, P_c, N, \tau_c) \tag{14.1}$$

where τ_c is the critical shear strength of the two similar material.
    Temperature rise at the interface may be represented in the functional form as

$$T_f = f(F_f, k_m, t_f) \tag{14.2}$$

where k_m is the material thermal conductivity and t_f the total rotation time of friction. For two dissimilar materials, the thermal conductivity and flow stress behavior at elevated temperature for both the materials needs to be considered. An example of RFW of cylindrical-shaped products is shown in Figure 14.4.

FIGURE 14.4
Example of RFW of cylindrical-shaped products. (From Thompson Friction Welding, UK.)

14.3.2 Linear Friction Welding

The concept of conventional RFW technology led to the development of linear friction welding (LFW). LFW [2] follows the same basic principle of taking advantage of heat generation from friction between two joining metal members in a linear reciprocating motion as illustrated in Figure 14.5. Friction welding is carried out by oscillating one joining metal object at high speed relative to the other stationary metal object along a common interface, with a compressive force across the joint. In the very beginning of the reciprocating motion, there is sliding friction, then a combination of sliding and sticking friction, and finally 100% sticking friction before forging takes place after completion of oscillation. Frictional heat is generated and conducted away from the interfacial area. Plastically deformed interface material is extruded out of the edges of the joint so that clean material from each component is left along the original interface. The relative motion is then stopped and a final compressive force is applied causing forging to take place within the clean material at the interface. This creates the frictional weld joint with flash as shown in Figure 14.5b. Flash is formed preferably in the direction of oscillation. LFW samples of metal blocks of different cross section are shown in Figure 14.6.

The major variables involved in LFW are:

- Reciprocating motion, $V_r = f(A_r, f_r)$, where A_r is the reciprocating amplitude, f_r is the frequency of oscillation
- Contact pressure, P_C
- Frictional force at the interface, F_f
- Joining/forging pressure, P_W

P_W can be equal to P_C or greater than P_C depending on the joining materials. Some materials need higher forging forces to achieve good quality weld. Frictional heat needs to be retained within the area for the final forge operation. Depending on the thermal conductivity of the

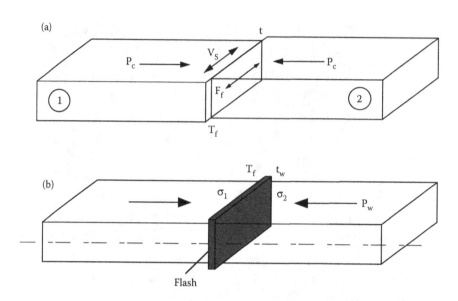

FIGURE 14.5
Schematic diagram of LFW: (a) Oscillating object 1 is brought in contact with object 2 and (b) welding takes place with flash.

FIGURE 14.6
Example of LFW of metal blocks. (From Thompson Linear Friction Welding, UK.)

material, it determines the forging decisions to be made at the end or during the reciprocating motion. Since aluminum is a highly conductive material, it transfers heat very quickly from the high temperature side (contact surface) to the other end of the material; the forging is performed at the end of oscillation. LFW has been successfully used to join materials including steel, intermetallic materials, aluminum, nickel, and titanium alloys.

LFW has been proven for industrial application in aircraft engine manufacturing. This process has proved to be an ideal process for joining turbine fan blades to discs or hubs as shown in Figure 14.7 [3]. LFW has been shown to be more cost effective than machining blade/disc (blisks) assemblies from a solid large metal block.

The specific benefits of LFW are the following:

- Solid-state joining process—reproducible process with high-quality weld
- Filler material or shielding gas is not normally required

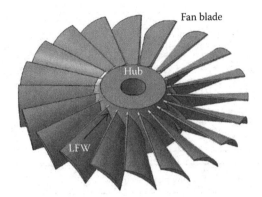

FIGURE 14.7
Linear friction welded aircraft engine fan blade. (From Thompson Linear Friction Welding, UK.)

- No fume or spatter and energy efficient—green process
- Relatively little loss of material
- Dissimilar materials can be joined

14.3.3 Friction-Stir Welding

FSW is another solid state joining process under the influence of frictional heat. There is no melting involved in the process unlike conventional fusion welding processes. The FSW process follows the principle of utilizing the frictional heat between a rotating tool and the two joining metal interfaces as shown schematically in Figure 14.8. The tool serves two basic purposes as follows:

1. Heating of the workpiece material due to friction and plastic deformation
2. Stir movement and containment of material to produce the joint

The FSW process was originally invented and experimentally verified by Wayne Thomas and his team at The Welding Institute (TWI) in UK in 1991. TWI holds number of patents on FSW process with the first one issued in 1991 [4]. FSW is a TWI-licensed process and is currently in use by many industries, institutions, and organizations all over the world. The primary features, advantages, and applications of FSW are summarized in the original introductory paper [5]. Since the introduction of FSW by TWI, the amount of research work conducted on this technology has increased tremendously and more than 1000 articles have been published since then. A review article [6] is recommended to the beginner to have over-all idea of this technology including process, tooling, microstructure, properties, and more.

FSW uses a nonconsumable rotating tool, with a specially designed shoulder with an attached pin. The length of the pin is usually equal to slightly less than the thickness of the workpiece materials being joined or welded. The tool is rotated with a vertical downward force (F_z) and slowly the pin of the tool is plunged into the abutting edges of the workpiece materials, which are butted together with clamping pressure P_C until the tool shoulder contacts the top surface of the workpiece. The tool with rotation (N_T) in clockwise or counterclockwise is traversed with a linear feed (f_T) along the weld or joint line as shown in Figure 14.8a. The relative motion between the tool and the workpiece materials generates frictional heat at the interfaces of the shoulder and to a lesser extent, the pin (also called

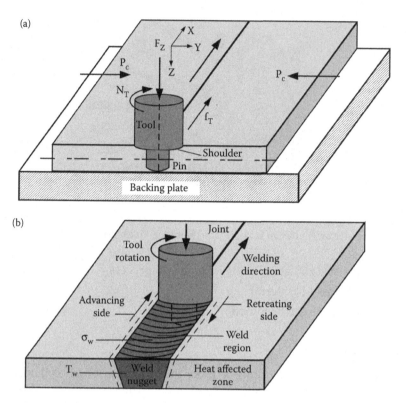

FIGURE 14.8
Schematic diagram of friction FSW: (a) Tool plunged into the work-piece and (b) FSW begins.

probe) with the workpiece materials. The rotational speed, feed, and vertical or axial load on the pin are determined to generate sufficient heat to soften or plasticize the material without reaching the melting point. The rotation of the tool results in stirring and mixing of material around the rotating pin, and the feed of the tool moves the stirred material from the front to the back of the pin and completes the welding.

The major variables associated with FSW are as follows:

- Tool geometry (shoulder and pin) and their combined contact area (Ac)
- Tool rotation speed, N_T
- Linear feed of the tool, f_T
- Axial force, F_Z
- Clamping pressure between two joining objects, P_C
- Flow stress of joining workpiece, σ_w

The amount of heat generated in the FSW process is mainly dependent on the following factors:

- Frictional force
- Thermal conductivity of the workpiece materials

The peripheral tool shoulder velocity is an important limiting factor in FSW as too high a rubbing speed can cause over softening (and in extreme cases melting) of the workpiece

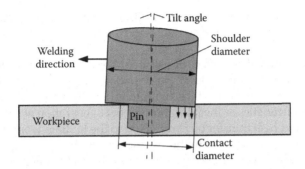

FIGURE 14.9
Schematic of tool tilt in FSW.

material, which leads to poor weld quality and/or breakdown of the process. Weld quality can also be improved by tilting the tool system with an angle of around 2° or 3° with respect to the normal flat position as shown in Figure 14.9 [7]. This increases the pressure at the back of the tool, where the weld is formed. A brief summary of major FSW terminology is shown in Table 14.1.

14.3.3.1 Tribology and Thermodynamics in FSW

Tribology and thermodynamics and their relationship play a tremendous role in the FSW process and influence the strength, quality, and the appearance of the FSW joints. Tribology is the science of surface topography and its effect on friction between two rubbing surfaces.

TABLE 14.1

Major FSW Terminology

Terminology	Function
Tool shoulder	Part of the rotating tool, which sits to come in contact with the workpiece surface either in flat or in tilted position to generate frictional heat
Tool pin/probe	Part of the tool that penetrates the workpiece
Advancing side	Side of weld where the direction of tool travel is the same as direction of rotation of shoulder
Retreating side	Side of weld where the direction of tool travel is the opposite as direction of rotation of shoulder
Axial force	The force applied to the FSW tool downward to plunge the probe into the workpiece and maintain shoulder in contact with the workpiece surface
Traverse force	The force required to feed the rotating FSW tool along the joint line through the workpiece material
Tool rotation	The tool rotation speed (can be represented by revolution per unit time)
Traverse speed	Speed of FSW tool movement along the joint line
Spindle torque	Torque required to rotate the FSW pin when plunging as well as feeding through the workpiece along the joint
Tool tilt angle	The angle at which the FSW pin is positioned relative to the workpiece surface
Heel plunge depth	The depth of the back of a tilted tool shoulder relative to the workpiece surface
Tool shoulder footprint	Area of shoulder in contact with the workpiece
Anvil/backing plate	The supporting plate used to prevent the forces applied to the workpiece
Weld nugget	Large volume of material processed in FSW
HAZ	HAZ is the area that experiences heat effects of welding on the workpiece material surrounding the weld

Friction will be controlled by the surface condition of the tool geometry at the tool and workpiece interfaces. The real area of contact will be a function of real geometry provided to the tool shoulder.

Thermodynamics in FSW is the effect of frictional heat, which brings the flow stress (σ_w) of the joining material to a proper plastic deformation stage to transfer the material from the leading edge to the trailing edge of the tool probe. Finally, thermodynamics may change the metallurgical characteristics of the joint through heat generation at the tool shoulder and workpiece interface and heat transfer by conduction to the tooling, workpiece, and weld and convection from tooling and the weld to the atmosphere as shown in Figure 14.10. Examples of FSW machines used by TWI for their continuous research in FSW are shown in Figure 14.11.

The major advantages of FSW are as follows:

- Solid state welding (no porosity, segregation, etc.)
- Excellent mechanical properties can be achieved

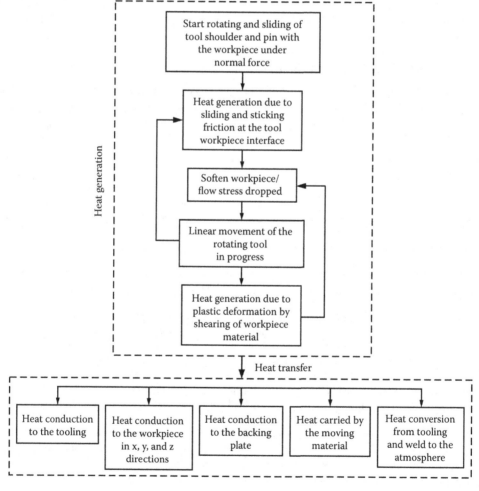

FIGURE 14.10
Tribology and thermodynamics in FSW.

(a)

(b)

(c)

FIGURE 14.11
FSW machines: (a) Precision spindle machine, (b) high-power FSW system, and (c) robot FSW system. (From TWI, UK.)

- Green process (low energy consumption)
- Less residual stress
- Automated machine tool process
- Opportunity to weld non-fusion-weldable alloys
- No shielding gas or filler wires needed

Proper design of the FSW tool and the selection of effective tool materials are critical factors that determine effective welding parameters for the production of good quality joints. It is understood from the principle of the FSW process that the shoulder and the pin perform different functions at the start to the end of the process as follows:

1. Shoulder
   a. Heating and softening the workpiece material
   b. Consolidating softened material to form solid phase joint

2. Pin
   a. Dispersing the oxide layers in the joint
   b. Extruding material from the front of the tool to back
   c. Extruding material from top to bottom of joint

Selection of proper tool material is an important consideration. The essential properties of the tool materials are as follows:

- High strength at service temperature (at ambient and elevated temperatures)
- Good wear resistance (at ambient and elevated temperatures)
- Good creep resistance (at ambient and elevated temperatures)
- Good fracture toughness (at ambient and elevated temperatures)
- Good manufacturability into complex shapes
- Inert with workpiece material

Besides the essential properties of the tool material, there are some desirable properties too:

1. Thermal conductivity and diffusivity
2. Environmentally stable
3. Reasonable price
4. Availability from multiple sources

AISI-H13 hot die steel and similar steels have proved to be satisfactory in FSW of most aluminum alloys. The hardness of the tool is usually around 45HRC. Tungsten alloys, such as Densimet grade D176, have proven satisfactory for some higher Zn concentration aluminum alloys.

14.3.3.2 Tool Design

Tool design is a continuously developing area in FSW process. Early work in TWI used a cylindrical (or slightly tapered) probe or pin beneath a cylindrical shoulder [4]. Early research on the FSW process reported similar simple tool designs and showed the tool probe as a small rounded protrusion beneath a flat tool shoulder [8]. The early developments at TWI in tool design for FSW concentrated on the development of improved FSW tool forms, and the assessment of their capabilities in a range of aluminum alloys and thicknesses. A wide range of possible tool designs for FSW of aluminum sheet of 0.25 in (6.4 mm) thickness were assessed, as illustrated in Figure 14.12.

The continuous development of tool designs brought an improved tool design consisting of a concave shoulder, and a parallel cylindrical pin-type probe with a screw thread feature as shown in Figure 14.13. This tool represented the state of the art for the FSW of aluminum in 1995. The thread features were designed to force material downward against the backing plate.

Further patent filing in 1997 by TWI [9] detailed a large number of possible FSW tool designs and features. The main tool designs from this selection were also presented at the 3rd World Congress on Aluminum in the same year [10]. The Whorl™ tool designed by TWI consists of a frustum probe with a coarse helical ridge profile, which acts to force

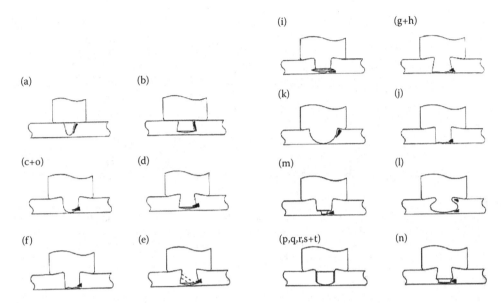

FIGURE 14.12
FSW tool profiles considered for 0.25 in (6.4 mm) thick aluminum sheet/plate product, the dark areas indicate the approximate positions of any weld flaws observed in each case. (a–t): Each tool profile (probe and shoulder geometry) having their combined frictional contact area generates certain amount of frictional heat in the joining material and allow the soften or plasticized material to flow along the weld direction that may lead to the weld quality. (From TWI, UK.)

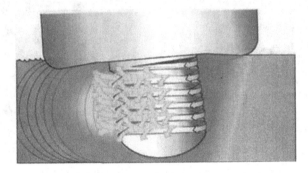

FIGURE 14.13
Schematic of tool design with screw thread features. (From TWI, UK.)

material downward during the welding process. A summary of the Whorl™ tool design was reported in 2000 [11,12] showing a number of possible variants as shown in Figure 14.14.

More recent developments in tool design in the industry have shown an enhancement of the variant in Figure 14.14d. This tool features three re-entrant flutes in a helix form on the probe, coupled with another smaller helix screw thread around the outside surface. In addition, tool shoulder features have been added, which assist in the development and containment of the material shear zone. This technology is designated as the multi-helix tri-flute or MXTriflute™ tool [13,14], the state-of-the-art FSW tool design. A typical example of MXTriflute™ tools is shown in Figure 14.15a,b. This design allowed improved welding speeds, and represents the current state-of-the art as of 2006. The geometry of the rotating tool in FSW is a cylindrical-shouldered tool with an extended probe or pin. The extension could be threaded or unthreaded.

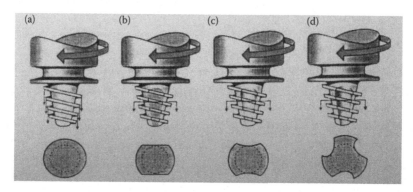

FIGURE 14.14

Basic variants of the Whorl™ tool: (a) Oval shape, (b) paddle shape, (c) two re-entrant, and (d) three sided probe. (Adapted from Nicholas, E.D. and Kallee, S.W., Friction stir welding—A decade on, *IIW Asian Pacific International Congress*, Sydney, Australia, October 2000; Russell, M.J., *Development and Modeling of Friction Stir Welding*, PhD thesis, University of Cambridge, UK, August, 2000.)

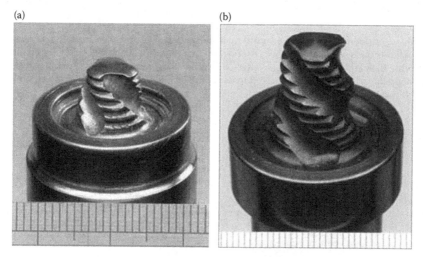

FIGURE 14.15

MXTriflute™ tools developed by TWI: (a) Shorter pin length and (b) longer pin length. (From TWI, UK.)

14.3.3.3 Microstructural Zones in FSW

Owing to the contribution of severe plastic deformation and high-temperature exposure within the stirred zone during FSW, three distinct zones have been identified within a typical micrograph as illustrated schematically in Figure 14.16:

1. Stir zone (weld nugget)
2. Thermo-mechanically affected zone (TMAZ)
3. Heat affected zone (HAZ)

The stir (nugget) zone or dynamically recrystallized zone is a heavily deformed material that roughly corresponds to the location of the pin during welding. Various shapes of nugget zones are observed depending on the FSW process parameter, tool geometry,

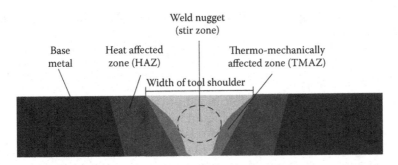

FIGURE 14.16
Schematic of different zones in FSW.

temperature rise in the workpiece, and the thermal conductivity of the material. The TMAZ zone is a transition zone between parent material and the nugget zone. In this region, the deformation and temperature are lower and as a result the effect of welding on the microstructure is correspondingly lower. The HAZ is the area that experiences the thermal effects of welding; the HAZ material surrounds the weld. The quality of weld would depend on the materials used and the heat input of the welding process used. The HAZ can be of varying size and strength. The HAZ is relatively small when the cooling rate is high due to higher conductivity of the material. On the other hand a slow cooling rate will typically provide a larger HAZ. HAZ size is process-dependent where both the amount and area of concentration of heat inputs are applied.

The heat from the FSW process and subsequent cooling causes microstructure and properties to be altered in the area surrounding the main FSW nugget. The extent and magnitude of property change depends primarily on the workpiece material, and the amount of heat input by the welding process. Close control of heat and its transfer toward joining material is very important to ensure the right temperature is achieved for the right material property.

The microstructure of the FSW joint is mainly dependent on

- Tool geometry
- Material characteristics of the workpiece material
- Welding parameters
  - Tool rotational speed
  - Tool feed
  - Applied pressure
  - Temperature of weld
  - Time of weld
- Backing plate material (e.g., tool steel, copper, tungsten, and ceramics)

14.4 Fusion Welding with External Heating Sources

External heat sources to generate localized heat at the joining interfaces between the metallic work-pieces could be from many different welding technologies [1,15,16] including high-energy beam welding (laser, EB) and arc welding (TIG, MIG), which will be discussed in this section.

14.4.1 Laser

LASER is an acronym for "light amplification by stimulated emission of radiation," and is defined as "any of several devices that emit highly amplified and coherent radiation of one or more discrete frequencies." LBW is a modern welding process, which is a high-energy beam process that continues to expand into modern industries and new applications because of its many advantages such as deep weld penetration and minimizing heat inputs. This technology can be considered to weld a variety of materials including steel, titanium, aluminum, and nickel alloys, generally used for the aerospace industry.

The fundamentals of LBW are illustrated in Figure 14.17. A laser beam is focused through lenses or mirrors. The focal spot is targeted on the workpiece surface, which will be welded. At the surface, the large concentration of light energy is converted into thermal energy. The surface of the workpiece starts melting and progresses through it by surface conductance. The role of focusing lenses in this process is really important because it concentrates the beam energy into a focal spot as small as 0.005 in (0.127 mm) diameters or even less.

A laser beam is generated either in a gas or in solid media. The most used industrial lasers are the carbon dioxide (CO_2) laser and the Nd-YAG laser:

1. Nd-YAG (neodymium–yttrium–aluminum–garnet) laser—uses a man-made crystal and produces a laser beam with 1.06-μm wavelength.
2. CO_2 laser—uses a mixture of gases including CO_2 as the active medium and produces light with 10.6-μm wavelength.

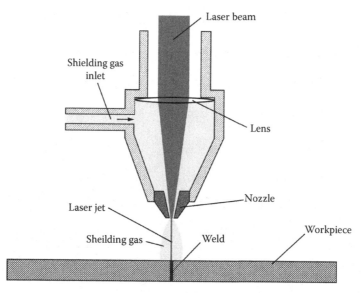

FIGURE 14.17
Fundamentals of laser welding principle.

The major process variables included:

- Type of joining materials (conductivity, reflectivity)
- Joint design
- Type and size of laser
- Position of focal point
- Focusing
- Welding speed
- Pulse frequency
- Duration of pulses
- Type and amount of shielding gas
- Nozzle design

Advantages of laser welding

1. Deep and narrow welds can be done
2. Absence of distortion in welds created
3. Minimal heat-affected zones in welds created
4. Excellent metallurgical quality will be established in welds
5. Ability to weld smaller, thinner components
6. Increased travel speeds
7. Non-contact welding

14.4.2 Electron Beam

In EB welding [1,16], heat is generated by high-velocity, narrow-beam electrons as shown in Figure 14.18. In heat-generation process, the kinetic energy of the electrons is converted into heat as the electrons strike the workpiece. The EB welding process requires special equipment to focus the beam on the workpiece and also a vacuum atmosphere. The higher the vacuum, the more the beam penetrates into the workpiece having higher depth-to-width ratio. The vacuum levels are commonly specified as high vacuum (HV), medium vacuum (MV), and no vacuum (NV). NV could be effective on some materials. Generally, any similar or dissimilar metals could be butt or lap welded using EB process, starting

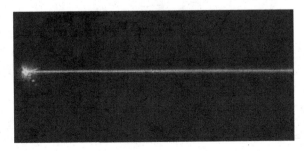

FIGURE 14.18
EB welding, gun on the left. (From TWI, UK.)

from foil to the plate thickness of 5.9 in (150 mm). Generally, no shielding gas, flux, or filler metal is required for this welding process. Capacities of EB guns range to generally 100 kW.

EB welding was developed in the 1960s, and continues producing high-quality, deep, and narrow welds with small HAZ. The results of EB welding depend on certain welding parameters that included:

- Beam power (kW)
- Power density (focusing of the beam)
- Welding speed
- Material properties
- Geometry of the joint

High-power EB welding has been performed successfully to join aluminum alloys including 2xxx, 5xxx, and 7xxx series to produce single pass welds and takes advantage of high joining rate with high weld quality:

- Deep, narrow fusion zone
- Low distortion and shrinkage

Typical applications of EB welding include joining of aircraft, missile, nuclear, and electronic components, as well as gears and shafts in the automotive industry.

14.4.3 TIG

Gas tungsten arc welding (GTAW) is frequently referred to as TIG welding and became an overnight success in the 1940s for joining magnesium and aluminum. TIG welding has become a popular choice of welding processes when high-quality, precision welding is required. In TIG welding an arc is formed between a nonconsumable tungsten electrode and the metal being welded. Gas is fed through the torch to shield the electrode and molten weld pool. If a filler rod is used, it is added to the weld pool separately. Figure 14.19 shows the schematic representation of TIG-welding process.

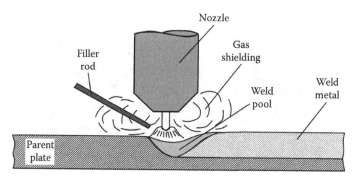

FIGURE 14.19
TIG welding process.

The use of an inert gas shield instead of a slag to protect the weld pool was a highly attractive replacement for gas and manual metal arc welding. TIG has played a major role in the acceptance of aluminum for high-quality welding and structural applications.

The major variables that influence weld quality include the following:

- Welding current and voltage
- Type of current used
- Gas coverage of weld zone
- Cleanliness of weld area
- Joint design
- Welding speed or deposition rate
- Electrode composition and end profile
- Conductivity of electrode metal
- Composition and properties of joining material

Orbital welding follows the same fundamentals of GTAW process in a controlled manner for welding tubing and pipe products. In the 1950s, orbital welding technology was developed by North American Aviation to weld hydraulic lines for the X-15 rocket vehicle when leaks developed. The aerospace industry recognized orbital welding technology in the 1960s as filling a need for a superior joining technique for aircraft hydraulic lines, fuel lines, and ducting. The orbital welding process has a mechanism to rotate a welding arc around a tube-weld joint. In orbital welding, the GTAW process is used as the source of the electric arc that melts the base material and forms the weld. During GTAW process, an electric arc forms between a tungsten electrode and the tube workpiece. Figure 14.20 shows an example of orbital tube weld process where an orbital weld head rotates an electrode and electric arc around the weld joint to make the required weld.

The major variables in the orbital welding process are:

- Tube material
- Tube edge—squareness and burrs
- Tube cleanliness
- Tungsten geometry and eccentricity

FIGURE 14.20
Example of orbital welding of tube. (From Techsouthinc http://www.techsouthinc.com/orbital/ehead.html)

- Power
- Purge gas quality and quantity

The advantages of orbital welding include (1) repeatable process, (2) smooth appearance, (3) less HAZ, and (4) very minimal defects such as porosity and cracks.

14.4.4 Metal Inert Gas

Metal inert gas (MIG) welding was first patented in the United States in 1949 for welding aluminum. The arc and weld pool formed using a bare wire electrode was protected by helium gas, readily available at that time. From about 1952 the process became popular in the United Kingdom for welding aluminum using argon as the shielding gas, and for carbon steels using CO_2. CO_2 and argon–CO_2 mixtures are known as metal active gas (MAG) processes.

Figure 14.21 illustrates the fundamentals of MIG welding process. MIG welding is a commonly used high-deposition-rate welding process. A consumable electrode is continuously fed into the weld pool. MIG welding is therefore referred to as a semiautomatic welding process. The shielding gas forms the arc plasma, stabilizes the arc on the metal being welded, shields the arc and molten weld pool, and allows smooth transfer of metal from the weld wire to the molten weld pool.

The primary shielding gasses are generally argon and helium. Argon gas is usually used for welding mild steel, aluminum, titanium, and alloy steels. Helium is generally used for high-speed welding of mild steel and titanium and also for stainless steel and copper. These gasses are often mixed with CO_2 to reduce cost. CO_2 is also used in its pure form in some MIG welding processes. CO_2 used as a component of a gas mixture is generally used for carbon and low alloy steels.

The major variables involved in MIG welding are:

- Welding current
- Type of current, AC or DC
- Heat impact/arc voltage/preheat
- Gas coverage of weld zone

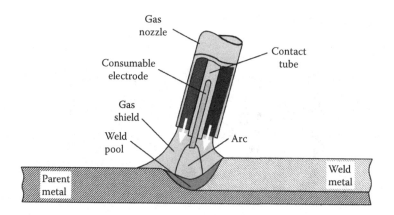

FIGURE 14.21
MIG welding process.

- Cleanliness of weld area
- Welding speed
- Electrode composition and end profile
- Metal conductivity
- Joint design
- Joining material composition and properties

References

1. http://www.twi-global.com
2. http://www.linearfrictionwelding.com
3. Jones, S., How to linear friction weld a blisk, *Presented at AeroMat Conference*, Long Beach, CA, 2011.
4. Thomas, W.M., Nicholas, E.D., Needham, J.C., Murch, M.G., Temple-smith, P., and Dawes, C.J. Friction-stir butt welding, GB Patent No. 9125978.8, 1991.
5. Dawes, C. and Thomas, W.M., Friction stir welding, *11th Annual North American Welding Research Conference*, Ohio, November, 1995.
6. Mishra, R.S. and Ma, Z.Y., Friction stir welding and processing, *Materials Science and Engineering* R50 2005.
7. Thomas, W.M., Friction stir welding and related friction process characteristics, *Proc. INALCO'98 7th Intl. Conference on Joints in Aluminum*, Cambridge, UK, 1998.
8. Barrett, J., *Energised Tool Stirs Up Welding*, Eureka on Campus, Spring Issue, UK, 1994.
9. Thomas, W.M., Nicholas, E.D., Needham, J.C., Temple-Smith, P., Kallee, S.W., and Dawes, C.J., Friction stir welding, UK Patent Application GB2306366A, 1997.
10. Thomas, W.M. and Nicholas, E.D., Friction stir welding and friction extrusion of aluminum and its alloys, *3rd World Conference on Aluminum*, Limassol, Cyprus, April 1997.
11. Nicholas, E.D. and Kallee, S.W., Friction stir welding—A decade on, *IIW Asian Pacific International Congress*, Sydney, Australia, October 2000.
12. Russell, M.J., *Development and Modeling of Friction Stir Welding*, PhD thesis, University of Cambridge, UK, August, 2000.
13. Thomas, W.M. and Andrews, R.E., High performance tools for friction stir welding (FSW), International Patent No. WO 99/52669, 1999.
14. Thomas, W.M., Nicholas, E.D., and Smith, S.D., Friction stir welding—Tool developments, *TMS Annual Meeting and Exhibition*, New Orleans, 2001.
15. Todd, R.H., Allen, D.K., and Alting, L., *Manufacturing Processes Reference Guide*, Industrial Press Inc., NY, 1994.
16. Kalpakjian, S. and Schmidt, S.R., *Manufacturing Processes for Engineering Materials*, Pearson, NY, 2008.
17. http://www.techsouthinc.com/orbital/ehead.html

15

Metal Cutting and Machining Technology

15.1 Introduction

Metal cutting and machining technology is the most common and necessary technology to enable the aerospace industry to manufacture series of components of an aircraft. There have been tremendous amount of improvements made, developing from simple metal cutting machines with manual controls to very special purpose numerically controlled (NC) machines to meet the requirements of the aerospace customers. For aerospace applications, NC machines have been developed ranging from small to very large sizes, which are hard to imagine, if not personally witnessed to manufacture mainly the wing components of large aircraft. The major aerospace input structural raw materials are produced mainly by rolling, extrusion, forging, and other forming operations as discussed in Chapter 4. The geometry and the dimensions of the initial raw stock materials are selected based on the final geometry of the part after machining to satisfy various engineering requirements; mainly structural strength, toughness, fatigue, and the aircraft weight and also the ease of manufacturing. For manufacturing major structural components, quite a substantial amount of machining is required to finish the part per geometry and dimensional tolerances from the original wrought products received from the metal manufacturers. The buy-to-fly ratio (weight ratio between input raw material used and the weight of the final component installed in the aircraft), a term commonly referred to within the aerospace community, is an important factor to be considered owing to various reasons including energy used, capital, tool, labor costs, etc. Generally, the lower the ratio, the more economical it is to produce the parts and finally reduce the overall manufacturing cost of the aircraft. The metal cutting processes used in manufacturing aircraft components are dependent on many factors including the type of material, the size and shape of the part, the dimensional tolerances, surface characteristics, and finally the production economics. The cost of machining is very much dependent on total machining time and the type of material being machined, which determines the life of a cutter and the capital cost of the machine. To meet the manufacturing and economic needs of the aircraft parts of various metals including aluminum, titanium, steels, and super alloys, various machining centers have been developed in the last few years. Similarly, huge varieties of cutting tools have been developed to satisfy the requirements of the machining centers as well as the life of the cutting tools. This chapter illustrates the fundamentals of metal-cutting principles and machining technology as the author learnt from his undergraduate course of Workshop Technology [1], and Principles of Machine Tools [2] in India during 1972. This chapter also illustrates machining of most common aircraft metals including aluminum, titanium, steel, and composite materials used for major structural components of an aircraft.

15.2 Fundamentals of Metal Cutting

This section describes the basic concepts of metal-cutting principles to help understand some important aspects of metal cutting including metal-cutting methods, cutting tool nomenclature, mechanics of chip formation, kinematics and dynamics of metal cutting, and thermodynamics of metal cutting.

15.2.1 Methods of Metal Cutting

The two basic methods of metal cutting using a single-point tool are (1) orthogonal or 2D, and (2) oblique or 3D. In orthogonal cutting, the cutting face of the tool stays at 0°, 90°, or 180° to the path of the tool. In oblique cutting, the cutting face of the tool stays less than $0° < \theta < 90°$, or $90° < \theta < 180°$ to the path of the tool. Orthogonal and oblique cutting principles are illustrated in Figure 15.1, which shows two workpieces having identical cutting parameters including depth of cut and feed. But in oblique cutting the cutting force acts on a larger area of contact at the tool–workpiece interface.

Figure 15.2 illustrates the flow of a chip in orthogonal and oblique cutting. In orthogonal cutting shown in Figure 15.2a, the cutting edge, AB of the tool is at right angle to the direction of relative velocity, V_w, of the workpiece. The chip curls into a tight flat spiral. In oblique cutting as shown in Figure 15.2b where the cutting edge, AB of the tool is inclined at an angle θ_n, the chip flows sidewise in a long curl. The inclined angle θ_n is defined as the angle between the cutting edge and the normal to the direction of the velocity, V_w, of

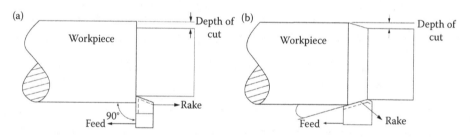

FIGURE 15.1
Basic method of metal cutting: (a) Orthogonal and (b) oblique.

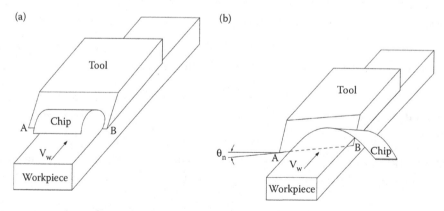

FIGURE 15.2
Chip flow direction: (a) Orthogonal and (b) oblique.

the workpiece. Orthogonal cutting is actually confined mainly to certain metal-cutting operations including knife turning, broaching, and slotting. The bulk of machining operations are done by oblique cutting.

15.2.2 Mechanics of Chip Formation

Figure 15.3 shows the schematic representation of a metal-cutting process with a single-point tool where the tool is moving toward the stationary workpiece. The metal in front of the tool is getting compressed with a high compressive stress. That stress brings metal into a plastic deformation stage. When the compressive stress in the workpiece reaches and exceeds the value of the ultimate strength of the workpiece material, the metal particle in front of the tool tip will shear to form a chip. The shearing movement of each successive element is arrested by work hardening, and the movement is then transferred to the next element. This process is repetitive, and a continuous chip is formed with a highly compressed and burnished underside and the top side is minutely serrated as caused by shearing action. The plane along shearing metal elements is called the shear plane. In summary, the chip is formed by the plastic deformation of the metal grain structure along the shear plane. In actual operation, the deformation does not occur sharply across the shear plane instead it occurs along a narrow band as shown in Figure 15.4.

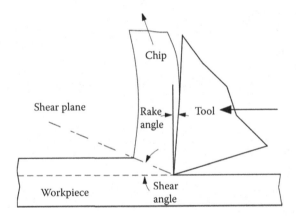

FIGURE 15.3
Tool and chip configuration.

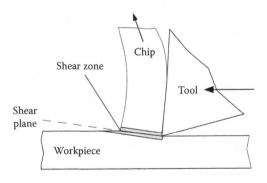

FIGURE 15.4
Shear zone in metal cutting.

15.2.3 Type of Chips

In a machining operation, the material is removed in the form of chips from the cutting tool–material interface. The dimension and the form of the chip will indicate the nature and quality of the machining process. The type of chip formed is mainly influenced by the properties of the individual material and the cutting variables. In machining of metals, three major chips are formed (Figure 15.5) as follows:

1. Continuous
2. Discontinuous
3. Continuous with build-up edge (BUE)

In continuous chip formation (Figure 15.5a), the metal chip coming out from the tool–metal interface is bonded firmly without any break. The metal flows continuously along the shear plane by plastic deformation and makes a continuous ribbon of metal chip from the machined surface at the metal–tool interface. The bottom part of the chip, which always slides over the tool face, becomes shinny. The continuous chip formation is considered most desirable for

- Low friction at the tool–chip interface
- Lower power consumption
- Long tool life
- Good surface finish

The major factors in favor of forming continuous chip are:

- Metal ductility
- Fine feed
- High cutting speed
- Large rake angle
- Sharp cutting edge
- Smooth tool face
- Effective lubrication

In discontinuous chip formation (Figure 15.5b), the chips consist of elements fractured into fairly small pieces ahead of the cutting tool. Discontinuous chips are generally obtained while machining brittle type material. These materials rupture during plastic

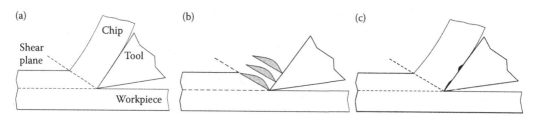

FIGURE 15.5
Three major type of chips: (a) Continuous, (b) discontinuous, and (c) continuous with BUE.

deformation along the shear plane, and form chips as separate small pieces as shown in Figure 15.5b. Discontinuous chips also can be formed on some ductile metals under certain conditions mainly at very low cutting speed and if the friction coefficient is low. This resulted in bad surface finish and shorter tool life with ductile metals. The conditions that tend to promote formation of discontinuous chips include:

- Brittle metal
- Larger depth of cut
- Low cutting speed
- Small rake angle

The building up of a ridge of metal on the top surface of the tool above the cutting edge is called the BUE. During cutting, a pile of compressed and highly stressed metals forms a build-up at the edge of the tool. Due to high heat and pressure generated at the cutting tip, the piled up metal is welded or bonded to the cutting tip and forms a false cutting edge to the tool as shown in Figure 15.5c. This welded or bonded metal is basically strain-hardened and very brittle in nature. As the chip moves along the tool, the weaker chip metals tear away from the weld. The build-up becomes unstable and breaks down as some of the fragments adhere to the workpiece surface to produce a rough surface finish.

The following conditions tend to promote forming BUE:

- Low cutting speed
- Low rake angle
- High feed

15.2.3.1 Chip Thickness Ratio

In the orthogonal cutting tool model shown in Figure 15.6, the tool with a positive rake angle β moves along the workpiece with a velocity, V_t, and a depth of cut, t_d. A chip of thickness t_c is generated ahead of the tool by shearing the metal along the shear plane making an angle with the horizontal surface of the workpiece. As the cutting continues, the chip is generated above the shear plane and climbing up the face of the cutting tool. Since there is no loss of material, the volume constancy relation holds good at the shear plane–tool interface to calculate the thickness of the chip. From the chip formation geometry

$$t_c = t_d \frac{\cos(\alpha - \beta)}{\sin \alpha} \tag{15.1}$$

$$t_c > t_d \tag{15.2}$$

The outward flow of metal causes the chip to be thicker. Chip compression ratio as defined by the ratio of chip thickness t_c and the depth of cut t_d can be written as

$$r_c = \frac{t_c}{t_d} = \frac{\cos(\alpha - \beta)}{\sin \alpha} \tag{15.3}$$

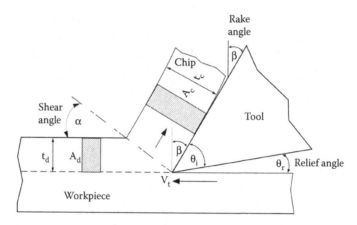

FIGURE 15.6
Chip formation in orthogonal cutting.

15.2.4 Kinematics of Metal Cutting

15.2.4.1 Velocity Relation

The velocity relationship for orthogonal cutting is shown in Figure 15.7. Figure 15.7a shows three velocity components to complete the velocity diagram (Figure 15.7b). The components are listed as follows:

1. V_t, the cutting velocity
2. V_s, the velocity of shearing material
3. V_c, the velocity of chip.

From the geometrical relationship, each velocity components can be driven as follows:

$$V_s = V_t \frac{\cos\beta}{\cos(\alpha - \beta)} \tag{15.4}$$

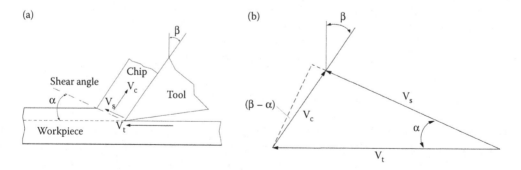

FIGURE 15.7
Velocity relationship: (a) Velocity components and (b) velocity diagram.

$$V_c = V_t \frac{\sin \alpha}{\cos(\alpha - \beta)} \tag{15.5}$$

$$V_t = V_s + V_c \tag{15.6}$$

15.2.4.2 Forces on Chip

The forces acting on the chip in an orthogonal cutting operation are shown in Figure 15.8. The shearing force F_s acts along the shear plane that is opposing to shear the metal in forming the chip. F_n is acting perpendicular to the shear plane provided by the workpiece. Force F_r is the frictional resistance acting on the tool–chip interface opposite to the direction of chip flow. The force, N, on the chip is acting normal to the cutting face of the tool.

15.2.4.3 Cutting Force

Figure 15.9 shows the Merchant's circle diagram showing the relationship between various forces and angles. In this diagram, two force triangles shown in Figure 15.8 have been combined. R and R_1 together have been replaced by R in the circle diagram. Force R has two components: F_c, which is the cutting force provided by tool on the workpiece and F_f is the feed force.

The following mathematical relationships are determined from the force system shown in Merchant's circle diagram.

$$F_r = F_f \cos \beta + F_c \sin \beta \tag{15.7}$$

$$N = F_c \cos \beta - F_f \sin \beta \tag{15.8}$$

where F_r is the frictional resistance at the rake surface and N is the normal force on the rake surface.

The average friction coefficient μ can be estimated as

$$\mu = \tan \gamma = \frac{F_r}{N} = \frac{F_f + F_c \tan \beta}{F_c - F_f \tan \beta} \tag{15.9}$$

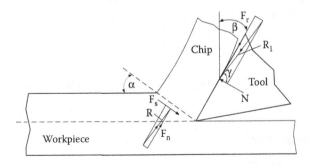

FIGURE 15.8
Forces on chip.

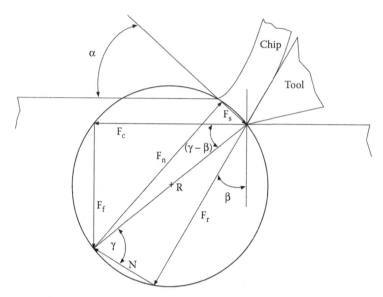

FIGURE 15.9
Merchant's force circle diagram.

where γ is the mean friction angle at the rake surface.

Similarly F_s and F_n can be determined by the following relationships:

$$F_s = F_c \cos\alpha - F_f \sin\alpha \qquad (15.10)$$

$$F_n = F_f \cos\alpha + F_c \sin\alpha \qquad (15.11)$$

F_n can be written in terms of shearing force, F_s as

$$F_n = F_s \tan(\alpha + \gamma - \beta) \qquad (15.12)$$

Merchant's developed relationship between the shear angle α, friction angle γ, and the rake angle β in the following form:

$$2\alpha + \gamma - \beta = C \qquad (15.13)$$

C is the machining constant related to the workpiece material, which is a dependent parameter.

15.2.4.4 Work Done in Metal Cutting

Total work done in metal cutting is given by

$$W_T = F_c V_t = W_S + W_F \qquad (15.14)$$

Similarly, work done in shear and friction are given by

$$W_S = F_s V_s \tag{15.15}$$

$$W_F = F_r V_c \tag{15.16}$$

where F_c is the cutting force, F_s the shear force, and F_r the friction force. V_t is the cutting speed, V_s the speed of shearing material, and V_c the speed of chip.

15.2.5 Thermodynamics in Metal Cutting

Figure 15.10 shows the sources of heat energy in metal cutting at three interface locations. Location 1 is at the shear zone where the plastic deformation is taking place. The major portion of the plastic deformation energy is converted into heat energy. Location 2 is another source of heat where the chip is leaving from the tool surface with the friction drag between the flowing chip and the tool surface. Location 3 has additional frictional force along the direction of the tool at the tool tip and the workpiece interface.

Heat generation and heat transfer in a single-point cutting operations is shown in the schematic representation of thermodynamics cycle, Figure 15.11. Thermodynamic cycle is a very critical aspect to determine the tool life and the maintenance of the tool as well as the surface quality of the machined surface. Both the factors are contributing toward the economics of machining.

15.3 Cutting Tool

As discussed earlier, a cutting tool has a unique role in metal-cutting operations since it is actually removing material under certain kinematic conditions of a metal-cutting

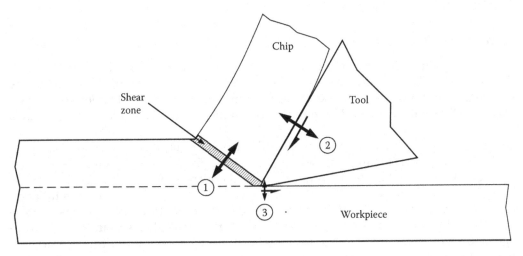

FIGURE 15.10
Sources of heat in metal cutting.

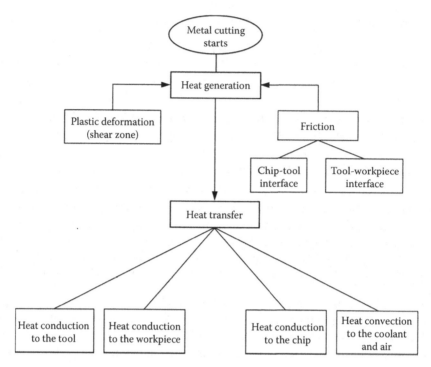

FIGURE 15.11
Thermodynamics in metal cutting.

operation. To understand how the tool is removing material, it is necessary to understand the various parts and angles of a cutting tool. The cutting tools are mainly divided into two groups as follows:

1. Single-point
2. Multipoint

15.3.1 Cutting Tool Parameters

The principles of various cutting tool angles as shown in Figure 15.12 are the same for either single-point or multipoint tool. Figure 15.12 shows the complete nomenclature of a single-point tool, which makes it easy to understand the fundamentals of a cutting tool. Different elements shown in the tool nomenclature can define the shape of a tool.

Single-point cutting tools having wedge-like action are utilized for a wide variety of applications in lathe, shaping, and slotting machines. Single-point tools are the simplest form of cutting tool, whereas in multipoint tools, there are two or more single-point tools arranged together as a unit. The milling cutter is one of the best examples of multipoint tool. The cutting processes involved with multipoint tools are closely looked at as machining by a tool having multiple single-point tools. Figure 15.13 shows examples of single-point tools in turning and multipoint tools in milling operations. In this chapter, the fundamentals of the cutting principles were discussed using a single-point cutting tool as an example.

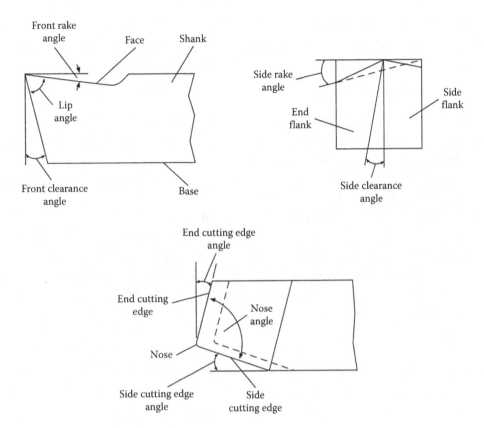

FIGURE 15.12
Single-point cutting tool nomenclatures.

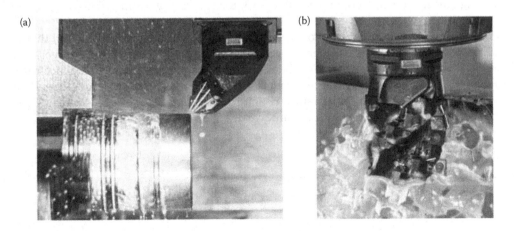

FIGURE 15.13
Examples of cutting tools in operation: (a) Single-point tool (Sandvik CoroTurn HP) in turning and (b) multi-point tool (Sandvik CoroMill 690) in milling. (From Sandvik Coromant AB.)

15.3.2 Cutting Tool Materials

The selection of the right tool material depends mainly on the machinability of the material to be machined. The major characteristics of cutting tool materials are as follows:

1. Wear resistance—must withstand excessive wear due to extreme machining conditions
2. Hot hardness—must maintain hardness at elevated operating temperature
3. Toughness—must have sufficient combination of strength and ductility to withstand shocks and vibrations and to prevent breakage
4. Friction—must remain low friction at the chip–tool interface for minimum wear and better surface finish
5. Fabrication cost—must maintain low cost in fabrication of tool

No single tool material is good for machining all materials and selection is also driven by the type of application. Based on the characteristics and performance of various tool materials, proper tool material is selected to perform a specific machining operation on a specific workpiece material. The principal cutting tool materials are:

- Carbon steel
- High-speed steel (HSS)
- Stellite alloys
- Cemented carbide
- Ceramics
- Diamond

15.3.3 Tool Wear and Tool Life

Every cutting tool has its definite tool life. Tool life is the time a tool performs satisfactorily until it is dulled or broken. A blunt tool normally can cause high cutting force and power, over heat of the tool and finally poor surface finish. The tool life is affected by many factors including cutting speed, feed, and depth of cut, tool geometry, cutting fluid, and rigidity of the machine.

In metal cutting operations, the tools are subjected to extreme mechanical and thermal conditions. During the process, there is a direct interaction at the tool and the workpiece material interface as well as the chip slides at high speed along the tool rake face resulting in high friction at the surfaces causing it to rise to high surface temperatures. A key factor in the tool wear is the temperature rise during operation. Figure 15.14 illustrates the common types of tool wear including flank wear, crater, and chipping.

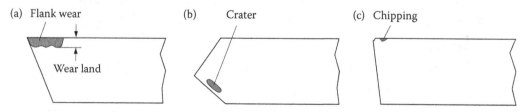

FIGURE 15.14
Common type tool wear: (a) Flank wear, (b) crater, and (c) chipping.

15.4 Cutting Fluid

The principle reason for using a cutting fluid is to reduce the friction at the cutting tool–workpiece material interface. Figure 15.15 shows the capillary passages allowing the cutting fluid to penetrate the interface and improve lubrication and cooling. When the chip moves up the tool face, it mainly touches the tip of the asperities of the tool surface and creates a capillary action between chip and tool. Those capillaries draw cutting fluid. The cutting fluid reacts chemically with the cutting tool surface to create low shear strength boundary lubrication film to reduce the friction at the tool–chip interface under the cutting condition at high pressure and elevated temperature.

15.4.1 Type of Fluids

Cutting fluids in the same manner as lubricants or coolants have a very major role in machining or metal-cutting operations. Cutting fluids are mainly used for

- Cooling tool
- Cooling work piece
- Reducing friction
- Improving surface finish
- Breaking the chips
- Washing the chips from the tool

Selection of cutting fluid is a very important task to satisfy many requirements including:

- Good lubrication to reduce friction
- High heat absorption
- High flash point
- Health issues (skin irritation, odorless, breathing, etc.)
- Noncorrosive to the workpiece or to the machine elements

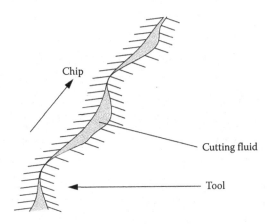

FIGURE 15.15
Cutting fluid at the chip–tool interface. (From M.E. Merchant.)

- Low in viscosity
- Price

The type of cutting fluid selected depends on the workpiece material and the type of the machining process. Depending on the machining process, cutting fluids may be a direct lubricant or an emulsified coolant. Cutting fluids are classified into several groups as follows:

- Water
- Soluble oils
- Direct oils
- Mixed oils
- Chemical additives
- Solid lubricants

15.5 Major Metal-Cutting and Machining Operations

Major metal-cutting operations involved in fabrication for producing various aircraft components are illustrated in Figure 15.16. Major metal-cutting operations are classified mainly as continuous and intermittent type cutting.

15.5.1 Turning

Turning is the metal cutting or removal process using a machine called a "Lathe" and is considered to be among the oldest machine tools, and was developed in the 1750s. Turning is accomplished by holding the workpiece securely and rigidly on the machine and then turning it against cutting tool, which will remove metal from the workpiece in the form of chips. The major operations accomplished by using lathe include:

- Straight turning
- Shoulder turning

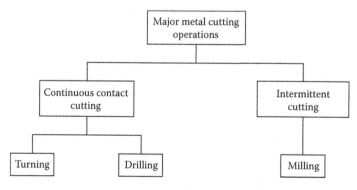

FIGURE 15.16
Major metal-cutting operations.

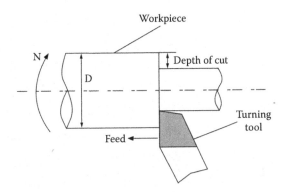

FIGURE 15.17
Principle of turning operation.

- Taper turning
- Eccentric turning
- Chamfering
- Grooving
- Polishing
- Tread cutting
- Facing
- Knurling and more

Some additional operations that can be performed by holding the workpiece by a chuck include:

- Drilling
- Reaming
- Boring
- Counter boring
- Internal thread cutting and more

The basic principle of straight turning operation as an example is illustrated in Figure 15.17. The workpiece is turned straight when it is made to rotate about the lathe axis, and the tool is fed parallel to the lathe axis. The straight turning produces a cylindrical surface by removing excess metal from the workpiece. There are normally two kinds of cuts utilized as follows:

1. Roughing cut or rough turning by applying high rate of feed and heavy depth of cut
2. Finishing cut or finish turning by applying high cutting speed, small feed, and a very small depth of cut

15.5.2 Drilling

Drilling machines are primarily designed to originate a hole in a workpiece. The hole is generated by the rotating edge of a cutting tool known as the drill bit, which exerts a

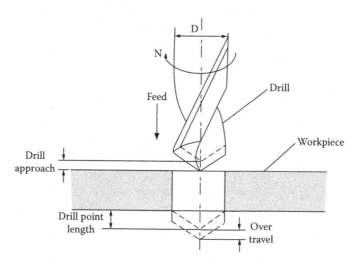

FIGURE 15.18
Fundamentals of drilling operation.

large force on the workpiece. The fundamentals of a drilling operation are illustrated in Figure 15.18.

Different operations can be performed using the drilling machines. Some major operations include:

- Drilling
- Reaming
- Boring
- Counter boring
- Counter sinking
- Spot facing and more

15.5.3 Milling

Milling is a faster metal-cutting operation using a machine tool called a "Milling Machine," which removes the metal as the workpiece is fed against a rotating multipoint cutter. The cutter rotates at a high speed and because of the multiple cutting edges it removes metal at a high rate. The machine may hold one or more number of cutters at a time. Based on the direction of spindle axis with respect to the machine table, the milling machine can be classified as (1) horizontal milling, (2) vertical milling, and also (3) universal milling. The principle of horizontal type milling (spindle axis is parallel to the table) is illustrated in Figure 15.19.

There are wide applications of milling operations in the aerospace industry due to large volumes of metal removal acquired from the primary metal stocks including extrusion, forging, rolled bar, and plate products. Figure 15.20 illustrates various types of milling operations generally used in a fabrication process. The major machining operations performed in a milling machine include:

- Plain milling
- Face milling

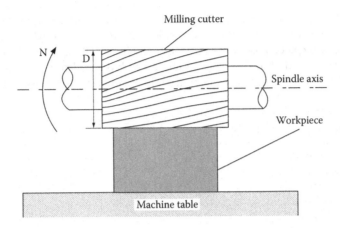

FIGURE 15.19
Principle of milling operation.

FIGURE 15.20
Various types of milling operations. (From Sandvik Coromant.)

- Side milling
- End milling
- Profile milling
- Form milling
- Gang milling
- Saw milling
- Milling grooves and slots and more

15.5.4 Cutting Speed, Feed, and Depth of Cut in Machining

In metal-cutting operations, the cutting speed may be defined as the speed difference between the cutting tool and the surface of the workpiece. Cutting speed is also considered as a surface speed that is defined as how fast the metal comes into contact with the tool at the cutting point. For an example, on a lathe, the workpiece is turning with a certain speed at which the surface of the workpiece passes the cutting tool. On a milling machine, cutting speed is considered to be the speed of the rotating cutter at which the end-mill moves past the stationary workpiece. The speed and feed are generally determined by the following major factors:

1. Type of material being cut
2. Type of tool material and tool life
3. Shape and tool geometry
4. Chip dimensions
5. Type of finish
6. Machine rigidity
7. Type of coolants

15.5.4.1 Cutting Speed

Cutting speed may be calculated for different metal cutting operations including lathe, drilling, and milling as

$$V_c = \pi DN \tag{15.17}$$

D is the diameter of the workpiece (lathe) or the diameter of the cutter (drilling/milling) and N is the turning of the workpiece (lathe) or the revolution of spindle (drilling and milling) per unit time.

15.5.4.2 Feed

The feed of a cutting tool in turning operation is the distance the tool advances for each revolution of the workpiece or per unit time.

Feed in a drilling operation is the linear distance the drill moves into the workpiece at each revolution of the spindle. In drilling, feed is also defined by the axial distance moved by the drill into the workpiece per unit time or for each revolution of the drill.

The feed in a milling operation is defined as the rate with which the workpiece advances under the cutter, expressed in three different methods as follows:

1. Feed per tooth—the distance per tooth of the cutter
2. Feed per cutter revolution—the distance per revolution of the cutter
3. Feed per unit time—the distance workpiece advances per unit time

15.5.4.3 Depth of Cut

The depth of cut is the distance measured from the machined surface to the uncut surface of the workpiece. In a lathe turning operation, the depth of cut is expressed as one-half of the difference between the diameter of the workpiece before machining and the diameter of the machined surface.

In a drilling operation, the depth of cut is equal to one-half of the drill diameter. The depth of cut in a milling operation is the thickness of the material removal in one pass of the workpiece under the cutter.

Besides speed, feed, and depth of cut, the metal removal rate (Q) is an important factor to be considered. Q is normally expressed in volume of metal removed per unit time. Q is established for each machining operation as expressed below:

$$Q_{milling} = \text{Width of cut} \times \text{Depth of cut} \times \text{Feed per unit time} \tag{15.18}$$

$$Q_{turning} = \pi D \times \text{Depth of cut} \times \text{Feed per unit time} \tag{15.19}$$

$$Q_{drilling} = \left(\frac{\pi D^2}{4} \right) \times \text{Feed per unit time} \tag{15.20}$$

15.6 High-Speed Machining Centers

A machining center is nothing more than a highly automated machine tool capable of performing multiple machining operations including turning, milling, drilling, tapping, and many more under CNC control. The term "machining center" describes almost any CNC machine that includes an automatic tool changer and a table that clamps the workpiece in place. The orientation of the spindle is the most fundamental defining characteristic of a machining center generally vertical or horizontal. Another common machining center type is the five-axis machining center, which is able to pivot the tool and/or the part in order to mill and drill at various orientations. Figure 15.21 shows an example of a 5-axis multitasking machining center, which can process an entire workpiece in a single setup from initial metal stock to the final machining per

FIGURE 15.21
Example of a 5-axis machining center from Mazak.

drawing. Metal cutting processes include turning, milling, drilling, tapping, and more. High-speed machining centers are widely used in manufacturing various aircraft components from various materials including aluminum, titanium, steels, and also composites.

15.7 Machinability

The machinability of a metal is defined as the ease with which the given metal is worked with the cutting tool. Good machinability means satisfactory performance in machining. The major factors influence the machinability of a metal including:

- Chemistry of the workpiece metal
- Grain structure
- Physical and mechanical properties
- Cutting conditions

The evaluation criteria of machinability of a metal include:

- Rate of metal removal
- Tool life
- Cutting forces
- Surface finish
- Type and size of chips
- Temperature rise in cutting

The machinability of different metals may be compared by the machinability index, which is defined as

$$\text{Machinability Index (\%)} = \frac{\text{Cutting Speed of Test Metal for Tool Life Time, t}}{\text{Cutting Speed of a Standard Metal for Tool Life Time, t}}$$

15.8 Machining of Various Metals Used in an Aircraft

The most common aircraft metals used to produce various shapes are mainly in the form of extrusion, rolled plates and bars, or forgings for machining to manufacture major structural components of an aircraft including:

- Aluminum
- Titanium
- Steel

15.8.1 Aluminum

High-strength aluminum alloys are in use for making major structural parts of a metal aircraft including fuselage skins, wing spar chords, wing spar webs, wing stringers, wing skins, wing-to-body joint chords, stabilizer skins, stringers, and chords. High-strength aluminum alloys mainly 2xxx and 7xxx series are used in manufacturing wrought products using various forming operations including extrusion, forging, and rolling as described in Chapter 4. Major structural components of fuselage, stabilizers, wing, and wing-to-body joining components for the aircraft are generally manufactured by extensive machining of initial input raw materials as received from the supplier, or are machined post-forming/processing of the initial wrought metal products. Due to high metal removal rate, high-speed machining has become a standard practice for producing aircraft parts. Figure 15.22 shows an example of a machined wing rib produced from an aluminum forging.

15.8.2 Titanium

Over the last 60 years, the aerospace industry is the single largest market for titanium products. Titanium is ideal for many aircraft applications due to its unique advantages such as high strength to weight ratio and exceptional corrosion resistance. The applications that are most significant include jet engine and airframe components that are subject to temperatures up to 1100°F and for other critical parts at the interfaces with composite aircraft structures. Figure 15.23 illustrates the application of titanium at various locations of an aircraft.

Machinability of titanium alloys has traditionally been considered to be poor. Its machining window is small, and careful consideration must be given to more factors than just selection of the tool. In order to secure successful machining in titanium, there are four areas that require special attention [3]:

1. Coolant pressure and volume
2. Programming techniques
3. Machine requirements
4. Tools with their tool holding

Milling operations are the dominant machining method, for titanium particularly in structural aerospace parts. An example of a structural part machined from titanium forging

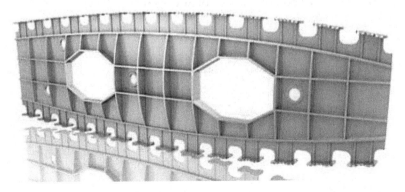

FIGURE 15.22
Example of machined wing rib from an aluminum forging. (From Sandvik Coromant.)

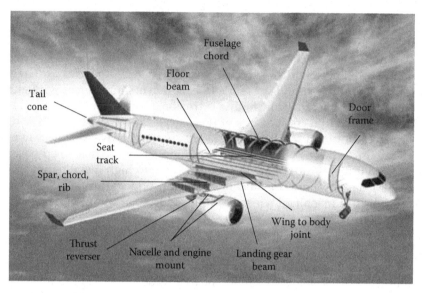

FIGURE 15.23
Usages of titanium at various location of an aircraft. (From RTI International Metals, Inc.)

is shown in Figure 15.24. There are a number of titanium alloys with different properties, and their machinability varies considerably—from traditional Ti6-4 to stronger alloys like Ti10-2-3 and now Ti5553. Among the main characteristics is a risk of rapid tool wear due to the cutting edge being exposed to higher temperatures because of its poor thermal conductivity. Tool wear or breakdown is also common due to the smearing tendency of titanium, which is reactive with tool materials. Other issues include material deflection/chatter tendency due to the elasticity of titanium. There are few general rules for titanium machining that can help overcome those issues:

- Use relatively low cutting speeds
- Use tool with sharp cutting edges
- Optimize feed rates
- Use large volume of coolant at high pressure
- Replace cutting edges at the first sign of any wear

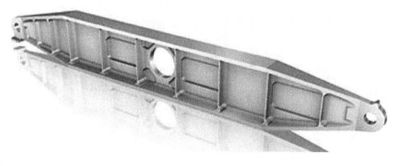

FIGURE 15.24
Example of a machined landing gear beam from a titanium forging. (From Sandvik Coromant.)

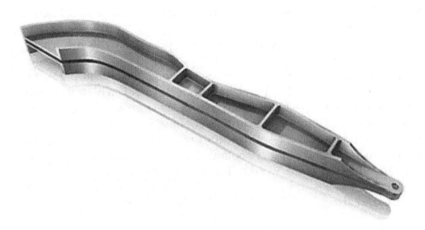

FIGURE 15.25
Example of a machined 15-5 stainless steel flap track. (From Sandvik Coromant.)

15.8.3 Steel

There are lot of applications of high-strength low-alloy steels such as 4340, 4330, and 4340 M and high-strength corrosion-resistant steels such as 15-5PH at various locations in an aircraft including steering yoke, truck position bell crank, wing landing gear lower side strut, torsion link, flap linkages, flap tracks and flap carriage, actuator arms, various components of landing gear, and more. Figure 15.25 shows an example of a 15-5 stainless-steel flap track machined from a forging.

15.9 Machining of Composite Materials

Machining of composite materials [3] has become an urgent need for the aerospace industry. In the present time, aircraft manufacturers are focusing on usage of more composite materials for new light weight fuel-efficient aircrafts. The Boeing 787 is the first example of a commercial aircrafts with its fuselage and wing structures that are made of composite materials. There are two types of fiber reinforcement that the aerospace industries use for their composite products—glass and carbon. Glass (fiber glass) is less expensive than carbon and is used for applications in which affordability and strength are required. Carbon is typically chosen for its stiffness—a carbon fiber part can be as stiff as steel at one-fifth of steel's weight. There are no major machinability differences between the two fibers. Glass is more abrasive than carbon, which causes tools to wear faster. Carbon is more brittle than glass, and it makes a finer dust that can pose larger cleanliness issues.

Depending on the circumstances, HSS, carbide, and diamond-coated cutting tools are all viable choices for machining composites. Regardless of the cutting tool's material or coating, its geometry should have a positive rake angle. A positive cutting edge tends to knife its way into the material. This shears the fibers more effectively than negative or neutral edge profiles that rub or drag across the material. When a tool begins to lose its sharpness, it tends to catch fibers without breaking them, pulling, and unwinding fibers from the part instead of cutting them.

The choice to machine wet or dry largely depends on the machining operation. Composite materials have low thermal conductivity, so coolant will be used for high-heat cutting operations to reduce the chance of degrading the part's resin. Coolant also keeps cutting tool edge temperatures under control, which is especially important for diamond-coated tools to prevent the coating from releasing the tool's substrate. Pure water is the most common machining coolant. If a machine will be cutting both composites and metals, a water-soluble fluid with 1%–2% of rust inhibitor is used. It is important to use a water-soluble coolant, because it will not adversely affect paint bonding as an oil-based fluid would. The porous surface of a machined composite part has numerous nooks and crannies where coolant oils could hide. Oil- or solvent-based coolants can also wick along exposed fibers, and they will not evaporate as pure water does.

With an operation such as single-point turning though, the dust is released in a finite area, which allows easy collection by a vacuum system installed at the machine. Those operations will be performed dry unless workpiece or tool frictional heating is an issue. The dust generated by machining composites is considered a nuisance dust (as is sawdust). While glass fiber is itchy, it is not like asbestos fiber, which can attach to a person's lungs. Carbon dust is electrically conductive, may affect electrical parts, and increase wear in a spindle. While machining, dust extraction is strongly recommended.

The major machining features that are involved for manufacturing typical structural components or an assembly of an aircraft section made with composite materials are illustrated in Figure 15.26:

- Hole making
- Edging
- Trimming
- Surface machining

Hole making (drilling and reaming) of the composite materials is very common for fastening the materials together, especially on carbon fiber reinforced plastic (CFRP) and also on the stacked materials including composites and metals such as aluminum and titanium products. The major requirements for making a hole are as follows:

- Surface roughness on both CFRP and aluminum and titanium
- Hole tolerance
- No delamination or chipping in the hole exit
- No splintering
- No chip erosion on the carbon fiber from the metallic-stacked metals

A couple of very common hole quality issues are illustrated in Figure 15.27. In splintering, there are residual fibers in the interior of the hole due to poor cutting action. In de-lamination, there is separation of the bottom layer(s) due to the thrust of the drill. The hole quality determines when the tool will need to be changed or indexed. In drilling of composite materials, there may be loss of hole quality before the tool failure.

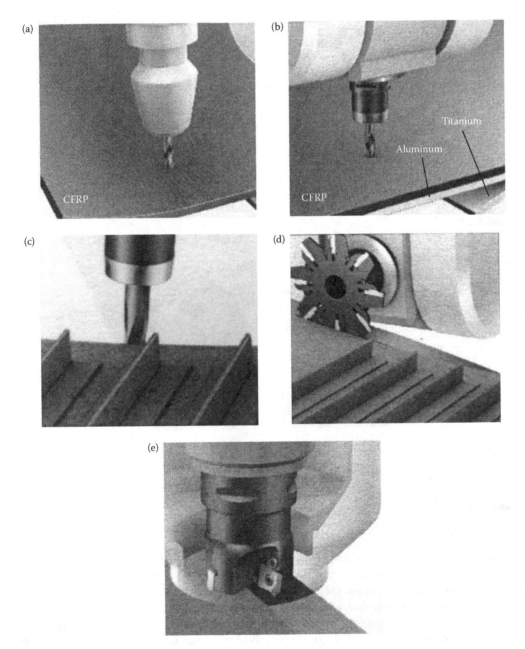

FIGURE 15.26
Typical component features of composite machining: (a) Hole making in CFRP, (b) hole making in stack, (c) edging, (d) trimming, and (e) surface machining. (From Sandvik Coromant.)

To resolve the issues of delamination and splintering in drilling composite materials or a stack of composite and other metals, the tool manufacturers are constantly bringing new solutions to mitigate the issues. Figure 15.28 shows a couple of drill tip geometries designed to prevent splintering (Figure 15.28a) and fraying and also to reduce delamination (Figure 15.28b) issues.

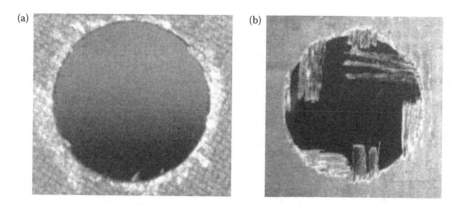

FIGURE 15.27
Hole quality issue: (a) Splintering and (b) delamination. (From Sandvik Coromant.)

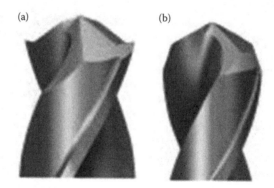

FIGURE 15.28
Drill tip geometries to prevent hole quality issues in composite drilling: (a) Coro Drill 854 and (b) Coro Drill 856. (From Sandvik Coromant.)

15.10 Surface Finish in Machining

Machining is basically a finishing process maintaining specified dimensions, tolerances, and surface finish of the part per engineering drawing. The type of surface finish that a machining operation generates is related mainly to the type of machining operation, workpiece materials, and also the type of tool, tool design, and the tool sharpness. The irregularities on the machined surface are classified into roughness or waviness depending on their size. Section 7.3 provided the fundamentals of tribology covering surface topography, friction, lubrication, and wear in major manufacturing processes. The factors that have influenced on surface finish and integrity in cutting operations have been provided [4].

15.11 Economics of Machining

Figure 15.29 illustrates the important factors influencing the cost of machining. The goal for any metal-cutting operations is to obtain the lowest possible unit cost and highest

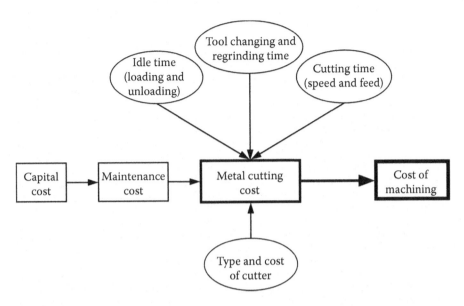

FIGURE 15.29
Economics of machining.

possible production rate. The total unit cost of machining is based on cutting cost, idle cost, tool changing cost, tool grinding cost, and maintenance cost.

Cutting cost depends on the time the tool is actually engaging with the metal blank. This time can be reduced by removing metal at a higher rate. The metal-cutting cost can always be optimized by precisely adjusting the cutting parameters. Idle time includes the time spent in loading and unloading the workpiece to and from the machine using material handling equipment. The tool changing cost includes an operator's time to remove the worn tool and install a new tool. Tool regrinding cost includes depreciation of the tool and cost for regrinding. This cost may become excessive at higher cutting speed. Maintenance cost varies with the type of machine.

References

1. Hajra Choudhury, S.K. and Bose, S.K., *Elements of Workshop Technology*, Vols. I and II, Asia Publishing House, Bombay, India, 1971.
2. Sen, G.C. and Bhattacharya, A., *Principles of Machine Tools*, Vol. I and II, New Central Book Agency, Delhi, India, 2009.
3. www.sandvik.coromant.com
4. Kalpakjian, S. and Schmidt, S.R., *Manufacturing Processes For Engineering Materials*, Pearson, NY, 2008.

16

Abrasive Metal Removal and Cutting Processes

16.1 Introduction

In Chapter 15, the conventional metal-cutting and machining technologies were discussed. Abrasive metal removal processes are generally used as post-machining operations to remove burs and sharp edges and even for polishing the surface and improving the aircraft metal component to meet the engineering surface finish requirements. Several standard processes are commonly used for final finishing. Generally, the metal parts are finished using abrasive materials in the form of sandpaper, emery cloth, disc, stone, wire brush, belts, and more. An abrasive material particle is a small sharp-edged hard element of irregular shape capable of removing a small amount of material from the interacting workpiece surface. This chapter illuminates the fundamentals that draw on the principles of tribology to explain how the abrasive particle interacts on the metal surface during metal-removal process.

In addition to abrasive metal removal using mainly metal finishing processes, there is also an abrasive cutting and machining technology known as "abrasive water-jet machining." This process, which was developed by integrating the cutting principles of sharp abrasive particles with the principles of water-jet, was also discussed. The abrasive water-jet process is generally used for cutting, trimming, profiling metal blanks, parts, nonmetallic, and composite material components. A water-jet provides the concentrated force (result of momentum change of the water stream) on the workpiece. This chapter will also cover an overview of abrasive water-jet cutting, which is widely used in the aerospace industry.

16.2 Fundamentals of Abrasive Metal Removal Processes

In abrasive metal removal processes, the tool or media carries the abrasive particles of regular or irregular geometry. An abrasive is a hard particle with sharp edges capable of removing soft material from the metal surface. An abrasive process is basically a wear mechanism created when two surfaces with a relative motion interact with each other. Wear is often thought of as a harmful phenomenon, but the wear mechanism in the abrasive process is considered beneficial to remove some thickness of material from the work-piece by mechanical action between the abrasive media and the workpiece surface to meet its surface finish requirements. The term abrasive wear covers two types of situation as follows:

1. A rough hard surface slides against the softer surface.
2. The abrasion is caused by loose hard particles sliding between two rubbing surfaces.

In both cases abrasive wear is caused by the ploughing out of the soft workpiece by a hard indenter/grit with applied pressure P and relative sliding velocity V_s as illustrated with a conical indenter in Figure 16.1.

In recent years, tribology (the science of interacting surfaces in relative motion) has received increasing attention from the industry to explain the interrelation between friction and wear principles and practices [1]. Without this science, an abrasive metal removal process would be impossible to characterize/quantify when the surfaces are in relative motion. In order to study and have a better understanding of wear, it is essential to recognize several mechanisms of wear. An excellent survey was made and the result listed four mechanisms including adhesive wear, abrasive wear, corrosive wear, and surface fatigue [2].

A useful review was provided of the wear process in metal-forming operations [3]. Abrasive wear covers generally two types of situations. In both cases, wear occurs by the plowing-out of softer material of a given volume by the harder indenters of an abrasive surface. In the first instance a rough, harder abrasive surface slides against a softer metal surface. In the second case, abrasion is caused by loose hard particles sliding between the rubbing abrasive and metal surfaces. An excellent review of abrasive processes and finishing operations was provided [4]. The size of an abrasive grain (grit) is identified by a number. The smaller the grain size, the larger the number is. A given surface is scored only if it is softer than the abrasive. It is therefore not surprising that resistance to abrasive wear is a function of a material's hardness. Many investigators have confirmed that hardness is the most important parameter in abrasive wear.

Quantifying and qualifying the abraded surface is an important consideration from the application point of view. Most of the previous research explored the abrasive process characteristics and mechanism by using surface roughness profiles. Surface profiles did not adequately characterize the behavior of abrasive cutting edges acting against a sheet surface to remove material during the process. In addition to the surface roughness measurements, the microscopic changes produced in the abraded surface texture were examined [5] to characterize the abrasive metal removal process.

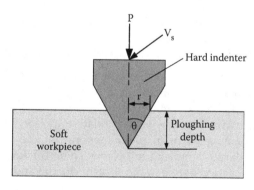

FIGURE 16.1
Fundamentals of abrasive wear mechanism.

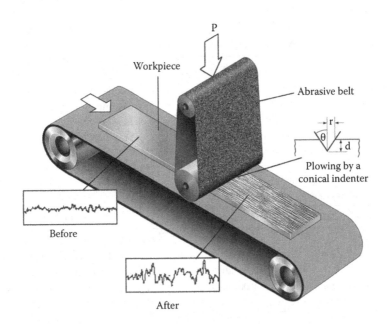

FIGURE 16.2
Tribological model of an abrasive wear process.

The Society of Manufacturing Engineers (SME) handbook [6] provides an overview of mechanical and abrasive and finishing processes including polishing abrasives with their standard grit sizes for various applications. Figure 16.2 represents the tribological model of the unidirectional abrasive wear process. In this model, the wear mechanism takes place due to plowing of material by a conical hard particle (indenter). As the flat sheet enters into the gap between the abrasive media and the conveyer belts, the surface finish of the top side of the material conforms to the size and distribution of indenters on the belt due to their plowing action on the soft workpiece material with the applied pressure P on the belt assembly. As the workpiece moves with the relative linear velocity between the conveyer and the abrasive belt, the abrasive process continues and generates a new surface finish as compared to the original surface roughness of the workpiece. Depending on the surface finish requirements of the workpiece, special attention is needed to select the types of abrasive belts according to the grit size and the distribution of the abrasives; otherwise, the penetration of abrasive particles into the soft material like an aluminum surface may cause quality issues. The depth of penetration is also a function of

- Geometry of hard indenters
- Applied pressure
- Hardness of the workpiece material

16.3 Abrasive Materials

The abrasive materials commonly used in the manufacturing processes include grinding, polishing, deburring, lapping, honing, pressure blasting, and more. The materials

TABLE 16.1

Abrasive Grit Size and Grade

Abrasive Grit Size [7]	Grade
8–24	Coarse
30–60	Medium
70–180	Fine
220–1200	Very fine

TABLE 16.2

Abrasive Materials

Abrasive Materials	Knoop Hardness [7]	General Used For
Silicon carbide	2100–3000	Nonferrous metals
Aluminum oxide	2000–3000	Ferrous alloys and high tensile metals
Diamond	7000–8000	Ceramic, titanium
Cubic boron nitride	4000–5000	Tool steel, high-strength steel, and titanium

used for abrasives generally have high hardness and moderate-to-high fracture toughness. Each abrasive particle acts like a single-point cutting tool. The total effect of abrasives is quite significant when hundreds or thousands of abrasive particles are integrated in a small area. Abrasives are designated in different particle or grit sizes as listed in Table 16.1 depending on the purpose and the amount of metal removal. Abrasives of coarser grades/grits are used when it is required to remove a high volume of materials, whereas fine grades are used to produce a smooth surface finish, and very fine for polishing metal surfaces. The most common materials used for abrasives are mainly silicon carbide, aluminum oxide (which are considered to be conventional abrasive), cubic boron nitride, and diamond as listed in Table 16.2. Abrasives are significantly harder than conventional cutting tool materials. Cubic boron nitride and diamond are considered to be super abrasives. Another important characteristic of an abrasive is friability, which is the ability of an abrasive grain to break down into smaller pieces. Friability provides abrasives self-sharpening characteristics that are important in maintaining the sharpness of the abrasives during use. High friability means low strength or low breakdown resistance. For an example, aluminum oxide has lower friability than silicon carbide. Depending on the application, abrasives are generally classified as

- Bonded
- Coated

16.3.1 Bonded Abrasive

Bonded abrasives consist of natural or synthetic abrasive grits that are attached to a matrix, which is called a binder and formed into useful shapes by pressing. The binder materials are generally vitrified or ceramic, rubber, resin, and metal. Bonded abrasive products include grinding and cut-off wheels, snagging wheels, segments, mounted wheels, plugs, and cones. Figure 16.3a shows an example of bonded abrasive wheel manufactured with flat grinding rims or faces and are designed for either side grinding, when used at

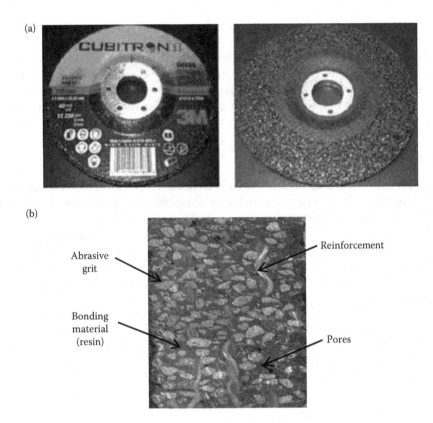

FIGURE 16.3
Example of bonded abrasive material technology: (a) Bonded abrasive wheel, (b) cross section of bonded abrasive wheel. (From 3M Corporation, MN.)

a slight angle to the workpiece, or peripheral grinding. Figure 16.3b shows the major elements of the bonded abrasive wheel structure. The total volume of the grinding wheel contained abrasive grain, the bonding material, and the porosity or otherwise called pore volume. Pore volume in the structure has an important role in the grinding process to form chip chambers and also beneficial for cooling during a grinding operation. Each grinding wheel has its natural porosity. Natural porosity is considered as standard structure. Grinding wheels with increased porosity can be produced artificially by the addition of a pore generating agent to the structure to make it a more porous structure. Abrasive grits are held together in proper place with a certain distribution by the bonding material. The type of bonding material and its percentage of volume of the wheel determine the following parameters of the wheel:

1. Strength
2. Hardness
3. Cutting performance

Resin-bonded grinding wheels are widely used in the industry.

The strength of the bonding material is a significant parameter of bonded abrasive media. If the bond strength is high, the used dull particles will remain in place while the remaining sharp particles come in contact with workpiece and remove materials eventually. If the bond strength is weak, the abrasive particles are easily disintegrated from the bonding material increasing the wear rate of the abrasive media in turn causing difficulty in maintaining the accuracy of surface roughness and the final dimensions. Selection of bonding materials are recommended based on the type of workpiece materials (hard or soft) and the functional use of the workpiece materials after the abrasive process. Standard marking systems for various bonded abrasives are identified by the manufacturer with certain formats [7].

16.3.2 Coated Abrasive

Coated abrasives are made up of abrasive grits bonded to a backing material such as cloth, paper, and rubber, polyester or even metal as illustrated in Figure 16.4. The bonding is done by adhesives including resins, glues, or combinations of the two.

Coated abrasives are formed by gluing natural and synthetic abrasive grains. Abrasive grains include mainly aluminum oxide, silicon carbide, zirconium, diamond, and hybrid. The coated abrasives are manufactured in "jumbo" rolls then cut into different shapes such as belts, discs, rolls, and sheets for surface-treatment and polishing applications of mainly metallic materials.

Coated abrasives usually have a much more open structure than the abrasives used in bonded grinding wheels. Coated abrasives are used extensively in finishing flat or curved surfaces of metallic and nonmetallic materials. The coated abrasive media are available in variety of shapes and forms (Figure 16.5) depending on the type abrasive equipments and their applications, which are mainly:

- Flexible discs
- Semi-rigid discs
- Flap discs
- Drum belts
- Small belts
- Large belts

Nonwoven nylon abrasives are open weave nylon materials coated with abrasives available in the form of pads, wheels, and brushes (Figure 16.6a). Examples of some backing materials are shown in Figure 16.6b. Flap discs (Figure 16.6c) have unique construction of overlapping

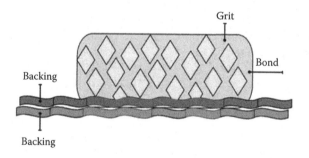

FIGURE 16.4
Schematic illustration of coated abrasive structure. (From KGS Precision Catalogue 2012, KGS Diamond International, The Netherlands.)

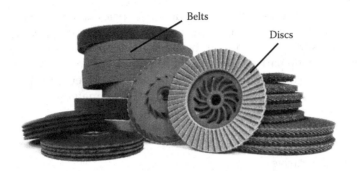

FIGURE 16.5
Example of different type of coated abrasive media. (From KGS Precision Catalogue 2012, KGS Diamond International, The Netherlands.)

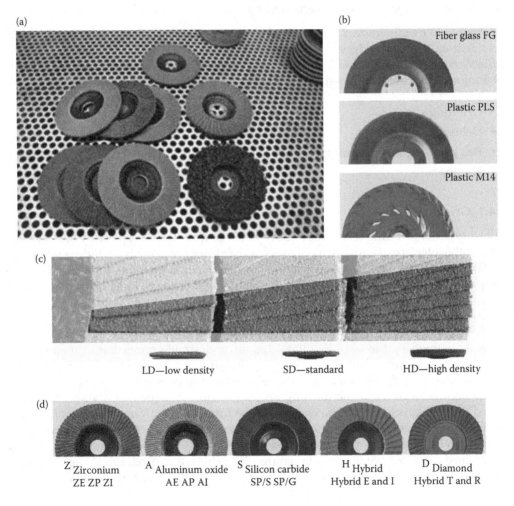

FIGURE 16.6
Selection of disc type products: (a) Type of abrasive discs, (b) backer materials, (c) examples of flap discs with different coated abrasives, and (d) different format. (From KGS Precision Catalogue 2012, KGS Diamond International, The Netherlands.)

layers of coated abrasive flaps with more abrasive particles attached to a central hub. Flap discs have a wide range of applications in the metal-working processes, which include mainly grinding, finishing, surface preparation, beveling, deburring, and edge chamfering. Flap discs are flexible to abrade parts with curves and contour with one-step grinding and blending. Figure 16.6d shows different format having low-density layer to the high-density layers.

The distribution of abrasive particles on the bonding materials is very important to provide as follows:

1. Clearance of fine chips to pass
2. Cooling of the abrasives

16.4 Major Abrasive Metal Removal Processes

The amount of metal removal, surface finish, and the dimensional accuracy are very much dependent on the sharp edges of the abrasives and the processes being used. All abrasive media starting from sandpaper and emery cloth to the abrasive belts are identified with the grit number. Major abrasive metal removal processes used in the aerospace industry include:

- Grinding
- Deburring
- Belt grinding and polishing
- Honing

16.4.1 Grinding

Various types of grinding machines remove material from the workpiece by abrasive processes at the interface of abrasive media such as hard-bonded abrasive grinding wheel and the relatively softer workpiece material by removing certain amount of material from the workpiece surface. Figure 16.7 illustrates the fundamentals of grinding operations.

16.4.2 Deburring

A burr is defined by the SME handbook, as an undesirable projection of material that results from a cutting, forming, blanking, or shearing operation. The process of removing burrs is commonly known as deburring. The need to deburr blanks prior to forming depends on the severity of the forming operation and subsequent operations to be performed on the blanks, and on the end use of the formed parts. Various types of abrasive deburring machines and processes are used to deburr the edges of the sheet metal blanks before applying forming processes to produce the aircraft part. Deburring is also applied to machined extrusions or forgings to remove the sharp edges produced in the machining operations. Major deburring processes include the type equipment used as follows:

- Grinding wheels
- Disc grinding
- Stoning

- Filing
- Wire brushing
- Belt type abrasive

Figure 16.8 shows the operating principles and kinematics of the belt-driven unidirectional deburring machine commonly used to deburr the edges of sheet metal blanks.

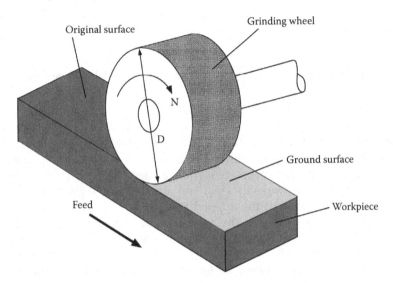

FIGURE 16.7
Fundamentals of grinding operation.

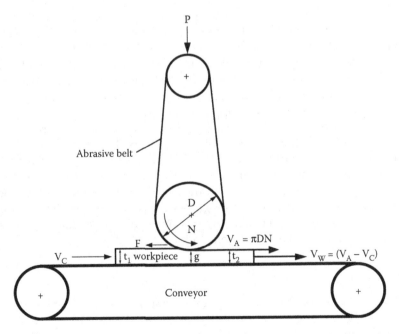

FIGURE 16.8
Operating principles of belt-type deburring machine.

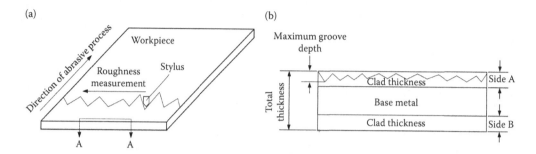

FIGURE 16.9
Tribology in clad sheet deburring: (a) Roughness after abrasive process and (b) Section AA.

A workpiece with initial thickness (t_1) is fed into the gap (g), between the abrasive belt and the conveyor that is moving with a velocity (V_C). The linear velocity of the belt, V_A (πDN), is acting in the same direction as the conveyor movement. The belt frame is applied with the induced pressure (P) caused by the difference between the workpiece original thickness (t_1) and the gap (g). Finally, the workpiece is dragged against the friction force (F) through the belt frame with a relative velocity (V_W) and abraded to the final thickness (t_2). More thickness variation may occur if the applied pressure P is kept higher to remove more material from the abraded surface. In this type of belt-driven process, the entire surface of the sheet metal blank is abrasively processed, removing a very thin skin of material along with deburring the edges.

A case study of interaction between different abrasive belt type media and aluminum flat sheet (both bare and clad) was investigated [8] to determine the best possible abrasive media including type and size of grit for unidirectional abrasive machines. A series of surface roughness measurements were conducted to study the surface finish of aluminum flat sheet after the abrasive process. Direct surface roughness measurements on aluminum clad sheet using a surface profilometer may not be sufficient to qualify and quantify the surface condition, due to a material's critical purpose of application. In the case of the clad aluminum flat sheet material, there is a definite layer (about 5% of the total sheet thickness) of nearly pure aluminum bonded metallurgically on the substrate (2024-T3). This clad material is produced in the rolling mill for the purpose of improved corrosion resistance and a shinier more visually appealing finish.

In roughness measurements, the diamond stylus could trace up to a certain depth within any groove (valley) due to the finite size of the stylus as shown in Figure 16.9. However, under no circumstances, should the clad layer be allowed to be penetrated by the abrasive particles. To find the maximum groove depth, especially in clad material, an optical microscopic analysis was performed for all the different processing combinations. In the microscopic measurements, the maximum groove depths as well as distribution of groove depths were compared for different abrasive grit sizes. Also, performance comparison was made between new and used abrasive belts in the case study.

16.4.2.1 Bare Sheet

Figure 16.10a and b shows some typical profilometer traces from the bare flat sheet test parts of 2024-T3 before and after abrasive processing. Figure 16.10a is typical of traces produced from the original surface of the flat sheet as received from the rolling mill, while Figure 16.10b shows the surface after an abrasive process (both in the rolling

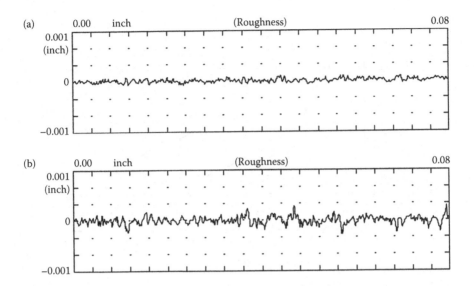

FIGURE 16.10
Typical profilometer traces before and after unidirectional abrasive process on 2024-T3 bare flat sheet: (a) Original mill finish surface (Ra = 25 μin = 0.635 μm) (measured perpendicular to the rolling direction) and (b) after abrasive process (Ra = 55 μin = 1.397 μm) (grit size 240) (measured perpendicular to the abrasive direction). (Adapted from Saha, P.K., *Wear*, 258, 13–17, 2005.)

direction and measured perpendicular to the abrasive direction). Figure 16.11a and b shows the variations in arithmetic mean value roughness (Ra) of the test parts as a function of equivalent grit size of three different abrasive grades, for two different alloys, and for three different abrasive directions. Typically, the roughness on the abraded surface generated at 45° to the rolling direction (measured perpendicular to 45°) was found to be much lower than that measured perpendicular to the longitudinal or transverse abrasive directions.

The original (mill) surface of the test pieces has a profile that resembles a series of peaks and valleys extending lengthwise in the rolling direction. If the abrasive belt interacts with this surface in a direction parallel to the rolling direction, the compliant abrasive belt can force the abrasive particles to penetrate to the full depth of the mill finished valleys between the peaks. This effectively allows the abrasive belt to deepen these valleys, thereby resulting in a larger measured difference between the peaks and valleys. This effect is lessened somewhat when processing in the angular direction (45 ± 10°) due to the act of shaving off the peaks from the original rolled finished material, while not inducing as much deepening of the mill finished valleys. In turn, the measured difference (roughness) will also be less.

16.4.2.2 Clad Sheet

A series of similar roughness measurements were performed on clad sheet material to quantify the surface finish. The measurements showed that the roughness values of the abraded clad material surface using the same abrasive grade were higher than those of the bare material of the same 2024-T3 alloy. This may be due to the softer surface of the clad layer of nearly pure aluminum. To quantify and qualify the clad integrity, after abrasive processing simple optical microscopic methods were used to measure the

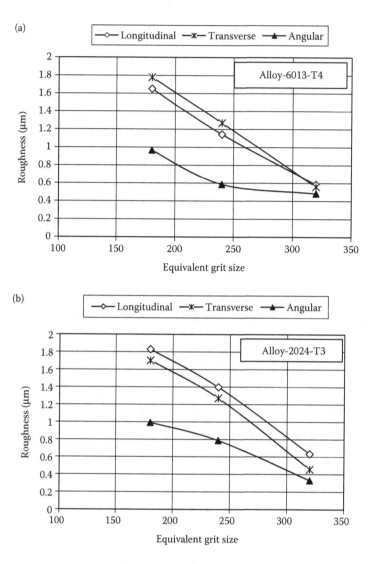

FIGURE 16.11
Variations in surface roughness on a flat sheet after abrasive processing in three different directions: (a) 6013-T4 (gage 0.100 inch [2.54 mm]) and (b) 2024-T3 (gage 0.100 inch [2.54 mm]). (Adapted from Saha, P.K., *Wear*, 258, 13–17, 2005.)

groove depth of the clad flat sheet test parts. A series of microscopic mountings were made to study various aspects of the abrasive processing including belts of different grit sizes varying from 180 to 320 and new and old belts of the same kind. Groove depths on 2024-T3 of 0.063 inch (1.6002 mm) thick clad sheet were measured for each of the various abrasives grades.

Figure 16.12 shows the comparison of the maximum groove (wear) depth within the clad thickness on 2024-T3 sheet due to the application of different grit size belts from the same manufacturer, varying from very fine to medium grade. Figure 16.12b shows that clad thickness is about to be compromised.

Figure 16.13 shows the comparison of maximum groove (wear) depth within the clad thickness on 2024-T3 sheet for two different belt conditions (brand new and used). Under

(a)   (b)

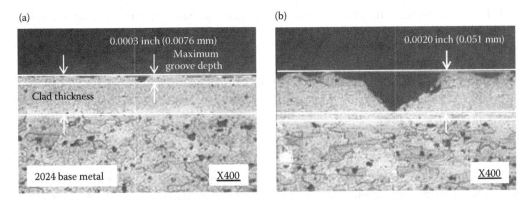

FIGURE 16.12
Maximum groove depth comparison of clad 2024-T3 sheet of thickness 0.063 inch (1.6002 mm): (a) 2024-T3 clad sheet processed with abrasive belt of grit size 320 and (b) 2024-T3 clad sheet processed with abrasive belt of grit size 240. (Adapted from Saha, P.K., *Wear*, 258, 13–17, 2005.)

(a)   (b)

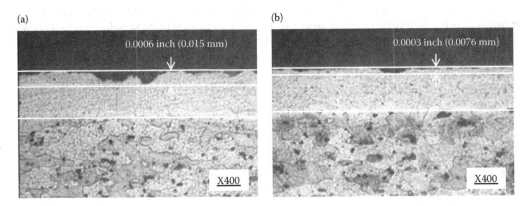

FIGURE 16.13
Maximum groove depth comparison of 2024-T3 clad sheet thickness 0.063 inch (1.6002 mm): (a) 2024-T3 clad sheet processed with abrasive belts of grit size 320 (new) and (b) 2024-T3 clad sheet processed with abrasive belts of grit size 320 (used). (Adapted from Saha, P.K., *Wear*, 258, 13–17, 2005.)

the same abrasive conditions, the new belt shows more metal removal compared to that of the used belt.

The following conclusions were made from this case study:

- The abrasive wear process on aluminum flat sheet is dependent on the type, size, and distribution pattern of the grit of the belt.

- The roughness on the abraded surface to the angular direction measured perpendicular to the abraded direction was found to be much lower than that measured perpendicular to the longitudinal or transverse directions.

- The interaction between an abrasive belt and the aluminum sheet may vary in a complicated manner when the clad sheet is processed using different equivalent grit-sized abrasive media.

- Optical microscopic investigation determined the relationship between the equivalent grit sizes of the abrasive media and the maximum penetration depth into the clad layer after the directional abrasive process.

- It was found that any abrasive belt media with equivalent grit size below 320 may be inappropriate to process clad aluminum sheet due to the potential for penetration of abrasive into the clad layer resulting in exposure of the substrate aluminum alloy to the atmosphere.

- A new abrasive belt removes more material than a belt that has been under use for some period of time. This loss of abrasiveness happens because the media loses some high peak abrasive particles during the abrasive process.

16.4.3 Belt Grinding and Polishing

Belt grinding is consistently demonstrated to be the most economical production process widely used for abrasive metal removal process in the industry where the abrasive media is an abrasive laden belt. Basically, sanding is a metal removal process by an "abrasive wear" mechanism on relatively softer material to generate a new surface finish after abrasion. The surface finish is very much dependent on the geometry of the abrasive particle and its depth of penetration under pressure and relative motion between two surfaces.

There are many different types of abrasive belts that are available from different manufacturers for the belt type sanding machines used for various metals. Selection of a correct abrasive belt for a specific application is very important for maintaining the part quality and cost. Different kinds of belts have been designated with varying abrasive particles including aluminum oxide, silicon carbide, cubic boron nitride, and diamond abrasive particles are bonded using generally resin or metallic compounds on the backing materials of the belt surface by many abrasives manufacturing companies. There are four basic factors to consider while selecting a type of abrasive belt as follows:

- Substrate material: alloy and temper
- Shape and condition: surface geometry and the amount of stock to be removed
- Equipment: design requires prefabricated belts
- Finish: type that is needed, which determines the type and grit size of abrasive

Figure 16.14 shows an example of diamond abrasive belts having few different patterns of the bonded abrasives on the backing materials, which have been successfully used in operations for belt sanding and polishing of harder alloys including steel and titanium alloy products.

(a)   (b)

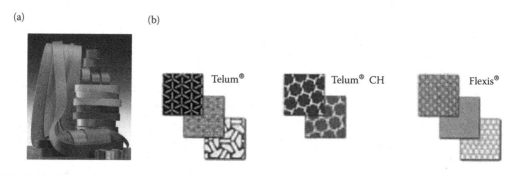

FIGURE 16.14
Diamond abrasive belts for belt sanders: (a) belts of different size and (b) example of few abrasive patterns. (From KGS Precision Catalogue 2012, KGS Diamond International, The Netherlands.)

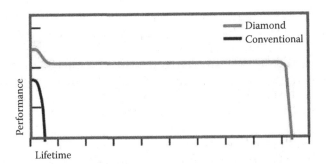

FIGURE 16.15
Performance comparisons between two types of belts. (From KGS Precision Catalogue 2012, KGS Diamond International, The Netherlands.)

Industrial study showed (Figure 16.15) life performance of diamond products versus conventional products for grinding and finishing applications including ceramics, composites, hard steel, titanium, and hard coatings [9]. Diamond abrasive belts proved their performance value due to

- Longer belt life
- Predictable, consistent, and repeatable finish
- Higher metal removal rate

Figure 16.16 shows a couple of examples of mainly belt grinding and polishing machines designed for grinding and polishing of harder alloy products of compound contour especially turbine blades, engine blades, and more. The MTS (Figure 16.16a) is a rigid heavy duty six-axis computerized numerical control (CNC) machine design for wet belt grinding. MTS stands for "Maschine zum Schleifen von Turbinenschaufeln," which translates into "machine for turbine blade grinding." Common applications of MTS machine are:

1. Wet belt grinding and polishing of blades in chromium-nickel steel, titanium, nickel-based alloys, titanium alloys, and nonferrous metals
2. Forged, milled, and cast blades for energy and aerospace industries

(a)

(b)

FIGURE 16.16
Belt type grinding, polishing, and deburring machines: (a) MTS-6NC and (b) SPE-6NC. (From IMM Maschinenbau GmbH, Germany.)

The MTS is designed for larger blades from approximately 400 mm (15.75 inch) length to 1800 mm (70.87 inch) length. In order to clamp the blade, a tailstock must be used. The blade is oriented in a horizontal position in the machine.

SPE (Figure 16.16b) is a rigid heavy duty six-axis CNC machine for dry or wet processing. SPE stands for "Schleifen Polieren Entgraten," which translates into "Grinding Polishing Deburring." The common applications of SPE machine-configured three tool units are:

1. Belt drive: Wet belt grinding and polishing of turbine blades in chromium–nickel steel, titanium, nickel-based alloys, titanium alloys, and nonferrous metals
2. Wheel unit: Processing including polishing, grinding, and brushing of complex parts, dressing of transition areas on turbine blades
3. Hard point Tool Unit: Chamfering and deburring of edges on parts such as gears, manifolds, and turbine blades

SPE is a vertical belt sander for smaller blades, even down to the size of helicopter engine blades.

Figure 16.17 shows an example of a turbine blade being ground and polished by a SPE type belt sanding machine. The finished roughness was measured in the order of, Ra—0.6 μm (23.62 μ inch).

Grinding and polishing machines as shown in Figure 16.16 are automated and CNC controlled. There are wide varieties of belt grinding and polishing machines for manual finishing also needed for finishing various aerospace parts. Figure 16.18 shows an example of a doubled-ended design enabling a choice of belt grinding and polishing on the same machine.

(a)   (b)

FIGURE 16.17
Turbine blade processed with SPE machine: (a) Before and (b) after. (From IMM Maschinenbau GmbH, Germany.)

FIGURE 16.18
Combined belt grinding and polishing machine. (From IMM Maschinenbau GmbH, Germany.)

16.4.4 Honing

Honing is a special type of polishing operation to provide holes with a very smooth surface finish. The honing tool consists of bonded abrasive stones, which are mounted on a mandrel that rotates with a given spindle speed in the hole applying a radial force with a reciprocating axial motion generating the finish in the hole. The stones are generally adjusted radially for different hole sizes. The surface finish of the hole can be controlled by

1. Type and size of abrasive
2. Speed of rotation
3. Pressure applied

Cutting fluid is used to remove very fine abraded chips and to compensate for the temperature rise during honing.

16.5 Abrasive Cutting

Abrasive cutting is a generalized industrial tool for cutting a wide variety of materials using the cutting power of abrasive particles. An abrasive water jet (AWJ) is an example of an abrasive cutting process and refers specifically to the use of a mixture of water and abrasive particles. AWJ cutting starts up with the principle of water jet machining (WJM), using the physics of a high-pressure water jet process, which was developed in 1930s. The major process steps for WJM are shown as a flow diagram in Figure 16.19. Water from the reservoir is pumped into the intensifier using a hydraulic pump. The intensifier increases the water pressure generally to the range of 200–400 MPa (29–58 Ksi). Pressurized water is sent to the accumulator, which temporarily stores the pressurized water. The pressurized water then enters the nozzle by passing through the control valve and flow regulator,

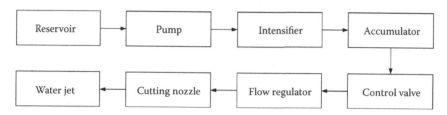

FIGURE 16.19
Major elements of WJM.

which regulates and controls the flow rate of water. A nozzle produces a very high pressure water jet directed to the workpiece. WJM only without the use of added abrasives is often used for cutting softer materials such as wood and rubber products.

Material cutting and profiling using the fundamental properties of abrasive particles has become popular in metal industries since the mid-1990s. Combining the principle of WJM with added abrasive particles in the water stream, the new term "AWJ" has been widely introduced in the manufacturing industry. Figure 16.20 shows a schematic representation of AWJ cutting [10]. High velocity water exiting the orifice or jewel creates a vacuum, which pulls abrasives from the abrasive inlet line mixing them with the water in the mixing tube supported by the guard. The beam of water accelerates abrasive particles to speeds fast enough to cut through the workpiece material. AWJ cutting is widely used in the aerospace industry for cutting and shaping various materials including aluminum, steels, titanium, inconel, ceramic and laminated materials.

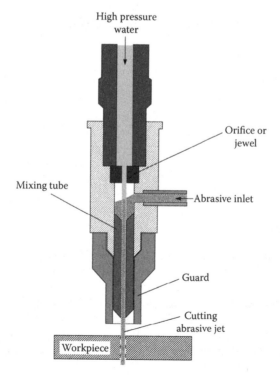

FIGURE 16.20
Schematic diagram of AWJ cutting.

The important process parameters of AWJ machining or cutting are as follows:

1. Mass flow rate of abrasive particles
2. Distance of cutting jet from the workpiece
3. Water pressure
4. Velocity of abrasive particles
5. Mixing ratio of abrasives with the water
6. Abrasive grain size

Major advantages of AWJ are listed below:

1. Relatively fast process and cut virtually any material including harder materials such as titanium, inconel, and stainless steel
2. Prevents forming heat-affected zone (HAZ)
3. No mechanical stresses introduced into the material
4. No start hole is required
5. Can machine thicker material
6. Safe to operate with less maintenance
7. Environmentally friendly

References

1. Halling, J., *Principles of Tribology*, Macmillan Education Limited, London, 1978.
2. Burwell, J.T., Survey of possible wear mechanisms, *Wear*, 1, 119–141, 1957.
3. Schey, J.A., *Tribology in Metal Working*, ASM, Metals Park, OH, 1983.
4. Kalpakjian, S., *Manufacturing Engineering and Technology*, Addison-Wesley, MA, 1992.
5. Yamaguchi, H. and Shinmura, T. Study of the surface modification resulting from an internal magnetic abrasive finishing process, *Wear*, Vol. 225–229, Part I, 246–255, 1999.
6. Wick, C. and Veilleux, R.F., *Tool and Manufacturing Engineers Handbook*, Society of Manufacturing Engineers (SME), Vol. III, MI, 1985.
7. Kalpakjian, S. and Schmidt, S.R., *Manufacturing Processes for Engineering Materials*, Pearson, NY, 2008.
8. Saha, P.K., A case study on the abrasive surface finishing of aluminum flat sheet, *Wear*, 258, 13–17, 2005.
9. www.kgsdiamond.com
10. www.watrejets.org

17

Chemical Metal Removal and Chemical Processing of Metals

17.1 Introduction

A chemical metal removal process known as chemical milling is commonly used by the aerospace industry for selectively removing metal to reduce weight and produce parts of varying gage. It is particularly beneficial on structure where the removal of small amounts of metal over large surface areas is desired, or where the part geometry causes difficult access for cutting tools. It can be used on all metal forms including sheet metal, plate, forging, and extrusion products in manufacturing aircraft components. The chemical machining process was developed based on the principles of chemical reaction between metals and chemical reagents or etchants like acids or alkaline solutions. Chemical milling is the oldest of the nontraditional machining processes and has been used for many years in the aerospace industry. Advances in high-speed mechanical milling are gradually reducing the usage of chemical milling in aerospace industries.

Chemical processing of metal parts is critical operations for aerospace applications in order to prepare the metal for subsequent processing and improve the base material's surface properties. An example is improving the base material's corrosion resistance, wear resistance, lubricity, and the adhesion of subsequent coatings to improve its protective and aesthetic properties. The most detrimental environmentally induced effect is the corrosion of metals exposed to the flying environments, and also the stress corrosion effect induced in the metal during service. Selection of metals and their corresponding chemical processes utilized to finish the metal part before it is installed in the aircraft are some of the primary concerns of design engineers. A number of chemical processing solutions are used to clean and prepare metal parts for the subsequent chemical-coating processes. This chapter will cover the fundamentals of the chemical milling process and also major chemical cleaning and coating processes required for fabricating aluminum and hard metal alloy parts, including plating of hard metals as illustrated in Figure 17.1.

17.2 Chemical Metal Removal Process

Chemical metal removal and machining processes [1,2] are among the oldest nontraditional machining processes used in the aerospace industry to manufacture complex shapes and thin gage metal parts. These include fuselage skin panels incorporating

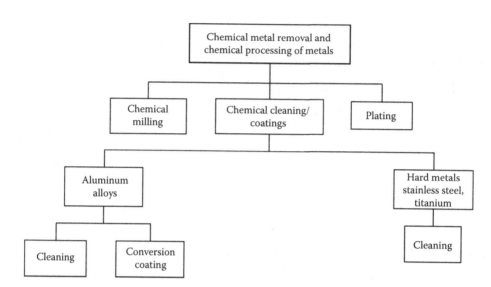

FIGURE 17.1
Major chemical metal removal and chemical processing of metals.

pockets mainly in the IML of the fuselage skin panels. These pockets provide weight reduction and failsafe structure. Chemical machining, commonly known as chemical milling in the aerospace industry, is the controlled chemical dissolution of the machined workpiece material by contact with an acidic or alkaline chemical reagent. The initial application of chemical machining in the industry was developed in 1953 by North American Aviation Inc. in California [3]. The process was used to etch aluminum components for rockets. The company named the process "chemical milling" and patented it (US Patent No. 2739047) in 1956 [4]. Figure 17.2 illustrates the fundamentals of a chemical milling process where the selective reaction by the chemical reagent on different areas of the workpiece surface is controlled by use of removable layers of masking material, which selectively protects nonetched areas when the part is fully immersed into the reagent tank. Figure 17.3 shows the major process flow chart of a chemical-milling process.

The workpiece material needs to be first cleaned by removing the oil, grease, dust, etc. and undesirable oxide to prepare the surface for subsequent masking. The cleaning method is determined based on the base material, the processing it has already been through, and the masking material being used. The cleaning process is selected to optimize the performance of the maskant without damaging the metal being protected. Some commonly used cleaning methods are hand solvent wiping, immersion in an alkaline cleaning solution, deoxidizing, and light etching.

In order to protect the metal from the etching process, the entire part is masked. The masking material is designed to protect the metal from the chemicals and temperature of the process it is exposed to. Masking materials generally include butadiene styrene (solvent based), latex (water based), and polyeurethane (92 K). It is optimized to be easily removable, provide enough adhesion to produce good definition between milled and unmilled surfaces, and easy to scribe. In order to ensure that the part is protected, the maskant is visually inspected and spark tested to detect defects. If defects are found, the maskant is repaired. Maskant commonly leaks on edges. If possible, excess should be allowed on the part edges and trimmed after chemical milling.

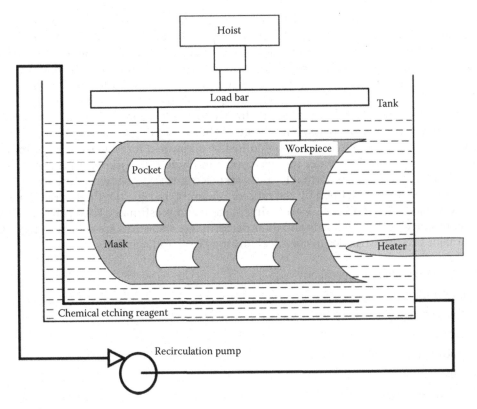

FIGURE 17.2
Fundamentals of chemical milling process.

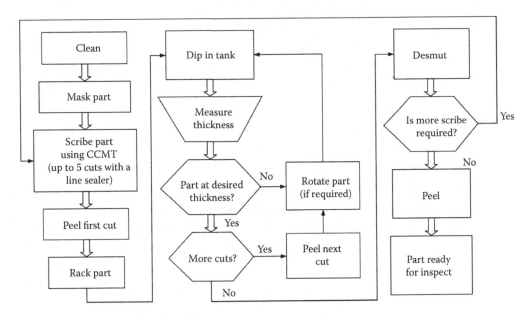

FIGURE 17.3
Major process steps of chemical milling.

Scribing is one of the most critical and undervalued tasks. It is typically done by hand or using a laser. When hand scribing, a tool is laid-up on the part and a sharp knife is used to trace the pattern. It is important that the scriber develop the skill to be able to sense when the maskant is cut all the way through without excessively cutting into the metal. This is important because if the maskant is not cut completely through a lift will occur during peeling, and it will have to be repaired. Multiple milling depths may be scribed at the same time using the same tool. The scribe lines of areas to be milled in later submersion steps are temporarily protected by a line sealer. When designing the tool or the laser scribe pattern, the undercut action of the etchant must be accounted for. When chemical milling, the etching solution etches under the maskant, parallel to the workpiece surface simultaneously with etching action normal to the workpiece surface as shown in Figure 17.4. The undercut ratio which is defined by the ratio of undercut and the depth of cut is impacted by the workpiece alloy and chemical etching reagent and is estimated based on experience. Some common etching chemicals and their performance measures are shown in Table 17.1.

The major process variables of chemical milling are as follows:

1. Chemistry of etching reagent
2. Etching rate
3. Temperature
4. Time
5. Undercut ratio
6. Part geometry
7. Workpiece alloy

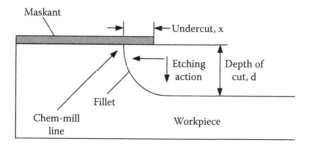

FIGURE 17.4
Chemical milling parameters.

TABLE 17.1

Common Etching Chemicals and Performance Measures

Alloy Etched	Typical Chemistry	Key Performance Property
Aluminum alloys	NaOH Na_2S Triethanolamine	Surface roughness
Titanium alloys	HNO_3 HF	Hydrogen pick-up Surface roughness
Steels	H_3PO_4 HCl HNO_3	Hydrogen pick-up Surface roughness

Once masked, scribed, and peeled the metal part is ready for the etching/milling process. Because there are many process variables as shown above that impact the etching of the part, etching is an interactive process that includes rinsing and often desmutting. Typically, solutions are maintained to obtain an etch rate between 0.001 inch (0.025 mm) and 0.002 inch (0.051 mm)/side/minute. Etch rate is controlled by chemistry and temperature of the etching reagent. Desmutting usually needs to be performed, and it is an acid-based process that removes remaining smut (metal sulfides) or oxides caused by alloy metals. Each level of metal removal takes a minimum of two times in the etching bath. Some of the factors that impact etch rate are remaining metal thickness, localized area being etched, relative location in the tank, and workpiece alloy. Once the targeted thickness is reached on an etching step, the next level of masking is removed. All metal removal is based on the thinnest area of the part (Figure 17.5). Because of all these variables, a milling recipe is developed to produce the needed thickness requirements for complex parts. Once the desired thickness is reached, the part goes through a final desmut, the maskant is removed, and the part is inspected for areas requiring manual rework and then is ready for final inspection.

Depending on the incoming material tolerance, final dimensional tolerances typically range from ±0.003 to ±0.005 inch (±0.127 to ±0.0254 mm). The surface finish produced usually ranges from 14 to 260 μ inch (0.35560–6.6040 μm), depending on the workpiece alloy and chemistry. It is typically independent of the initial surface properties once a minimum of around 0.01 inches is removed.

Figure 17.6 shows an example of a chemical milled fuselage skin of an aircraft, which is removed from the chemical etch tank with the maskant still intact.

The chemical milling process is also being used for titanium alloy products and is generally used to

- Remove metal from the surface of complex shapes and thin gage titanium metal parts

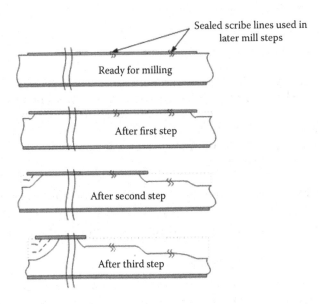

FIGURE 17.5
Chemical milling steps.

FIGURE 17.6
Example of an aluminum fuselage skin having chemical mill pockets. (From Spirit AeroSystems, Inc.)

- Remove oxygen-enriched α-stabilized, surface in titanium, resulting from elevated temperature exposure (usually called alpha case) from superplastically formed (SPF) parts
- Removing excess metal to lighten and/or shape the part and gage reduction

17.3 Chemical Cleaning

As highlighted in Figure 17.1, the major chemical cleaning processes for both aluminum alloys and hard metals including titanium and stainless steel will be briefly described in the following two sections.

17.3.1 Cleaning of Aluminum Parts

An aqueous degreasing process is widely used for general cleaning of aluminum parts, mainly to remove grease and oil before and between fabrication steps. It is also used to remove grease and oil before processing the parts in other tank-line processes, mainly for application of conversion and anodic coatings on the aluminum part's surfaces. Hard-to-remove contamination and marking inks may have to be removed by hand using a solvent. The degreasing step is often eliminated if there is no visible oil or grease. The major process steps for aqueous cleaning are:

1. Workpiece is immersed in hot cleaning solution with agitation (spray rinsing at the exit of the tank is often used to prevent the cleaner from drying on the part)
2. Immersed in first rinse tank (a heated rinse is often used to aid in removing the cleaner)
3. Immersed in the second rinse tank
4. If parts are not proceeding to other chemical processes in the same tank line, they may be dried in an elevated temperature high airflow dryer to accelerate the evaporation of water

Alkaline clean is used for general cleaning to remove surface soils, such as fingerprints, coolant, perspiration, machining, and aluminum or other metal dusts before other tank-line processing. Generally, alkaline cleaners are nonetching, and only remove aluminum surface contaminations. The process is usually validated using a water-break-free test. This is done by inspecting the part while still wet. The part should not have breaks in the water sheen in the middle of the part. Water-break-free can only identify if a part is still dirty, and cannot ensure the part is totally clean.

Etching is an aggressive cleaning process used for removing surface contamination, mainly from aluminum alloy wrought products during fabrication of aircraft parts. The most common etchants for aluminum are alkaline based. Etching is commonly used for specific applications that include:

- Removal of oxide films produced during heat treatment and other thermal operations in preparation for the application of anodic/conversion coatings and also before some mechanical operations including welding
- Removal of smeared metal from post-mechanical or machining processes prior to penetrant inspection
- Stripping of anodic or conversion coatings for rework

Etching is usually preceded by alkaline cleaning or aqueous degreasing to ensure uniform etching and followed by a deoxidizer to remove alloy metal oxides; the black-colored smut generated during etching from a part's surface. If the production facilities do not have an etchant solution, a deoxidizer may be used in its place. This is not desired because the etch rate in deoxidizers is very slow and more prone to selective etching.

Deoxidize/desmut is another chemical cleaning process that follows alkaline clean and/ or etching and is commonly used for the following major purposes:

1. Removal of surface oxides that may be of varying thickness and compositions before some mechanical operations including welding and also for the application of a conversion coating or anodize
2. Removal of heat treatment scale and smeared metal before penetrant or chemical finishing operations (not as effective as alkaline etching)
3. Removal of smut that is deposited on the surface during alkaline etching and chemical milling operations
4. Activate the metal surface for subsequent conversion coating processes
5. Stripping of anodic or conversion coatings for rework

17.3.2 Cleaning of Hard Metal Parts

Aqueous degrease and alkaline cleaning processes used for aluminum part cleaning are also applicable for processing hard metal parts. In addition, a scale conditioner may be used for softening and breaking up heat treatment scale on these materials. Parts are placed in a hot, alkaline oxidizing solution, which is a scale conditioner followed by a rinse process at room temperature. Due to the change of temperature from the hot scale conditioner to the cooler rinse, the scale is caused to crack, which facilitates attacking the scale by the acid when the parts are placed in an acidic solution.

Etching of titanium parts and descaling of steel parts are generally used to

1. Remove heat treat scale and oxides from titanium and stainless steel before some mechanical operations including welding and finishing like painting and assembly
2. Removal of smeared metal to enable penetrant inspection

Cleaning of titanium parts and passivation of stainless parts are commonly used to

1. Oxidize the surface to form a passive layer
2. Remove fluoride residue after nitric-hydrofluoric acid etch
3. Remove iron residue from the titanium surface left from previous operation
4. To change the surface stainless steel from a chemically active surface to a much less reactive surface

17.4 Conversion Coatings

Conversion coatings are processes to change the surface of the metal or a coating applied to the metal to improve the properties of the metal. The desired property depends on the metal and the application it is being used for, but the most common are to improve adhesion of subsequent coating, improve resistance to corrosion, and wear. The two most common types of conversion coating used on metal substrate in aerospace are:

1. Chemical film
2. Anodize

17.4.1 Chemical Film

The most common chemical film used in aerospace is chrome conversion coating. It is most commonly used on aluminum and plated surfaces. Chrome conversion coating is a protective surface layer on a metal that contains hexavalent chrome. Chrome conversion coating acts as an adhesion promoter and provides sacrificial corrosion resistance. The chrome conversion coating is applied to an aluminum surface; the natural aluminum oxide (Al_2O_3) on the surface is converted to thicker oxide containing chromium (Cr_2O_3). Chrome conversion coating provides the following:

1. Improves the corrosion resistance of aluminum due to incorporation of chrome
2. Excellent electrically conductive surface for good electrical bonding in an airframe
3. Paint adhesion to the alodine surface is significantly better than the natural oxide layer

The major process steps of chrome conversion coating are as follows:

1. Degrease followed by rinsing
2. Alkaline clean or etch followed by rinsing

3. Checking for water-break-free

4. Immersion or spray apply chrome conversion coating, typically 30–90 s

5. Immerse in the rinse tank

6. Air dry or blow dry

17.4.2 Anodize

Anodize is an electrochemical process in which the process forms a structured metal oxide surface layer. The process converts the metal surface into a decorative, durable, corrosion resistant, anodic oxide finish. The oxide layer formed is tough and often porous. The porosity helps to form a mechanical bond with subsequent coating to improve their adhesion. A seal solution may be used to hydrate the oxide, which helps to reduce the porosity and increase its corrosion resistance. Aluminum is an ideal material to anodize although other nonferrous metals including magnesium and titanium alloys can be anodized. The anodic oxide structure is generated from the original aluminum substrate and is composed entirely of aluminum oxide and is fully integrated with the underlying aluminum substrate. The fundamentals of an anodizing process are:

- Immersing the aluminum workpiece (anode) into an acid electrolyte bath (usually acidic) and passing an electric current through the bath and a cathode.
- The cathode is either metal tank structure or metal plates inside the anodizing tank.
- Oxygen ions released from the electrolyte combines with the aluminum atoms at the surface of the part being anodized. Hydrogen gas is released from the bath.

In anodic coatings, the thickness increases due to the surface oxides which are substantially thicker than the conversion coatings. Anodic coatings will also have better capability of withstanding abrasion. Conversion coatings are generally less costly than anodic coatings and can normally provide sufficient corrosion protection and paint adhesion also. Anodic coatings do provide certain advantages:

1. When sealed can provide more corrosion protection from the external environment and internal environmental control system of an aircraft

2. Increase in hardness, wear resistance, abrasion resistance, electrical insulation, and radiant heat absorption

3. Enhanced surface condition mainly for paint, adhesive, and sealant

4. Improved decorative texture, color, and appearance

Boric sulfuric acid anodize (BSAA), a type of anodize process widely used in the aerospace industry improves paint adhesion and provides substantial corrosion and abrasion resistance. BSAA is generally used for

- Corrosion protection and paint adhesion
- Adhesive bonding applications for nonstructural components
- Corrosion and erosion protection for aerodynamic leading edges, where normal paint finishes would fail because of rain erosion. Polished clad is preferred though

Other anodic processes like phosphoric acid anodize are used for providing a paint base for structural adhesive bonding applications and sulfuric acid anodize (SAA) is used to provide a finish on interior decorative parts and sometimes on exterior high erosion areas (sealed SAA). It is colorless in appearance.

17.5 Plating of Hard Metals

Electroplating processing is generally used to deposit a metallic coating onto another metal surface by electro deposition. Plated metals provide corrosion protection for the metal on which they are deposited and also for a decorative coating and for other important functional needs. Various types of plating materials used in aerospace are:

- Bright cadmium—used to provide corrosion protection or resistance on low alloy or carbon steel, threaded fasteners, and improve the compatibility of corrosion resistant stainless steel with other metals, such as aluminum. Applications are for gears, screws, shafts, fasteners, bushings, springs, pins, etc.
- Copper—used to protect steel parts during hardening processes including nitriding, carburizing as needed for gears and flap tracks, etc.
- Chrome—used to increase abrasive wear, sliding, and rolling-wear resistance. Used for flap carriage and landing gear.
- Nickel sulfamate—used to provide an undercoating for the noble plating alloys, such as chrome or cadmium plating on steel, improve sliding wear resistance of steel and bearing properties and provide barrier protection from corrosion or impact. A typical application is for seal surfaces for landing gear.
- Ti-Cad—used for paint adhesion on high strength, low alloy steel, corrosion resistance of high strength, low alloy steel, and to produce a porous plate; and is also used to minimize hydrogen embrittlement of high strength steel.
- Zinc-nickel—used to provide a sacrificial coating for corrosion protection on low alloy or carbon steel, nickel strike is performed first to improve the adhesion of the plating, to improve the compatibility of stainless steel (corrosion resistant steels) with other metal including aluminum, and to provide an alternative to cadmium plating in many low-strength steel applications.

References

1. Harris, W.T., *Chemical Milling: The Technology of Cutting Materials by Etching*, Oxford University Press, UK, 1976.
2. Todd, R.H., Allen, D.K., and Alting, L., *Manufacturing Processes Reference Guide*, Industrial Press Inc., NY, 1994.
3. Cakir, O., Yardimeden, A., and Ozben, T., Chemical machining, *Archives of Materials Science and Engineering*, 28(8), 499–502, August, 2007.
4. Sanz, M.C., Process of chemically milling structural shapes and resultant article, US Patent No. 2739047, 1956.

18

Manufacturing Processes of Composite Materials for Aerospace

18.1 Introduction

Composites with carbon and glass fibers have been steadily gaining a very high-level of importance within the aerospace industry since the 1970s. Continued technology development of these materials brought a new revolution to the commercial aircraft industry including Airbus and Boeing to allow introduction of wide body aircraft to the world's airline customers in the twenty-first century. Worldwide research and development on composite materials and their manufacturing processes for producing aircraft components with these materials mitigated to some extent the increased cost of materials. Reduction of manufacturing costs helped the aircraft industry to introduce new fuel-efficient long-range aircraft. Research is continuing performing exercises for the reduction of the cost of manufacturing by developing new processes. Manufacturing processes from the very labor-intensive prepreg hand lay-up process to newer processes like automated tape laying and or automated fiber placement have brought cost-efficient airframe composite structures to reality. Thermoplastic composites with their short manufacturing cycle time compared with thermoset composites have brought further improvement in manufacturing costs and thus increased usage of composites.

Engineered composite materials must be formed to shape certain geometry to meet the role of the various components of an aircraft. The proper selection of the type of composite materials and the corresponding manufacturing processes is one of the most important steps in producing composite aircraft components. A variety of fabrication processes suitable for various applications in the aircraft are available in the current high tech manufacturing industry. The composite cure processes, which follow different molding technologies, include mainly vacuum bag, pressure bag, autoclave, and resin transfer molding (RTM). Other molding processes like press molding, injection molding, pultrusion molding, and filament winding are also used. Large structural parts consisting of different materials or different components can be cured and bonded together in one step by the so-called co-curing process. Some components can be pre-cured while others become cured in the co-bond process. Thus, a number of structural or nonstructural components can be joined together with a minimum number of fasteners. The composite fabrication process is still dependent to some extent on skilled hand work that requires manufacturing and material costs compared to metallic structure, including cost of quality control. As seen in the case of conventional metal structure fabrication, there are two separate processes that include complex tooling and elaborate assembly along with multiple elements and joints. Much of the research on composites is focused on the development of affordable

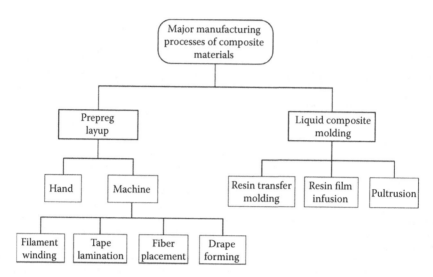

FIGURE 18.1
Major manufacturing processes of composite materials.

composite materials and manufacturing techniques, which may not require extensive use of an autoclave. Another important area of research is the development of automated manufacturing technologies for prepreg layup of large structural components of an aircraft.

In Chapter 5, the introduction of composite materials for aerospace and the manufacturing technology of input raw materials for composite fabrication have been discussed. Chapter 18 covers major manufacturing processes as illustrated in Figure 18.1 for producing aircraft components from composite materials.

18.2 Role of Composites in Major Aircraft Components

The major driving force of introducing composite materials for aerospace structures is the significant savings of weight compared to the traditional aluminum aircraft structures due to higher specific strength and stiffness of composite materials. More efficient structural design can be made by selecting the correct advanced composite materials to meet the structural requirements at different locations within an aircraft structure. The major drawback of composite materials is the higher materials and manufacturing cost of making aircraft components as compared to the metal components replaced. Research continues focusing on the development of cost-effective composite materials and manufacturing techniques. Industry focus is on the development of automation in composite manufacturing.

There are various applications of nonmetallic materials within the aircraft. Composite materials are realizing a growing number of applications in new aircraft design in the twenty-first century. Composites are replacing more common high-strength aluminum alloys to reduce gross weight of the aircraft. Figure 18.2 shows within the shaded areas the locations of typical applications of carbon fiber reinforced plastic, graphite, graphite/fiberglass, and fiberglass in an aircraft.

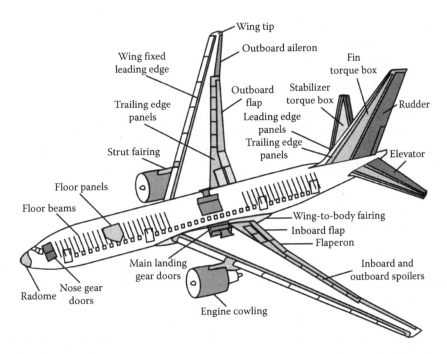

FIGURE 18.2
Typical applications of composites in an aircraft.

Some of the airplane structure is made of composite materials to improve resistance to damage and corrosion, and reduce airplane weight. Composite materials are layers or plies of high-strength fibers (carbon or fiberglass) in a mixture of plastic resin. Components made of composite materials use lamination or collation or stacking combine layers of the composite materials. Some components incorporate honeycomb core in between composite plies to form a sandwich type construction after curing. As much as 30% weight savings have been realized by the aircraft manufacturer on airplane components using composite materials. Wing leading- and trailing-edge panels, control surfaces, and wing-to-body fairings are constructed using laminated technology.

18.3 Manufacturing Processes of Composite Materials

Demands of composite materials in aerospace are reaching new levels of sophistication every year. There is wide variety of processes available to the composites manufacturers to produce cost-effective products for aerospace applications. Each of the manufacturing processes has characteristics that define the type of products to be produced. Developments of new technologies are proving the success of extensive usage of composite materials in manufacturing of modern aircrafts for the airline customers. In order to select the right manufacturing technology for a given part, several important factors are being considered:

1. User need and performance requirements
2. Material selection

3. Part geometry and overall size
4. Surface contour and physical appearance
5. Tooling and assembly cost
6. Production rate and volume of production

As discussed in Section 5.10, prepreg is the input raw material for fabrication of any particular composite part made by stacking and shaping layers of prepreg materials. Prepreg tows, tapes, or fabrics are used for many different types of lay-up operations. The major manufacturing processes of composite materials [1–3] for aerospace are mainly classified as

1. Prepreg layup
2. Liquid composite molding

18.4 Prepreg Layup

In a prepreg lay-up process, tows or fabrics are impregnated with controlled quantities of resin before being placed on the mold. Prepreg layup is commonly classified into two different types:

1. Hand layup
2. Machine layup

18.4.1 Hand Layup

Engineered composite parts for aerospace are generally formed to shape by lay-up processes using prepreg materials. Prepreg hand layup is one of the oldest techniques for manufacturing composite parts and is still one of the most used composite manufacturing processes in the aerospace industry. The process is considered to be labor-intensive, but on the other hand, it is simple to set up. The major process steps include:

1. The thermoset prepreg rolls are removed from the freezer and slowly brought to room temperature to prevent condensation
2. The mold is pretreated with a release agent
3. Individual plies are cut to shape from the prepreg roll by hand or using automated computerized numerical control (CNC) machines
4. Plies are laid up on the mold into a laminate maintaining the right fiber orientation of the plies, in a room usually called clean room with controlled contaminants, humidity, and temperature
5. The whole stack of laid up plies is then vacuum bagged and placed into an autoclave to cure resin at elevated temperature and pressure
6. The layup is held at often a few hundred degree Fahrenheit and several atmospheres pressure for over an hour
7. After cooling the laminate the product is removed from the mold for inspection

FIGURE 18.3
Prepreg hand layup. (Adapted from Kisch, R.A., Composites processing technology and innovation at Boeing, *Key Note Presentation for the Society of Plastics Engineers Automotive Composites Conference & Exhibition*, MI, September, 2007.)

Prepreg hand layup has design flexibility for producing large and complex parts. Minimum equipment investment and start-up lead time are required, but the hand layup process is labor-intensive since the quality is related to the skill of the operator, so it is typically a low-volume process. There is difficulty to maintain the product uniformity from one part to another or within a single part, which can result in a higher waste factor (Figure 18.3).

Many components are produced by prepreg hand lay-up processes, but it is more suited to highly complex parts that will fit on a work bench. The examples include fairings, vertical stabilizer skin panels, and wing control surfaces.

18.4.2 Machine Layup

Increasing demand for faster production rates has induced the aerospace industry to replace hand layup with automated machine lay-up processes. As is understood from the hand lay-up process it is very labor-intensive, in addition there are chances to have irregular placement of the tape that could cause inconsistency of quality of the cured laminates. The major machine lay-up processes are generally used to cover wide variety of parts of an aircraft:

- Filament winding
- Tape lamination
- Fiber placement
- Drape forming

18.4.2.1 Filament Winding

Filament winding is a process [1] in which continuous rovings or tows are wound over a rotating male mandrel. These rovings or tows can be resin impregnated before, during, or after the winding process. They are wound on various types of mandrels including cylindrical, spherical, or conical hollow configurations. Typical aerospace products include mainly rocket motor cases, pressure vessels, launch tubes, and drive shafts. The technique has the capability of varying winding tension, wind angle, or resin content in each layer of

(a)   (b)

FIGURE 18.4
Filament winding of large cylindrical structure. (a) Filament winding process and (b) fibers are feeding through the resin bath. (From Magnum Venus Plastech.)

reinforcement until the desired thickness and resin content of the composite are obtained with the required direction of strength. However, since filament winding requires constant tension, the tows cannot be cut or restarted during the process and concave surfaces are not possible. Properties of the finished composite can be varied by the type of filament winding pattern selected, which could be polar, helical, or modified helical, or hoop.

Figure 18.4a shows an example of a filament-winding process producing a large cylindrical structure where fiberglass roving strands are impregnated with liquid thermoset resins (Figure 18.4b) and wrapped onto a rotating mandrel in a specific pattern. When the winding operation is completed, the resin is cured or polymerized and the composite part is removed from the mandrel (Figure 18.5a). The mandrels are generally designed to have collapsible features with an open position (Figure 18.5b) and also an extracting position (Figure 18.5c) for releasing the part from the mandrel. Washout sand or salt mandrels are also used. Figure 18.6 shows an example of a filament-winding process for producing a smaller diameter tube.

The major limitations of the filament-winding process include:

- A complex and or expensive mandrel is required
- Limited component shape for removal of mandrel
- Convex structures can only be made
- Low fiber angles are difficult to produce
- External surface quality may not meet aerodynamic surface requirements

The applications of filament winding are generally limited to bodies of revolution like tanks and pressure vessels.

18.4.2.2 Tape Lamination

Automated tape lamination, generally called ATL process [2], is an automated process in which prepreg tape, instead of single tows, is laid down continuously to form parts. It is often used for making parts with low complexity and gentle surface contours. Tape layup is a versatile process allowing breaks in the process and easy directional change, and it can

FIGURE 18.5
Collapsible type mandrel used in filament winding machine. (a) Mandrel, (b) open, and (c) extract.
(From Magnum Venus Plastech.)

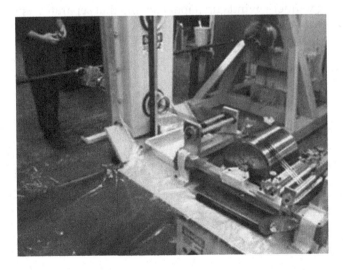

FIGURE 18.6
Producing smaller diameter tube by filament winding. (From Magnum Venus Plastech.)

FIGURE 18.7
Gantry type ATL machines are in operation for mass production. (Adapted from Kisch, R.A., Composites processing technology and innovation at Boeing, *Key Note Presentation for the Society of Plastics Engineers Automotive Composites Conference & Exhibition*, MI, September, 2007.)

be adapted for both thermoset and thermoplastic materials. ATL machines (Figure 18.7) were widely developed for aerospace applications including stabilizer skins, wing skins, and wing spars. The machines are usually gantry type with a cross feed bar containing the end effector delivery head for the rolls of tapes with widths from 3 inch (76.2 mm) to 12 inch (304.8 mm) can be laid down. The width of tapes selected depends on the required curvature of the part. Figure 18.8 shows an example of the ATL process of producing an airplane part using a gantry-type machine. The machine head includes the major elements:

1. Supply roll—raw material for layup, usually in the form of carbon tape preimpregnated with epoxy resin
2. Paper backing winder—winds up paper backing that separates tape on the supply roll and stores it until disposal

FIGURE 18.8
Example of tape lamination of horizontal stabilizer skin of an aircraft. (Adapted from Kisch, R.A., Composites processing technology and innovation at Boeing, *Key Note Presentation for the Society of Plastics Engineers Automotive Composites Conference & Exhibition*, MI, September, 2007.)

3. Edge guides—keep tape properly aligned with head

4. Knives—cut tape ends to desired shape prior to compaction

5. Compaction shoe—compacts tape onto lay-up mandrel

6. Tail compactor—compacts ends of tape where compaction shoe cannot do so adequately

The head may be located on the end of a multiaxis articulating robot that moves around the tool or mandrel to which the prepreg tape is being applied, or the head may be located on a gantry suspended above the tool. Alternatively, the tool or mandrel can be moved or rotated to provide the head access to different locations of the tool or mandrel to cover the entire geometry of the part. Tape is usually applied on the tool or mandrel in courses, which consist of one row of prepreg of any length at any angle. Multiple courses are usually applied together over the defined part geometry and controlled using NC software programmed with the input derived from the digital 3D part model.

The applications of ATL process are typically for flat or nearly flat shapes, like wing and horizontal and vertical stabilizer skins.

18.4.2.3 Fiber Placement

Automated fiber placement (AFP) process [1,2] was developed specifically for the aerospace industry. The development started in the beginning of the 1980s and the aerospace industry started using it from the early 1990s. In a fiber placement process, multiple individual prepreg tows are placed automatically on a mandrel at high-speed using an NC, articulating placement head. The fiber placement heads can be attached to a five axis gantry, post-style machinery, horizontal style machinery, or a robot as a turnkey system. Fiber placement combines the advantages of high precision fiber layup and differential tow payout capability of filament winding with the cut-restart and compaction capabilities of tape laying. The individual tows are fed through the delivery head and rolled onto the tool surface. These individual tows are typically 0.125 inch (3.175 mm), 0.25 inch (6.35 mm), or 0.5 inch (12.7 mm) wide. Tow cutters, clamps, and pinch rollers in the head enable the machine to start and stop individual tows at any location along a straight or curved programmed path. Heat and pressure are applied to ensure proper adhesion and compaction of the material during layup.

Multiple tows, up to 32, are placed during a course. The tows are generally placed in angular orientations like 0°, ±45°, and 90° depending on the mechanical properties designed. Every single tow can be dispensed at its own speed and cut independently, thus allowing each tow to conform independently to the surface. Consequently, the fibers are not restricted to geodesic paths like in filament winding. They can be steered to meet specific design goals.

The AFP process has been used successfully with both thermoset and thermoplastic matrix composite prepreg. When thermoplastic prepreg is utilized, the materials can be fusion (or *in situ*) bonded. This technique can eliminate the need for a post-process autoclave cycle and allows for integral attachment of internal composite components such as stiffeners. *In situ* fiber placed thermoplastic matrix composites can offer an attractive alternative for many applications because of reduced processing costs. Figure 18.9a,b, and c shows types of AFP machines used for making various airframe structures for the aircraft including fuselage barrels and fuselage skin panels. Figure 18.10 shows an example of a one piece fuselage forward section produced by AFP process.

The fiber placement process automatically places multiple individual prepreg tows onto a mandrel at high-speed using an NC articulating placement head. The fiber placement

FIGURE 18.9
Examples of AFP machines with different tooling systems for aerospace parts. (a) Large horizontal AFP with headstock/tailstock for tool handling, (b) medium, vertical Mongoose AFP, and (c) medium, vertical AFP. (From Ingersoll Machine Tools, Inc.)

heads are generally attached to a 5, 6, or 7-axis gantry. Machines are also available with dual mandrel stations to increase productivity. Advantages of fiber placement machines include:

1. Reduced flow time
2. Minimal material scrap
3. Reduced labor cost

FIGURE 18.10
Composite fuselage structure made by fiber placement technology. (From Spirit AeroSytems, Inc.).

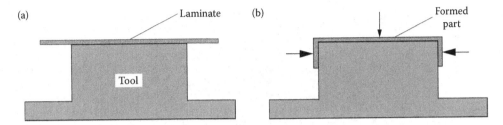

FIGURE 18.11
Schematic of drape forming. (a) Before and (b) after.

4. Opens design space
5. Facilitates integrated structure
6. Improved safety and ergonomics
7. Reproducible parts with high quality

The applications suitable for a fiber placement process typically include making a wide variety of parts' shapes and sizes; from small flat charges to large round structures like fuselage sections, and including fuselage skins.

18.4.2.4 Drape Forming

Hot drape forming (HDF) is an automated process of draping flat composite laminate onto a male tool to fabricate structural stiffeners for the wing, strake, fuselage, and empennage applications. The formed shapes are often C, L, or omega (hat)-shaped cross sections. HDF process is derived from the metal-forming technology of forming metal sheet on a male die. Figure 18.11 shows the schematic diagram of a drape-forming process where the laminate is placed on the male tool (Figure 18.11a). The flat laminate can be partially or totally heated to the right processing temperature before being pressed onto the mold by mechanical means or by a hydrostatic pressure. After forming (Figure 18.11b), the laminate is cooled under pressure to a certain temperature as specified for the specific resin/matrix material and then the part is either cured on the same mold or removed from the mold for further processing. HDF reduces fabrication costs by avoiding the intensive labor of hand layup and the repetitive stress injuries associated with it. Figure 18.12 shows a drape-forming equipment producing a stabilizer spar of an aircraft.

The applications of drape forming are generally suitable for straight or nearly straight bends like L- or C-section spars and stringers.

18.5 Liquid Composite Molding

Unlike prepreg process, liquid molding uses separate resin matrix and dry (unimpregnated) fibers combined during molding. The matrix material can be introduced before or after the reinforcement fibers are placed into the mold cavity or onto the mold surface. The matrix material experiences a melding process before the part shape is essentially set. Based on the nature of matrix materials, melding generally occurs in various ways

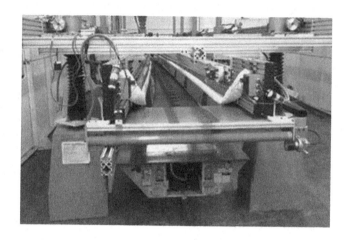

FIGURE 18.12
Drape forming of spar of an aircraft. (Adapted from Kisch, R.A., Composites processing technology and innovation at Boeing, *Key Note Presentation for the Society of Plastics Engineers Automotive Composites Conference & Exhibition*, MI, September, 2007.)

including chemical polymerization or solidification from the molten state. A large variety of molding processes can be used according to the design requirements of the end item. The molded products are generally referred to as panels such as wing skin, stabilizer skin, and many more. The molding process is very much dependent on the principal factors:

1. Type of matrix and fiber materials
2. Quantity of produced material

Large production volumes can justify high capital investments to introduce rapid and automated manufacturing technology. Whereas, lower capital investments with higher labor and tooling costs are suited more for small quantity production. The molding tool can be a single one-sided mold, or a two-piece mold with lower and upper molds. Panel fabrication begins by placing fibers and resin matrix into the lower mold. Here is the list of major types of molding methods of producing polymer matrix composites for aerospace applications:

1. RTM
2. Vacuum-assisted RTM (VARTM)/vacuum infusion
3. Resin film infusion (RFI)
4. Pultrusion

In liquid composite molding [1], there are four distinguished phases:

1. Manufacturing of preform by fiber reinforcement in the shape of the finished part
2. Injection or transfer of resin into the preform
3. Curing of the resin starting during or after the second phase
4. Demolding of the composite part after completion of curing

The preforming technology is an important factor of the liquid composite molding. Woven and stitched type fabrics can be used for layer-by-layer lay-up preform process. Complex 3D preforms can be made by using braiding, knitting, or stitching processes.

18.5.1 Resin Transfer Molding

In an RTM process, both lower and upper molds are used to form both surfaces of the panel in between the gap or cavity created by the molds. The reinforcement materials are placed into this cavity and the mold set is closed prior to the introduction of the resin matrix material. Figure 18.13 shows a schematic representation of RTM process. The resin is transferred to the reinforcement in the mold cavity using vacuum and pressure. This process can be performed at either room or at elevated temperature conditions. Figure 18.14 shows an example of RTM manufacturing process for production parts. The mold surface can produce high-quality finish at a fast rate while maintaining tight dimensional tolerances. The size of the part is limited by the mold geometry.

The applications of RTM process are generally suitable for small complex parts like ribs.

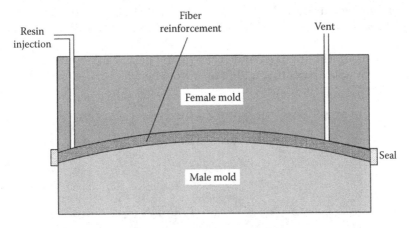

FIGURE 18.13
Fundamental of RTM.

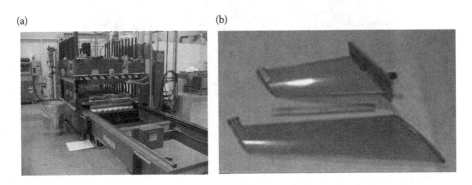

FIGURE 18.14
Example of RTM process. (a) RTM process equipment and (b) RTM products. (Adapted from Kisch, R.A., Composites processing technology and innovation at Boeing, *Key Note Presentation for the Society of Plastics Engineers Automotive Composites Conference & Exhibition*, MI, September, 2007.)

18.5.2 Vacuum-Assisted RTM

VARTM process has a fundamental difference with RTM of drawing resin into a preform through the use of a vacuum only. In the VARTM process, fiber reinforcements are placed on a single open mold with a flexible cover placed on the top of the mold to form a vacuum tight seal. The resin generally enters the mold where the fiber reinforcements are placed through the ports in place. The resin is drawn by vacuum through the reinforcements by means of a series of channels in the mold. This process is generally used to make both thin and very thick aerospace quality laminates. There are many names for VARTM depending on how vacuum and transport media are handled. Some other names include controlled atmospheric pressure resin infusion.

18.5.3 Resin Film Infusion

RFI is a process where a dry preform is placed in a mold on top of a layer or interleaved with layers of high-viscosity resin film. The vacuum is applied to the mold cavity at elevated temperature with ambient atmospheric pressure acting upon the vacuum bag. Under elevated temperature and pressure in the autoclave, the resin is drawn into preform and results in having a uniform high viscous resin distribution throughout the fibers due to the short flow distance. Figure 18.15 shows an example of an RFI process of making angle-shaped part.

The applications of RFI are like hand layup, many parts can be infused, but are more suited to highly complex parts that will fit on a work bench.

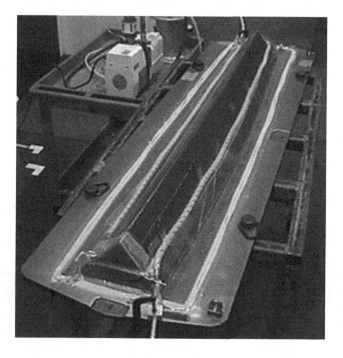

FIGURE 18.15
Example of RFI. (Adapted from Kisch, R.A., Composites processing technology and innovation at Boeing, *Key Note Presentation for the Society of Plastics Engineers Automotive Composites Conference & Exhibition*, MI, September, 2007.)

18.5.4 Pultrusion

Pultrusion is a continuous molding process that combines fiber reinforcements and thermosetting resins. Pultrusion processing is used to fabricate composite profiles with constant cross sections, including various rods, bars, and beam sections. Figure 18.16 shows the schematic representation of a pultrusion process. The reinforcements are drawn through a resin bath. The resin saturated fiber reinforcements enter through the heated die with a given shape of a finished part being fabricated. The heated die activates the curing or polymerization of the thermoset resin to the solid shape. The pultrusion shape solidifies when rolled and continuously emerges through the pull rollers. Figure 18.17 shows an example of a pultrusion process used in production of a "C" channel for aerospace applications.

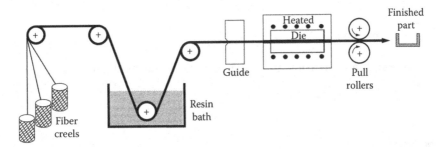

FIGURE 18.16
Fundamental of pultrusion process.

FIGURE 18.17
Example of pultrusion process in production of composite C-channel. (Adapted from Kisch, R.A., Composites processing technology and innovation at Boeing, *Key Note Presentation for the Society of Plastics Engineers Automotive Composites Conference & Exhibition*, MI, September, 2007.)

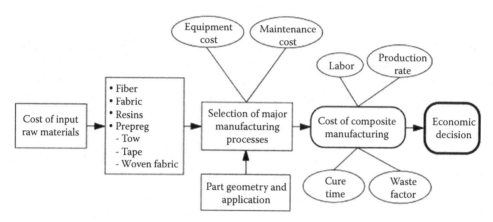

FIGURE 18.18
Economics of composite manufacturing.

The application of a pultrusion process is generally limited to constant cross sections, like cargo floor stanchions.

18.6 Economics of Composite Manufacturing

Figure 18.18 illustrates the important factors influencing the cost of composite manufacturing. The goal of composite manufacturing operations is to obtain the lowest possible unit cost and highest possible production rate with the highest quality. The cost of composite manufacturing is based on the initial cost of input raw materials and selection of major manufacturing processes, which is dependent on part geometry and application, and the equipment cost and its maintenance cost are also dependent on the selection of manufacturing processes.

The overall cost of composite manufacturing is related to labor cost. This is based on the skill of the operator, rate, or volume of production, which is a function of automation of the equipment, cure time of the composite, and also the waste of materials due to processing. The final economic decision is being taken based on the total cost of manufacturing.

References

1. Bersee, H.E.N., Composite aerospace manufacturing processes, *Encyclopedia of Aerospace Engineering*, John Wiley & Sons Ltd., NJ, 2010.
2. www.compositesworld.com
3. Campbell, F., *Manufacturing Processes for Advanced Composites*, Elsevier Advanced Technology, UK, 2004.
4. Kisch, R.A., Composites processing technology and innovation at boeing, *Key Note Presentation for the Society of Plastics Engineers Automotive Composites Conference & Exhibition*, MI, September, 2007.

19

Measurement and Inspection Methods
in Aerospace Manufacturing

19.1 Introduction

In manufacturing, the success or failure, and the next action plan are derived from the measurement and inspection methods applied in every step of producing aerospace components. Various manufacturing processes for making aircraft components from metallic products and also composite materials have been discussed in the Chapters 7–18. Measurement is the fundamental approach used to control, verify production performance, and as a starting point for process improvement. Measurement techniques play a very important role in every step of manufacturing processes starting from the input of raw materials to finished products. Measurement techniques are used to ensure engineering geometry requirements are met, from simple to complex contours. This is done to assure maintaining the highest quality of the aerospace products. Individual part measurements can be done directly using various measuring tools, gages, and also by using check fixtures for contoured parts. For all manufactured parts of an aircraft, there is a need to measure some fundamental dimensional elements including length, width, thickness, height, depth, and radius. Besides measurement methods to measure the engineering definition of a part, several testing/inspection methods are very critical to verify material soundness and integrity of the part, which ensures it is free from any surface and subsurface flaws including cracks. This chapter will cover the fundamentals of measurement and inspection methods required for manufacturing aerospace parts as illustrated in Figure 19.1.

19.2 Manual Bench

Engineering measurement of an object is performed to represent the physical dimensions in standard units within acceptance tolerances. The science of any type of measurement performed with appropriate units is generally called metrology. Standard measurement practices are based on measuring physical dimensions of an object using standard precision measuring tools available in the world market.

19.2.1 Measurement Tools

Manufacturing shop personnel use a vast range of measuring tools or instruments to check the dimensions and other geometry features of the initial raw materials received from

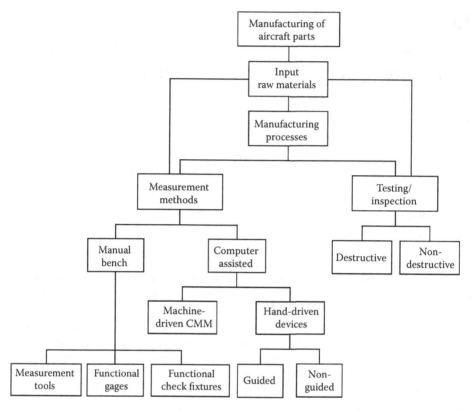

FIGURE 19.1
Measurement and inspection methods.

the primary metal producers, and also the subsequent manufactured parts. These instruments may range from simple rulers to precision measuring tools as shown in Figure 19.2. The precision measuring tools for dimensional checks are mainly:

1. Caliper
2. Micrometer
3. Square
4. Protractor
5. Height gage
6. Depth gage

19.2.2 Functional Gages

Figure 19.3 shows a few examples of standard functional gages used in day-to-day production processes to check engineering dimensions of the parts:

1. Feeler gage
2. Radius gage
3. Taper gage

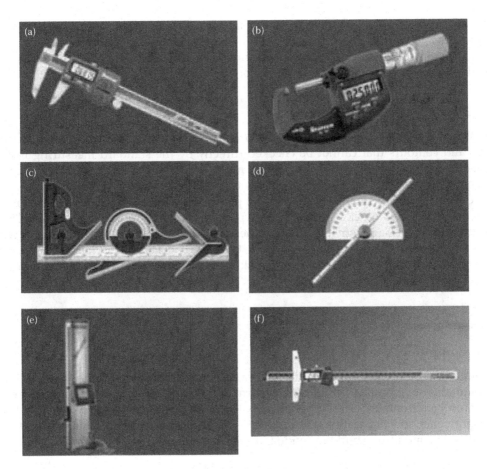

FIGURE 19.2
Example of precision measuring tools. (a) Caliper, (b) micrometer, (c) combination square, (d) protractor, (e) height gage, and (f) depth gage. (From Starrett, www.starrett.com)

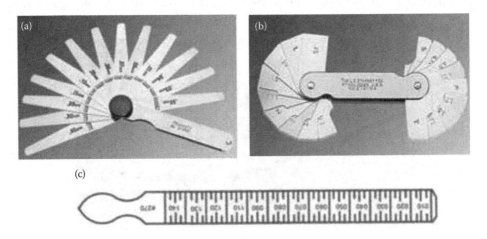

FIGURE 19.3
Example of few standard gages. (a) Filler gage, (b) radius gage, and (c) taper gage. (From Starrett, www.starrett.com)

There are applications of standard template for checking parts with certain contour geometries and various gages, including go/no go gages, which are referred to an inspection tool and used to check a workpiece against its upper and/or lower limits of dimensional tolerances.

19.2.3 Functional Check Fixtures

Templates and gages are specially made to check certain parts having critical dimensions, and contours that cannot be checked directly by the standard measuring instruments. Functional check fixtures are commonly used in the aerospace industry to check the conformity of the part contour geometry during production as shown in Figure 19.4. This example shows a small-size sheet metal part being inspected after a stretch-forming operation. It is a very quick way to check the part contour after forming, and also helps in adjusting the forming parameters to bring the part closer to the designed contour. A wide variety of sheet metal and extrusion parts are typically checked by check fixtures, starting from small to very large parts like fuselage skins, wing skins, spars, chords, and more.

19.3 Computer Assisted

Continuous developments of computer software technologies have become the foundation of automation in manufacturing measurement and control systems. Modern software in a graphical environment improves the ability to increase productivity in conducting tests, measurements, and automation solutions through data acquisition, data analysis, and data visualization and also to build individual measurement and control systems.

19.3.1 Machine-Driven Coordinate Measuring Machine

The coordinate measuring machine (CMM) plays an important role in the mechanization of the measurement process. CMMs have been widely used in the industry since the first was developed in the 1950s. Measuring platforms on which the workpiece may be

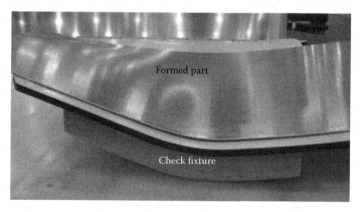

FIGURE 19.4
Example of a check fixture. (From The Boeing Company.)

measured used to be made from granite, steel, and most recently out of aluminum alloys and also ceramic materials. Platforms can be moved linearly or rotated. The machine could be manually controlled or computer controlled, as applicable, to the size and complexity of the part to be measured. CMMs are built rigidly and consist of three major functional components:

1. Measuring platform—three axes of motion
2. Probing system—mechanical, optical, laser, or white light
3. Data acquisition system—machine controller, computer, and application software

The machine follows the basic concept employing three coordinate axes to get precise movement in three coordinates: X, Y, and Z directions. Each axis movement is fitted with a linear transducer. Each transducer senses the direction of movement and shows a digital dimensional display. Based on the required direction of movements, a CMM is constructed with different types of structural arrangements:

- Bridge
- Cantilever
- Gantry
- Horizontal arm

Figure 19.5 shows an example of the most popular bridge-type CMM construction that provides better accuracy.

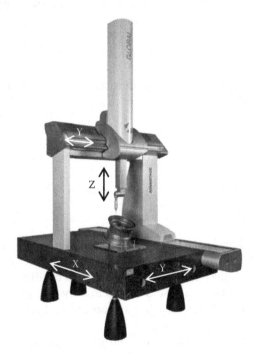

FIGURE 19.5
Example of bridge-type CMM. (From Hexagon Metrology, www.hexagonmetrology.us)

The precision and accuracy provided by a CMM is significantly high. Major advantages of CMM are highlighted below:

1. Flexibility
2. Reduced setup time
3. Single setup
4. Improved accuracy
5. Reduced operator influence
6. Improved productivity
7. Able to measure large parts

For increased accuracy and repeatability, thermocouples are incorporated throughout the machine and interfaced with the data acquisition system to compensate the temperature gradients during measurement.

19.3.2 Hand-Driven Devices

As seen in the enhanced benched method, a traditional CMM uses a probe that moves on the three Cartesian axes (X, Y, and Z) to measure the physical geometry of an object. In the case of the portable method of measurement, systems like CMMs use either articulated arms or arm-free scanning systems to have total freedom of movement around the object. Hand-driven portable CMMs equipped with the arm systems have six or seven axes with rotary encoders used instead of the linear axes in bench-type CMM. Portable arms are typically light weight and can be carried and used in any place within the manufacturing or assembly environments.

19.3.2.1 Guided Method

Portable CMMs have been developed in the industries that allow aerospace manufacturers easy verification of product quality mainly by

- Performing 3D inspection
- Tool certifications
- Point cloud-to-CAD comparison
- Dimensional analysis
- Reverse engineering

Figure 19.6 shows an example of a hand-driven guided state-of-the-art metrology system "FaroArm" introduced in the manufacturing industry. This system is equipped with both a hard probe and the hand-held laser line probe that can digitize interchangeably without having to remove either component. Users can accurately measure prismatic features with the hard probe (contact), and also laser scan (noncontact) sections requiring larger volumes of data.

Figure 19.7 shows another example of a hand-driven guided CMM device called The ROMER® Absolute Arm with hard probe with scanning pack that allows connection of laser scanning systems.

(a)   (b)

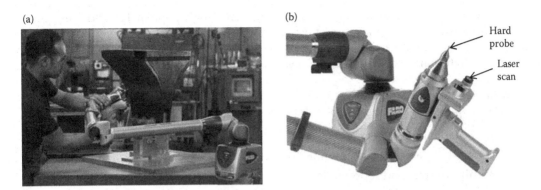

Hard probe

Laser scan

FIGURE 19.6
Hand-driven metrology. (a) Scanning the part geometry and (b) Faro edge and ScanArm HD. (From FARO, www.faro.com/edge)

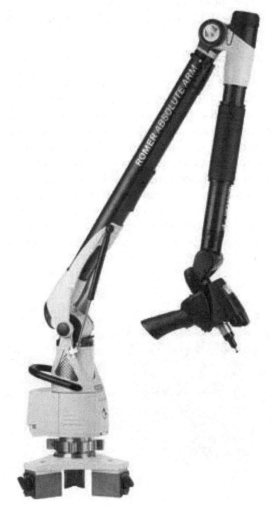

FIGURE 19.7
Example of another hand-driven guided measuring arm. (From Hexagon Metrology, www.hexagonmetrology.us)

Noncontact measurement devices are becoming increasingly popular. Hand-driven guided laser scanners provide a quick and effective way to inspect and reverse engineer complex parts and surfaces. Parts of complex shapes can be easily inspected without ever coming in contact with the part. The major advantages of scanArm-type portable CMMs are listed as

1. Rapid scanning speed
2. High definition data
3. Scan challenging part geometry
4. Compact, light weight, and simple to use
5. Highly accurate and repeatable

In the past, machine shop inspection has required the removal of parts from a CNC machine to a temperature-controlled inspection area for surface inspection or dimensional verification of the part. It was a time-consuming and expensive process. The portability of tools like the measuring arm enables efficient on the machine and in-process inspections. This delivers consistent, accurate measurements directly on or at the machine used for producing the parts.

19.3.2.2 Non-Guided Laser Tracker

Arm free scanning systems like the laser tracking system were introduced in the manufacturing world sometime during the 1990s. Since then this technology has established itself as a noncontact metrology tool for in-place inspection of large parts in the manufacturing process as well as in the assembly and integration processes. These are used in the aerospace industries which require accuracies within a couple of thousands of an inch. A laser tracker is fundamentally a portable coordinate measuring system that uses a laser beam to accurately measure and inspect over a very large spherical volume of a part as shown in Figure 19.8. The laser tracker basically measures two angles and a distance. It works as follows:

1. Distance measurement—the tracker sends a laser beam to a retro-reflective target held against the object to be measured. The reflective beam returns to the tracker and its precise distance is recorded.

(a)

(b)

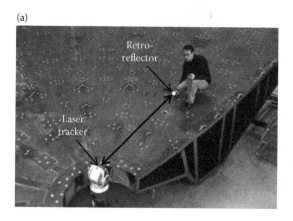

FIGURE 19.8
Application of laser tracker to inspect a very large part. (a) Measuring and (b) laser tracker. (From FARO, www.faro.com/edge)

2. 3D measurement—two angle encoders measure the elevation and rotational angles while a highly accurate absolute distance meter is used to determine the 3D position of the target. This position is shown in the software as an X, Y, and Z value.

3. Tracking—the target returns the reflected laser beam parallel but offset back to the tracker where it hits the position detector that calculates the offset between the outgoing and incoming beams. Servo motors continuously (thousands of times per second) steer the tracker's head to minimize the offset between the two beams, resulting in high-speed accurate measurements.

In aerospace industry, large-scale measurements are a common requirement to maintain consistency and accuracy of the part manufacturing and assembly integration for the aircraft design, and thereby maintain the highest quality products for the airlines customers. Laser trackers are widely used for various applications including:

- Large part inspection
- In-process inspection
- Tool building and setup
- CAD-based inspection
- Dimensional analysis
- Assembly integration
- Reverse engineering

A laser tracker can take precision measurements at any scale providing the highest accuracy with least amount of user variability. Measuring tool, check fixture building, and moldings of composite manufacturing for the aerospace industries all require a high degree of accuracy and repeatability. A laser tracker and also the measuring arm CMMs can perform full volumetric accuracy tests on tools, check fixtures, and molds to ensure parts are being manufactured to the highest standards. With these metrology instruments, tool makers can also identify or predict defects by evaluating the dimensional integrity and repeatability of tools, fixtures, and molds.

Portable CMMs simplify the implementation of tight requirements, drive the need to use CMM's, and provide solutions for CAD-based 3D inspections and nominal comparisons. Scan arm tools and laser tracker provide both contact and non-contact measurement capabilities. They can therefore be utilized with CAD overlays to check complex geometry against design, or provide CAD comparison to evaluate deviations in surface form, and thereby ensure every part is measured to an exacting tolerance.

Many machine shops are now equipped with portable CMMs from FARO, such as measuring arms and laser trackers. These meet a variety of dimensional measurement needs, such as in-process inspection, and CAD comparison to ensure proper tolerances are achieved.

Reverse engineering is becoming a more popular method of creating a 3D model of a part without existing CAD files. Reverse engineering reconstructs classic designs and implements new ones. It can generate lost or absent design documentation and update or create as-built documentation. Using noncontact 3D laser scanning, portable CMM, and 3D documentation solutions integrate 3D modeling processes into a single step. They can support rapid prototyping in plastic manufacturing and inspect composite surfaces in aerospace assembly.

19.4 Destructive Testing

Destructive testing includes methods of determining mechanical properties including strength, elongation, material flow stress, strain hardening index, fracture, and toughness parameters related to the input raw materials received for manufacturing. This testing occurs before the parts are being fabricated using different manufacturing technologies. Most common destructive testing of metals include:

- Tensile
- Compression
- Hardness
- Impact
- Fatigue
- Bend
- Macro- and microstructure
- Grain size

Destructive testing provides

1. Properties of raw materials per specification
2. Material's formability as it relates to part fabrication
3. Ensures compliance with regulations

19.5 Nondestructive Inspection/Testing

Nondestructive testing/inspection (NDT/NDI) is a wide group of inspection methods used in the industry to evaluate a material component or system without breaking or cutting the component for small-scale laboratory analysis. The major NDI methods used in aerospace manufacturing include:

- Dye penetrant
- Eddy current
- Ultrasonic

19.5.1 Dye Penetrant

Widely used in the aerospace industry, dye penetrant inspection, also known as liquid penetrant or penetrant inspection [4] developed in early 1940s using fluorescent or visible dye added to an oil, has since been used to detect flaws on the surface of the materials. The flaws include porosity, cracks, fractures, laps, seams, blisters, and other types that are open to the surface of the test part or production parts. These defect features may be caused by service load, applications, manufacturing processes including machining, forming and joining, casting, extrusion, forging, sheet metal and tubular products and also the effect of heat treatment quench, and more.

Varieties of penetrant materials are available, and the proper selection is done mainly on the basis of required sensitivity levels of the tests and the equipment available to conduct the test. There are two types of penetrant liquids:

1. Visible dye (colored red)
2. Fluorescent dye (colored green-yellow)

Fluorescent penetrants are generally classified by sensitivity levels ranging from 1 to 4 with 4 being the most sensitive, and used to detect the finest flaws when the penetrant (low surface tension fluid) penetrates by capillary action into a clean and dry surface to show the discontinuities.

The major process steps of penetrant inspection are as follows:

1. Pre-cleaning
2. Chemical etching
3. Application of penetrant
4. Removal of excess penetrant
5. Application of developer
6. Inspection
7. Post-cleaning

The workpiece surface is thoroughly cleaned by removing any dirt, oil, grease, heat treat scale, paint, or any other surface contamination prior to application of penetrant. Cleaning processes may include those used with ultrasonic waves, solvents, alkaline etch, or vapor degreasing. Vapor degreasing processes are discontinued in many industries due to inherent health hazards and safety issues. Post-machined, sanded, or grit-blasted parts may require chemical etching to remove a very thin film of smeared material that could block the opening of the flaw to prevent penetrant from entering.

Penetrant liquids are applied on the workpiece by spraying, brushing, or immersing into the penetrant bath. The choice of application is usually a matter of preference but could be influenced by the size and shape of the test or production parts, as well as the equipment used for the test requirements according to specifications applied to the part. The liquid penetrant is left on the workpiece surface for a definite time period, which is called a dwell time to allow the liquid to penetrate into any surface flaw openings. Dwell time mainly depends on the type of the penetrant, type of test material, and the size of flaws desired to be detected.

The excess penetrant is removed carefully after the dwell time is elapsed. In the case of a visible dye, excess penetrant from the workpiece is usually removed with a solvent. In the case of a fluorescent dye, excess penetrant is generally removed by rinsing with water. After rinse, the workpiece is dried before applying developer.

A thin layer of developer is applied either by dusting or spraying with a dry powder. The indications on the workpiece have higher contrast level between penetrant and developer to make the flaws more visible in inspection. Dwell time for the developer is generally stated in the test specification by the manufacturer.

The inspection process is carried out by the trained and certified inspectors using a visual examination. High-intensity UV lamp or black light is used when working with fluorescent penetrant, whereas sufficient white light is used when visible dye penetrant is used. The workpiece surface is cleaned after inspection.

19.5.2 Eddy Current

Eddy current inspection/testing [5,6] is widely used in a variety of industries including aerospace to find defects and make measurements. Aircraft components are inspected at various manufacturing stages before they are qualified to be assembled in the aircraft, and also periodic inspections are made throughout the life of the aircraft. One of the primary applications of eddy current inspection is to detect surface and near-surface defects in a relatively small area. The major type of measurements that can be done are:

1. Thickness of metal tube and sheet stocks, foils, and metallic coatings on metallic and nonmetallic substrates. This makes it a useful tool for detection and assessment of corrosion or other damage, which causes thinning of materials.
2. Cross-sectional dimensions of cylindrical tubes and rods.
3. Thickness of paints and other nonmetallic coatings on metallic substrates.

Since the eddy current is affected by the electrical conductivity and magnetic permeability of the materials, eddy current inspection can be widely used to check the conductivity of aluminum parts after heat treatment where the conductivity is a critical factor for aerospace parts.

In eddy current testing, an electrical current (eddy current) is generated in conductive material by changing the magnetic field. Meanwhile, the strength of generated eddy current is being measured. Material defects cause interruptions in the flow of eddy current, which may detect the presence of a defect or other change in the material. The eddy current working principles as illustrated in Figure 19.9 are explained below:

1. A magnetic field around the coil is generated when the alternating current is flowing through the coil at a certain frequency.
2. Eddy current is induced in the material, when the coil is placed close to an electrically conductive workpiece material.
3. If there is a flaw in the conductive workpiece material, it disturbs the eddy current circulation and the magnetic coupling with the probe is changed and a defect signal can be read by measuring the coil impedance variation.

Changes in metal thickness or defects like near-surface cracking will interrupt or alter the amplitude and pattern of the eddy current and the resulting magnetic field. This in turn affects the movement of electrons in the coil by varying the electrical impedance of the coil. The eddy current instrument plot changes in the impedance amplitude and phase angle, which can be used by a trained operator to identify changes in the test piece.

A wide variety of tests can be conducted using Eddy current instruments depending on the type of probe used. Measuring performance can also be optimized by careful selection of a probe. Major types of probes commonly used are:

- Surface probe—used to identify flaws on and below metal surfaces having large diameter to accommodate lower frequencies for deeper penetration and to cover larger areas

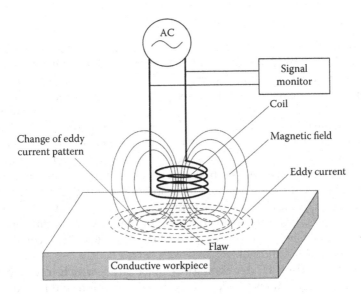

FIGURE 19.9
Eddy current inspection principles.

- Pencil probe—smaller diameter probes used for high frequencies and resolution to detect near-surface flaws
- Bolt hole probe—used to inspect the inside of a bolt hole
- Donut probe—used to inspect fastener holes with fasteners in place for aerospace
- Sliding probe—used also for aircraft fastener holes with higher scanning rate than donut probe
- ID probe—used to inspect metal tubing from the inside
- OD probe—used to inspect metal tubing and bars from the outside

Major factors affecting the capabilities of an eddy current inspection:

1. Eddy currents induced in the workpiece material with higher conductivity will be more sensitive to detect surface defects but will have less penetration into the material. Depth of penetration is also dependent on test frequency.
2. Higher test frequencies increase near-surface resolution but limit the depth of penetration, whereas lower test frequencies increase penetration.
3. Large coils can inspect a bigger volume of material from any given position whereas smaller coils are more sensitive to small defects.
4. Variations in permeability of a workpiece material generate noise, which can limit flaw resolution because of greater background variations.

At the start of a test, an eddy current instrument and the probe must be always calibrated with appropriate reference standards. The calibration involves identifying the baseline display from a given test piece and observing the changes under the conditions for the intended test. In the case of thickness measurement applications, the reference standards would consist of various samples of known thickness.

19.5.3 Ultrasonic

Ultrasonic testing (UT) [5,6] uses high frequency sound waves transmitted to detect imperfections including hidden cracks, voids, porosity, and other internal discontinuities into a workpiece material, as well as to measure thickness and analyze material properties. The materials include metals, composites, plastics, and ceramics. Since the early work in Europe and the United States in the 1930s, there has been tremendous growth in the use of ultrasonic measurement systems in the manufacturing, process, and service industries and also in medical science. In the 1940s, Japan pioneered the use of UT in medical diagnostics. In 1945, Floyd Firestone, a US researcher, patented an instrument called the "Supersonic Reflectoscope" as the first practical commercial ultrasonic flaw detector, which used the physics of the pulse/echo technique fundamentally used in today's technology. During the 1960s and 1970s, many companies developed ultrasonic flaw detectors, gages, and transducers. From the 1980s onward, availability of digital signal processing technology and inexpensive microprocessors brought the latest advancement of ultrasonic instruments in the world of industry.

The most commonly used UT technology is pulse/echo, which introduces sound into the test workpiece and generates a reflection (echo) from an internal flaw or the geometrical surface imperfection, which are returned to the receiver. The fundamentals of a pulse/echo inspection configuration are illustrated in Figure 19.10. A typical ultrasonic inspection system consists of three major functional units:

1. Pulser/receiver
2. Transducer
3. Display monitor

A pulser/receiver is an electronic device that can produce high-voltage electrical pulses. A transducer is the device that converts one form of energy to another. The ultrasonic transducer converts electrical energy into mechanical vibrations, sound energy, and sound wave into electrical energy. Driven by the pulser, the transducer generates high frequency ultrasonic energy. The sound energy is introduced and propagates through the materials in the form of waves. When there is a flaw (such as a crack, void) in the wave path, part of the energy will be reflected back from the flaw surface. The reflected wave signal is transformed

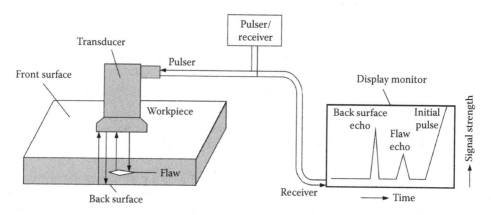

FIGURE 19.10
Fundamentals of ultrasonic measurement.

into an electrical signal by the transducer and is received by the display monitor. In the display monitor, the reflected signal strength is displayed versus the time from initial pulse generation to an echo was received. Signal travel time can be directly related to the distance that the signal travelled. The following information could be obtained from the signal:

- Reflector location
- Size
- Orientation
- Other features

Ultrasonic technology is a very useful and versatile inspection method and the inspections can be accomplished in a number of ways:

1. Pulse echo or through transmission energy used
2. Normal or angle beam enters the workpiece
3. Contact or immersion method of transducer to the workpiece
4. Noncontact generation and detection of ultrasound

Some of the advantages of ultrasonic inspection technology include:

- Sensitive to both surface and subsurface discontinuities
- The depth of penetration for flaw detection or measurement is superior to other NDT methods
- Single-sided access is only needed when using the pulse-echo technique
- High accuracy in determining flaw position and estimating size and shape
- Minimal part preparation
- Electronic equipment provides instantaneous results
- Detailed images can be produced with automated systems
- Additional advantage of thickness measurement in addition to flaw detection.

Ultrasonic inspection requires a trained operator to set up a test with the help of appropriate reference standards and properly interprets the results. Inspection of some complex geometries including rough, irregular in shape, very small, and exceptionally thin may be difficult. Ultrasonic thickness gages are more expensive than mechanical measurement devices.

Several noncontact type ultrasonic techniques [7] are available in various stages of development, namely capacitive pick-ups, electromagnetic acoustic transducers, laser beam optical generators and detectors, and air-coupled ultrasonic systems are widely used for process control of composite fabrication.

References

1. www.starrett.com
2. www.hexagonmetrology.us

3. www.faro.com/edge
4. Bentzley, S.M. and Goodwin, R., Basic principles of liquid penetrant inspection, *Quality Magazine*, 16–18, November 1, 2010.
5. www.nde-ed.org
6. www.olympus-ims.com/en/eddycurrenttesting
7. Djordjevic, B.B. and Green, R.E. Jr., Non-contact ultrasonic techniques for process control of composite fabrication, *NDE Conference Applied to Composite Fabrication*, St. Louis, Missouri, October, 1994.

20

Research and Development

20.1 Introduction

Research and development (R&D) plays a very critical role in aircraft manufacturing industries, enabling the more successful companies to stay in business in the highly competitive market place. Continuous improvement in productivity, reduced cost of manufacturing, quality, and safety performance of an aircraft all benefit from the successful application of R&D in aircraft manufacturing. This chapter highlights the major elements of aircraft design configurations, which will drive materials selection and the need for manufacturing R&D. Both will be applied to satisfy engineering design needs, drive down manufacturing costs, and improve the quality and safety of an aircraft as illustrated in Figure 20.1. Airline customers look for an aircraft that can fly longer distances with greater passenger capacities and also burn less fuel over the same flying range. Fuel savings enable flying extra miles/kilometers and thereby increase the flying range. Based on the aircraft design configuration, an aircraft manufacturer develops their engineering design needs, which in turn are supported by materials and manufacturing R&D. These are the two most important elements to be applied in satisfying all the design configuration elements of the airline customers. Successful achievement of these goals will lower aircraft manufacturing cost and provide highest quality and safety of an aircraft.

20.2 Aircraft Materials

Application of the right material for the right component of an aircraft derives from the properly researched selection of available materials. Materials could be metallic or nonmetallic, including composites. R&D starts from the houses of primary material producers who work together with the aircraft manufacturers to produce lower density, higher allowable material strength, and other properties that allow them to be considered for design and manufacturing. An aircraft component must fulfill the engineering design needs and also finally satisfy the performance needs of the airline customers. Figure 20.2 illustrates an idealized example of a materials R&D model.

The objective of introducing new materials (metallic or nonmetallic) to produce the needed light-weight airframe structure of an aircraft is to satisfy all the required material properties including strength, fracture toughness, fatigue, corrosion resistance, and compatibility with other materials at the interface surface during assembly or joining. These performance requirements must be balanced with consideration of cost of input raw

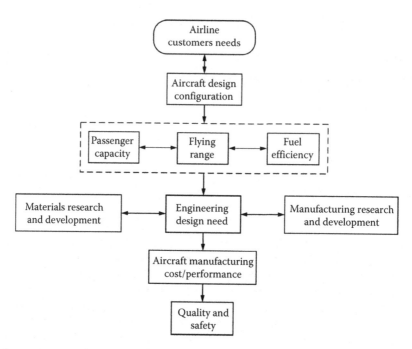

FIGURE 20.1
Major elements for aircraft design and manufacturing.

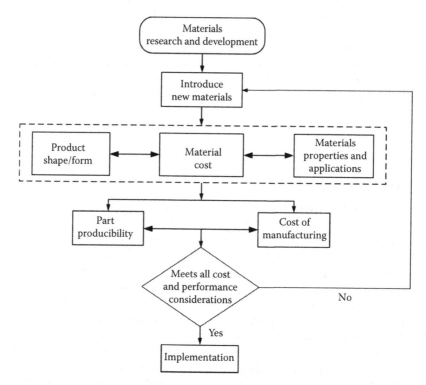

FIGURE 20.2
R&D model of aerospace material development.

materials, which in turn will determine the overall cost of the product. Part producibility studies of new materials and their corresponding product shape are very critical to verify that the part can be manufactured with minimum process flow time. Efficient processing will help to maintain the manufacturing cost targets, but must also assure the engineering requirements are met prior to being implemented in the production system.

20.3 Manufacturing Process Technologies

Once the proper input material is selected, manufacturing R&D plays a very important role to apply the right manufacturing technology for the right material for the right application. Figure 20.3 shows an outline of the manufacturing R&D model. Manufacturing technology development involves a continuous process of bringing new technology to the manufacturing world. This is especially true in the composite manufacturing facilities due to surging demand from the aircraft manufacturers. Once the part geometry is defined

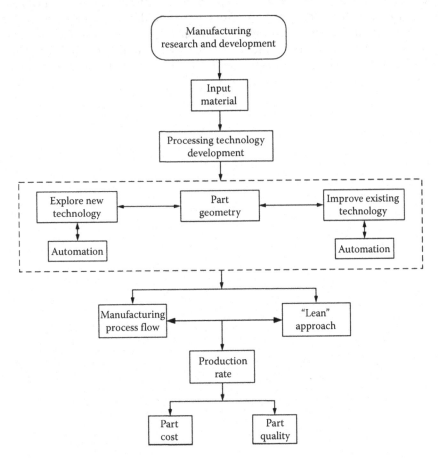

FIGURE 20.3
Manufacturing R&D model.

with the right input material, manufacturing facilities work on processing technology development by

1. Exploring new technology
2. Improving existing technology

Technology development's success determines the manufacturing process flow for either utilizing new technology or for the continued use of existing technology. New technology is being increasingly offered within integrated automation systems. The integration of automation to the existing manufacturing technology is often recommended to increase the production rate and help lower manufacturing costs. It is also important to apply a "Lean" approach to optimize the manufacturing process flow, which will also finally determine the part production rate, part cost, and part quality.

In summary, manufacturing of modern-day commercial aircraft is continuously evolving. This is driven by a complex combination of real-world needs. More and more people are able to utilize the aircraft mode of transportation as time goes by. This is, in large part, due to the efficiencies provided in the operation of modern-day aircraft. These efficiencies are made possible by improvements in aircraft technology and design, as well as the continued enhancement of the aircraft transportation infrastructure. But, another equally important reason for the wider use of air transport is that the real cost of the production of these aircraft is also decreasing, making the acquisition of these newer technology aircraft more affordable. These are all trends that will likely continue in the future as the aircraft industry seeks to improve all aspects of the business, including the cost of travel for the passengers.

Index

Printed in the United States
by Baker & Taylor Publisher Services